ADVANCES IN ONCOBIOLOGY

Volume 1 • 1996

SOME ASPECTS OF ONCOLOGY

ADVANCES IN ONCOBIOLOGY

SOME ASPECTS OF ONCOLOGY

Editors: GLORIA HEPPNER
Breast Cancer Program
Karmanos Cancer Institute
Detroit, Michigan

E. EDWARD BITTAR
Department of Physiology
University of Wisconsin Medical School
Madison, Wisconsin

VOLUME 1 • 1996

JAI PRESS INC.

Greenwich, Connecticut *London, England*

55 Old Post Road No. 2
Greenwich, Connecticut 06836

JAI PRESS LTD.
38 Tavistock Street
Covent Garden
London, England WC2E 7PB

ISBN: 0-7623-0146-5

Manufactured in the United States of America

Transferred to digital printing in 2008.

CONTENTS

LIST OF CONTRIBUTORS

Bharat B. Aggarwal	Cytokine Research Laboratory Department of Clinical Immunology and Biological Therapy The University of Texas M.D. Anderson Cancer Center Houston, Texas
Mark Bandyk	Department of Urology Danville Urologic Clinic Danville, Virginia
Leland Chung	Department of Urology University of Texas M.D. Anderson Cancer Center Houston, Texas
Bryant G. Darnay	Cytokine Research Laboratory Department of Clinical Immunology and Biological Therapy The University of Texas M.D. Anderson Cancer Center Houston, Texas
P.M. Doskoch	Memorial Sloan-Kettering Cancer Center New York, New York
Clive W. Evans	School of Biological Sciences University of Auckland Auckland, New Zealand
R.K. Ganjoo	Department of Medical Oncology St. Bartholomew's Hospital London, England

Martin Gleave — Division of Urology
University of British Columbia
Vancouver, British Columbia, Canada

Gloria Heppner — Breast Cancer Program
Karmanos Cancer Institute
Detroit, Michigan

A.N. Houghton — Department of Medicine and Immunology
Memorial Sloan-Kettering Cancer Center
New York, New York

S.A. Lynch — Memorial Sloan-Kettering Cancer Center
New York, New York

J.S. Malpas — Department of Medical Oncology
St. Bartholomew's Hospital
London, England

Kenneth Mossman — Department of Microbiology
Arizona State University
Tempe, Arizona

Elizabeth A. Musgrove — Cancer Biology Division
Garvan Institute of Medical Research
St. Vincent's Hospital
Darlinghurst, Australia

Peter J. Polverini — Department of Oral Medicine, Pathology, and Surgery
Laboratory of Molecular Pathology
University of Michigan School of Dentistry
Ann Arbor, Michigan

Jerome Ritz — Department of Medicine
Harvard Medical School
Division of Hematologic Malignancies
Dana-Farber Cancer Institute
Boston, Massachusetts

Michael J. Robertson	Department of Medicine Harvard Medical School Division of Hematologic Malignancies, Dana-Farber Cancer Institute Boston, Massachusetts
A. Rohatiner	Department of Medical Oncology St. Bartholomew's Hospital London, England
Randall Ruch	Department of Pathology Medical College of Ohio Toledo, Ohio
Flemming Brandt Sørenson	University Institute of Pathology University of Aarhus Aarhus, Denmark
Dennis W. Stacey	Department of Molecular Biology Cleveland Clinic Research Institute Cleveland, Ohio
Robert L. Sutherland	Cancer Biology Division Garvan Institute of Medical Research St. Vincent's Hospital Darlinghurst, Australia
S. Vijayasaradhi	Rockefeller University New York, New York
D.W. Vischer	Department of Internal Medicine Wayne State University Detroit, Michigan
Crystal M. Weyman	Department of Cell Biology Cleveland Clinic Research Institute Cleveland, Ohio

PREFACE

In the last few years considerable progress has been made in our understanding of molecular oncology and we are beginning to see how this understanding can be translated into strategies for treatment or prevention. It is now widely recognized that tumor suppressor genes play an important role in the development of tumors and that oncogene mutation represents only one mechanism which leads to cancer. In their seminal study of colon cancer, Volgelstein and coworkers were able to identify several genetic events, notably activation of a dominant oncogene (ras) and inactivation of several tumor suppressor genes including p53 and DCC that contribute to cancer development and progression. These events occur in clonal populations of adenoma cells. Whether products of tumor suppressor genes and oncogenes interact with each other is not yet known with certainty.

The present volume is the first in the advances in oncobiology series. It is meant to be useful not only to clinical and non-clinical oncologists but also to graduate students and medical students. The individual chapters are presented as self-contained summaries of current knowledge rather than as reviews. The last chapter deals with the subject of chemotherapy.

To the various authors are due our warmest thanks for their scholarly contributions and patience. We wish to thank Ms. Lauren Manjoney and the staff of JAI Press for their skill and courtesy.

Gloria Heppner
E. Edward Bittar
Editors

Chapter 1

Pathobiology of Neoplasia

D.W. VISSCHER and G.H. HEPPNER

Advances in Oncobiology
Volume 1, pages 1–41.

ISBN: 1-55938-146-5

INTRODUCTION

In order to understand the pathology of neoplasia, one must first understand that human tissues are dynamic, interactive and heterogeneous "cell societies." By the term heterogeneous, we are referring to the combined presence of multiple, embryologically distinct cell types which, in normal tissues, achieve and maintain a functional microanatomical distribution. By the terms dynamic and interactive, we are referring both to the cellular interactions which sustain this architecture and to the process of continuous cell proliferation by which a tissue balances physiologic or pathologic cell losses. The organization of heterogeneous cells into tissues, the specific interactions which maintain that organization, and the cellular replicative activity which renew it are, taken together, called differentiation. Neoplasia, in all of its clinical and pathologic diversity, is best viewed as a somatic (i.e., postembryonic) disease of differentiation.

CLASSICAL ASPECTS OF NEOPLASIA

Neoplasia as a Disease of Tissue Differentiation

Neoplasia usually results in autonomous growth of an abnormal mass. Thus, growth of the neoplasm not only exceeds and is uncoordinated with that of the surrounding tissue, but persists despite being dysfunctional or unfavorable to the host. Other diseases may result in an abnormal tissue mass without being neoplasms, including inflammatory lesions such as granulomas, as well as non-neoplastic disorders of histogenesis such as choristoma (an ectopic nest of normal tissue) or hamartoma (a mass of disorganized but otherwise normal cells and tissue components which are indigenous to that particular organ). Some descriptive terms are often used to refer to neoplasms which do not form solid masses within parenchymal tissues. These include: (a) polyp, which is any mass of tissue which visibly protrudes or projects above a mucosal surface or into the lumen of a duct or viscous. Sessile polyps are attached via a broad base whereas pedunculated polyps are attached by a thin, elongated stalk; and (b) cyst, which is any fluid-filled cavity lined by epithelial or mesothelial cells. Some neoplasms form cysts or polyps but not all cysts or polyps are neoplasms.

Table 1 is an abbreviated summary of human neoplasia classification and nomenclature. Note that neoplasms are grouped into categories (epithelial, stromal, hematopoietic, and miscellaneous) based on the embryologic lineage of the presumed neoplastic cell type. Inherent to this classification strategy is the concept of histogenesis; that different types of neoplasms arise from different types of somatic cells. Neoplasms are classified secondarily by their aggressiveness (i.e., benign versus malignant). This inference is based on the observed degree of histologic differentiation abnormality. For reasons which shall soon become apparent, the neoplastic cells in benign tumors, such as adenomas or lipomas, bear a striking

Table 1. Simplified Classification Scheme of Human Neoplasia

Neoplastic Cell Type	*Benign*	*Malignant*
Epithelium		
squamous	papilloma	squamous carcinoma
gland/duct lining	adenoma	adenocarcinoma
	papilloma	papillary carcinoma
	cystadenoma	cystadenocarcinoma
transitional	papilloma	transitional carcinoma
Connective Tissue		
fibrocytes	fibroma	fibrosarcoma
fat	lipoma	liposarcoma
cartilage	chondroma	chondrosarcoma
bone	osteoma	osteosarcoma
endothelium	hemangioma	angiosarcoma
smooth muscle	leiomyoma	leiomyosarcoma
striated muscle	rhabdomyoma	rhabdomyosarcoma
Blood Cells		
hematopoietic	—	leukemia (myelogenous, monocytic...)
lymphoid	—	lymphoma
		lymphocytic leukemia
		plasma cell myeloma
Miscellaneous		
mesothelium	benign mesothelioma	malignant mesothelioma
melanocytes	nevus	malignant melanoma
germ cells*	teratoma (♀)	seminoma
		embryonal carcinoma
		yolk sac tumor
trophoblast	hydatidiform mole	choriocarcinoma
glial cells	ependymoma	glioblastoma multiforme
neuroendocrine	carcinoid tumor	small cell carcinoma

Note: *Neoplasms derived from germ cells are special. As a totipotent (predifferentiated) cell, a neoplastic germ cell may differentiate toward germ cells, trophoblast, extraembryonic mesoderm (yolk sac), or any somatic tissue. Germ cell neoplasms, accordingly, often contain mixtures of these tissues. The stem cells of other organs, in contrast, are always tissue determined, as reflected in a more limited spectrum of neoplastic differentiation.

similarity to their putative normal (i.e., mature) counterparts. In carcinomas or sarcomas, however, the tumor cells resemble the "immature," or incompletely differentiated, precursors of their physiologic counterparts.

The clinical aggressiveness of human neoplasia constitutes a broad spectrum. Even the behavior of a given type of neoplasm—say breast carcinoma—may vary tremendously from individual to individual. This in great part reflects the considerable variability in degree of differentiation abnormality between neoplasms.

Thus, the concept of benign versus malignant sometimes conveys an overly simplified, dichotomous, view of neoplasia biology. In essence, though, a benign neoplasm is one which lacks the capacity to spread beyond its tissue of origin and, generally, grows at a slow rate. A malignant neoplasm, in contrast, is one which is capable of progressive, often rapid, growth, extension into surrounding structures (invasion) and, in some cases, dissemination to other organs (metastasis).

There are several nuances to be appreciated about the classification of neoplasia. First, due to the cellular heterogeneity of tissues, a given organ may give rise to a variety of histologically distinctive adenomas, carcinomas, or sarcomas. Second, due to metaplastic differentiation, an organ may also develop a neoplasm exhibiting cellular features which are not physiologically observed in that site. Squamous carcinoma of the lung, for example, is a common human neoplasm despite the absence of a squamous epithelium in normal pulmonary airways or alveolar parenchyma. Third, the neoplastic population may display morphologic features of more than one somatic cell type. These so-called biphasic neoplasms generally contain a neoplastic epithelial cell component admixed with a neoplastic stromal cell component. Examples include Wilm's tumor (a malignant kidney neoplasm), and mixed Mullerian tumor (a malignant neoplasm of the uterus or ovary). Fourth, not all of the cells within a neoplasm are neoplastic. On average, in fact, only 50% of them are. The non-neoplastic cells represent host-derived blood vessels, fibroblasts, and inflammatory cells. As we shall see, these constituents are critical to the biology of neoplasia. Finally, neoplasms are, in a sense, like snowflakes—no two are exactly alike and they occur in a near infinite variety. Thus, histologic classification of spontaneous human neoplasia can, at times, be subjective and controversial. This problem is compounded by the anecdotal and sometimes misleading diagnostic terms for neoplastic diseases which are often employed in clinical settings.

The development of neoplasia is profoundly influenced by nutritional, environmental, and genetic factors as well as by contact between humans and other living organisms. It should come as no surprise, therefore, that the incidence of various forms of neoplasia is in constant flux as humankind adapts to and alters lifestyles and environments. Although there are countless exotic tumors, most neoplasia mortality results from "garden variety" carcinomas of the lung, large intestine, prostate, and breast along with the hematopoietic malignancies (leukemia and lymphoma). Skin cancer cases numerically dominate these visceral neoplasms; however, most forms of skin cancer are readily curable. Common forms of human neoplasia become progressively more prevalent with increasing age. Neoplasia nevertheless accounts for 10% of total mortality in the pediatric age group, where the tissues involved by neoplasia and their histologic patterns are very different from mature adults. The most common forms of neoplasia in younger age groups are summarized in Table 2.

Table 2. Common Malignant Neoplasms in Younger Age Groups

Infants/Toddlers	*Children*	*Young Adults*
nephroblastoma*	leukemia (acute)	germ cell neoplasia
neuroblastoma	sarcomas	Hodgkin's disease (a lymphoma)
retinoblastoma	glial neoplasms	malignant melanoma

Note: *Also known as Wilm's tumor. The suffix blastoma is commonly used in neoplasia nomenclature. It connotes a resemblance between the neoplasm and embryonic tissue from that site. These neoplasms are therefore primitive in appearance and generally have a rapid growth rate and an aggressive clinical course if left untreated.

The Relationship Between Differentiation and the Development and Behavior of Neoplasia

The wide spectrum of clinical behavior and histologic patterns inherent to neoplasia are in many ways explained by the maturation arrest/blocked ontogeny model proposed by Sell and Pierce (1994). This theory holds that neoplasia results from delayed maturation, or differentiation, of organ-specific stem cell progeny. Maturation thus becomes arrested at a stage between the stem cell and its terminally differentiated daughter cells. As a consequence, incompletely differentiated cells accumulate out of proportion to mature, or terminally differentiated, cells. The stage of arrest defines the degree of differentiation abnormality which, in turn, determines whether a given neoplasm will behave in a clinically benign or malignant fashion.

The term differentiation is sometimes confusing because it may refer to divergent, but related, concepts. In one sense, it denotes the lineage of a cell or tissue (e.g., squamous epithelial cell versus lymphocyte versus fibrocyte). Differentiation also refers to the process by which tissues maintain normal architecture and function through continuous renewal of cell populations from an indigenous supply of "stem cells" which are supplied during embryonic histogenesis. In the postembryonic organism, these stem cells are tissue determined; that is, their progeny are capable of maturing in only one anatomic site and only into a certain type of somatic cell. Hematopoietic stem cells, for example, can not differentiate into epithelium. In non-neoplastic tissue, stem cells replace functioning somatic cells as needed, either in response to physiologic losses of mature cells which have a finite life span, or to replenish areas of tissue lost or damaged by disease. They do so by proliferating, thereby giving rise to daughter cells which differentiate into mature cells. All stem cell progeny eventually undergo terminal differentiation, thus maintaining an equilibrium between cell renewal and loss. Tissue-determined stem cells are "predifferentiated" and are thus lacking in functional morphologic characteristics which normally define mature cells from that tissue type. Epithelial stem cells also occupy a specific microanatomic site, the basal cell layer, thereby maintaining contact with the basal lamina.

Maturation of stem cell progeny occurs through orderly, stepwise acquisition of specific morphologic characteristics which define the nature of terminally differentiated somatic cells of that particular tissue. Terminal differentiation is associated with establishment and preservation of appropriate intercellular relationships and other cell orientation properties (e.g., polarity) acquired during the process of maturing. It is often associated with migration through layers of epithelium, along ducts, or into the blood stream (i.e., hematopoietic cells). Importantly, the progeny of stem cells are endowed with limited proliferative capacity such that, as functional differentiation proceeds, there is a progressive decline in the number of cell generations which may be spawned by an incompletely differentiated cell. Complete, or terminal, differentiation of stem cell progeny is both irreversible and associated with irrevocable loss of proliferative capacity. In fact, only predifferentiated stem cells retain the ability to proliferate indefinitely.

Finally, it is important to reiterate that tissues are actually "societies" of cells. Thus, growth and differentiation of various cell types must be interdependent. Through an elaborate system of growth factors, hormones, and molecular contacts between various cell types and surrounding extracellular matrix, tissues are able to coordinate the growth and maturation of functional parenchymal cells to the growth and differentiation of supportive connective-tissue, or stromal, elements.

According to the blocked ontogeny model of neoplasia, the inevitable consequence of arrested differentiation at the tissue level is that the ratio of precursor cells to terminally-differentiated cells, which is normally very low, will be increased. Due to greater proliferative capacity of precursor cells, an abnormal mass will result. Further, overall level of cell maturity, or differentiation, will be lessened. The cells within a given neoplasm tend to become arrested at a similar stage of maturation, defining the degree of maturation abnormality and thereby the histologic features, growth potential, and clinical behavior of the neoplasm. This is believed to reflect the stage of maturation at which neoplastic transformation of the initial cell occurred. Transformation of a nearly mature precursor cell would result in a "mild" differentiation abnormality. The neoplastic cells would resemble, for the most part, terminally differentiated normal counterparts and growth capacity would be limited, resulting in a benign neoplasm. Transformation of an immature, predifferentiated, precursor cell would result in a "severe" differentiation abnormality. Most neoplastic cells would resemble predifferentiated stem cell progeny and growth capacity would be substantially increased, resulting in a malignant neoplasm.

To quote Pierce again, neoplasia thus represents a "caricature" of physiologic tissue growth and differentiation. The abnormalities inherent to maturation delay account for the classical histologic alterations of malignant neoplasia outlined in Table 3, including the immature appearance of the cells and the presence of mitotic activity. The classically described disorganized growth patterns of neoplastic

Table 3. Histologic Features of Abnormal Differentiation in Malignant Neoplasia

Micro architecture	disorganized, non-physiologic, variable from area to area, secondary ulcer or infarction common
Host cells	proliferative response, inflammatory infiltrates
Functional maturation	resemblance to partially differentiated cells with increased nuclear to cytoplasmic ratio
Proliferation (mitoses)	increased, often morphologically abnormal (suggesting aneuploidy)
Nuclear morphology	nucleoli prominent, hyperchromasia, pleomorphism, enlarged (often greatly)
Invasion	present (always)
Metastasis	often (but not always)

tissues derive from the failure to establish appropriate intercellular relationships, which requires a normal ratio of immature to mature cells.

Abnormalities of differentiation, as we have defined them, do not account in a straightforward way for the following three features of malignant neoplasia listed in Table 3: invasion, metastasis, and nuclear alterations (enlargement, abnormal chromatin patterns, and pleomorphism). But invasion and metastasis are, in many respects, abnormalities of differentiation and we shall discuss them as such. Nuclear alterations are not necessarily a form of differentiation abnormality per se, but rather a consequence of structural genetic pathology which likely represents the underlying cause of arrested differentiation abnormalities (see later).

Dysplasia and Tumor Progression

All malignant neoplasm derived from epithelium (i.e., carcinomas) evolve from pre-invasive neoplasias, called dysplasias. These lesions exhibit classical abnormalities of growth and differentiation, corresponding to the blocked ontogeny theory of neoplasia. Dysplasias are thus clinically, as well as pathologically, relevant lesions. The prototype for this genre is cervical dysplasia, which is detected clinically with the Papanicolaou (Pap) smear.

Figure 1 compares dysplastic cervical mucosa to a normal sample. Note the morphologic abnormalities which are characteristic of dysplastic epithelial surfaces. First, the average nuclear-to-cytoplasmic ratio of the dysplastic cells is increased. This manifestation of cellular immaturity is a direct consequence of arrested differentiation. Normal squamous cells progressively acquire abundant cytoplasm, through the terminal differentiation cornification sequence, as they migrate from the basal layer to the superficial layers. Many dysplastic cells, though, resemble the predifferentiated cells in the suprabasal layer of the normal tissue. The dysplastic cells are also crowded, jumbled-appearing, and disorganized since they

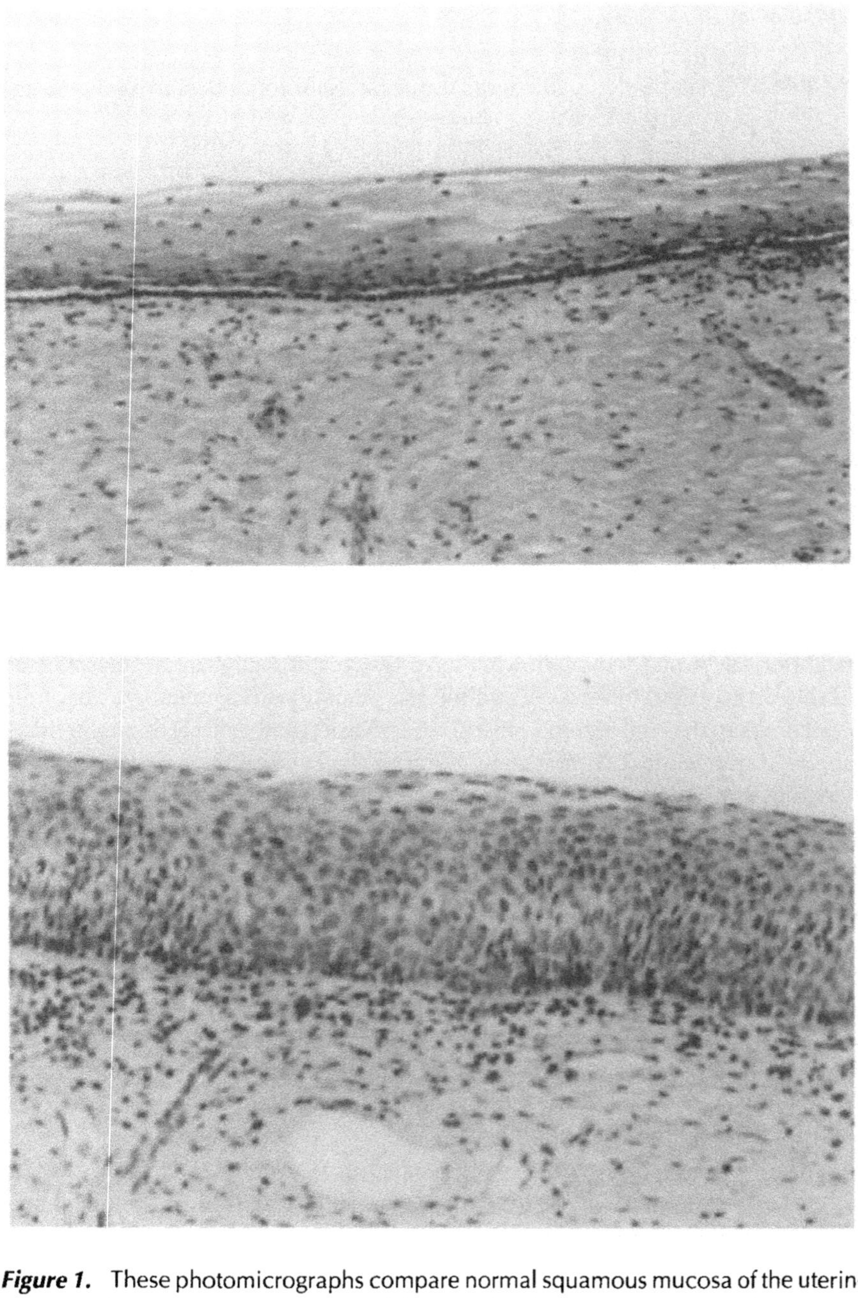

Figure 1. These photomicrographs compare normal squamous mucosa of the uterine cervix (top) to squamous mucosa exhibiting moderate to severe dysplasia (bottom). In the normal specimen, functional maturation of predifferentiated basal layer stem cells proceeds in an orderly manner from deep to superficial portions of the epithelium. Also, there is uniformity of nuclear size and shape. In contrast, the dysplastic mucosa exhibits significant maturation delay. Predifferentiated cells exhibiting increased nuclear-to-cytoplasmic ratio are present near the surface of the epithelium.

are unable to establish appropriate intracellular relationships or polarity due to incomplete differentiation. Mitotic figures are not only increased but present outside the normal proliferative zone of the suprabasal cell layer. Thus, proliferation and maturation are uncoupled. As a consequence of greater cell proliferation, there are more layers of cells, resulting in a thicker epithelium. Finally, nuclei are enlarged and contain dark, granular chromatin (hyperchromatism). These changes are caused by abnormal cellular DNA content and/or structure. Note the manner in which dysplasia represents a "caricature" of normal differentiation.

Dysplasias of the uterine cervix may vary in degree of differentiation abnormality. The most severe lesions, referred to as carcinoma-*in-situ*, show an epithelium comprised entirely of predifferentiated basaloid cells with virtually no maturation. The mildest lesions are characterized by subtle maturation delay within cells of the lower (i.e., more basal) third of the epithelium. Importantly, these morphologic distinctions have great clinical relevance. Severe dysplasias are clinically unstable and often progress to invasive carcinomas, whereas mild dysplasias are generally indolent and may even regress. It should also be emphasized that, in general, the severity of differentiation abnormality closely parallels the degree of nuclear enlargement and chromatin staining intensity. This is presumably due to differences in the extent of structural genetic pathology among these lesions.

In other tumor systems dysplasias are sometimes designated by misleading terms, including colorectal adenoma, and endometrial hyperplasia. These unfortunate historical artifacts do not negate the tremendous clinical relevance of dysplastic lesions, the detection and treatment of which represent a mainstay of cancer prevention. There are even some epithelial tumor systems in which all pre-invasive neoplasias are referred to as carcinomas-*in-situ*. This is the case, for example, in the breast. Pathologists have described morphologically-distinctive forms of mammary carcinoma *in situ* based on divergent features of differentiation. So-called low grade breast carcinomas *in situ*, which retain features of ductal architectural differentiation, have an indolent clinical behavior. In contrast, high grade ductal carcinomas *in situ* lack architectural differentiation and are characterized by more frequent and rapid progression to invasion.

Dysplasia differs from hyperplasia since, in the latter, proliferation is increased (as it is in dysplasia) but differentiation is normal (i.e., complete). Dysplasia also differs from metaplasia, which is defined as complete differentiation which is inappropriate for an anatomic site. Keep in mind, however, that dysplasias may themselves be metaplastic. This reflects the multipotent differentiation capacity of tissue-determined stem cells. Dysplasias may also evolve from hyperplastic lesions. In both hyperplasia and metaplasia nuclear morphology and staining characteristics are normal since there is theoretically no structural genetic pathology.

Invasive Growth and Metastasis

As a feature of malignancy, invasive growth has traditionally been taught from a surgical point of view as, for example, the proclivity of some lung carcinomas to invade the mediastinum, or the tendency of breast cancers to invade skin and chest wall. It is highly uncharacteristic for a benign neoplasm to breach functional and histologic tissue barriers. Another surgical concept of invasive growth refers to the tendency of many malignancies to have a permeative, crab-grasslike growth pattern. Thus, the advancing front of malignant neoplasms may be difficult to define precisely by palpation or visual examination. Benign neoplasms, in contrast, usually have a circumscribed interface with the host often resulting in a cleavage plane which may be blunt dissected. The term encapsulation is sometimes applied to benign neoplasms due to their circumscribed character. However, these neoplasms do not make a connective tissue capsule in the sense of, say, the liver. The fibrous rim which surrounds some benign neoplasms represents residual supportive tissues left behind by atrophy of parenchymal tissues, presumably due to interruption of vascular supply from pressure created by the mass. The concept of invasive growth is really of greatest relevance to epithelial neoplasias. This is because many sarcomas are deceptively well-circumscribed and also because leukemias and some lymphomas are circulatory (i.e., liquid) malignancies.

The permeative growth character of carcinomas is most readily appreciated at the microscopic level where it also becomes apparent that invasion may be viewed as a caricature of tissue repair and regeneration. This may be illustrated by briefly returning to basic concepts of tissue architecture and growth homeostasis. Epithelium and stromal extracellular matrix are always separated by a continuous layer of basal lamina, which is comprised principally of Type IV collagen, laminin, and various proteoglycans. Basal lamina determines the shape and configuration of epithelial units (i.e., glands and ducts) by providing a scaffolding for attachment. Stem cells are always attached to basal lamina since this structure transduces growth and differentiation signals, enabling coordination of epithelial and stromal differentiation. Following trauma or inflammation, tissues restore normal microarchitecture with a coordinated proliferative response involving both epithelial and stromal cells. To heal an ulcer, for example, proliferation of epithelial stem cells is required to resurface denuded areas. Proliferation of stromal stem cells, however, is also required for revascularization, remodeling of extracellular matrix, and re-establishment of the basal lamina. Cell division is combined with migratory activity of stem cell progeny to repopulate previously damaged areas. Following reconstitution of normal microanatomy, the repair process ceases through terminal differentiation. In a sense, then, the architectural and functional homeostasis between epithelium and supportive tissues, both under normal conditions and during states of inflammation or repair, represents a form of tissue differentiation as we have previously defined it.

There are numerous mechanistic and histologic analogies between tissue repair and tissue invasion by neoplasms. To initiate invasive growth, neoplastic epithelial cells traverse defects in the basal lamina and migrate into underlying stroma (see Figure 2). This event is accompanied by simultaneous growth of new blood vessels (angiogenesis/neovascularization) (Folkman, 1985) and remodeling/resynthesis of extracellular matrix by endogenous fibroblasts. Morphologically the invading neoplastic epithelial cells exhibit a primitive, or predifferentiated, phenotype. At some point, however, invading epithelial cells are believed to cease migration and proliferate, thereby forming a nest of cells. Sometimes these nests re-constitute a basal lamina and undergo architectural differentiation, such as gland formation, in a manner akin to tissue repair (Figure 2). This cycle of invasion followed by differentiation then repeats itself with other cells, either from the dysplastic surface or from the invasive nest, until a mass forms.

Invading epithelial populations require vascular supply as well as an extracellular matrix which favors their state of differentiation (Iozzo, 1995). In order to sustain invasive growth, a carcinoma must be capable of indefinitely maintaining a wound repairlike reaction from the surrounding host supportive tissues. Tissue sections from the invasive front of a carcinoma, accordingly, will show proliferating fibroblasts, new blood vessels, and an immature extracellular matrix resembling granulation tissue (see Figure 2). In a sense, then, invasive neoplasms may be viewed from a functional standpoint as persistent wound-like lesions.

Acquisition of an invasive phenotype is a seminal event in the natural history of an epithelial neoplasm and, for this type of neoplasia, reproducibly distinguishes benign from malignant. It may seem that by this criterion, all connective tissue-derived neoplasms, even benign ones, are invasive since by definition they permeate homologous supportive tissues from inception. However, connective tissues represent the physiologic environment for stromal cells, and thus growth within them is not necessarily a manifestation of invasion for neoplasms of stromal cell derivation. Thus, the histologic distinction between benign and malignant for stromal cell neoplasms is based on differentiation features intrinsic to the neoplastic cells per se as opposed to morphologic features of invasion.

One important ramification of evolution to the invasive phenotype is the potential for eventual growth into neighboring tissues or body cavities. Invasion also affords tumor cells the opportunity to grow through the walls of blood and lymphatic channels which course through supportive tissues (Figure 3). This event is a necessary, but not sufficient, condition for metastasis. Metastasis is defined as noncontiguous spread of a neoplasm, via vasculature, to a secondary organ with formation of a colony. The term colony was chosen to imply a self-sustaining mass, capable of eliciting vascular and stromal support in a secondary site. Thus, isolated neoplastic cells observed outside of their primary organ do not necessarily represent metastases. Malignant cells have been shown to exit their primary organ in large numbers via the efferent vasculature, without always having the phenotypic prop-

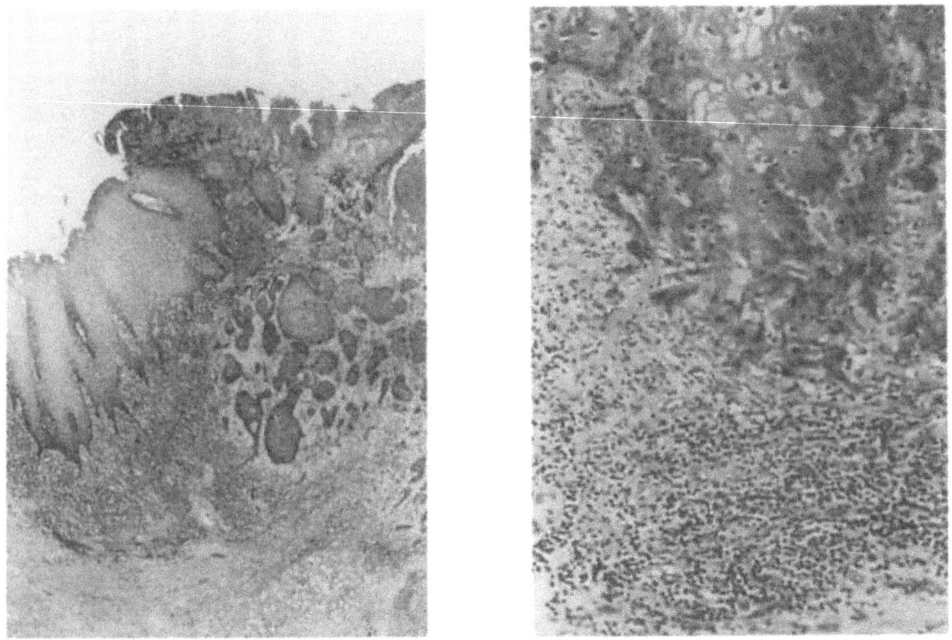

Figure 2. These photomicrographs illustrate an invasive squamous cell carcinoma at low magnification (right panel). It is possible to distinguish the invasive component of the neoplasm, characterized by infiltrating nests of neoplastic cells, from residual benign epithelium. At high magnification (left panel) small groups of neoplastic cells appear to detach from the overlying mass of neoplastic cells. Note that this is accompanied by a host response, consisting of an edematous extracellular matrix as well as accumulation of inflammatory cells.

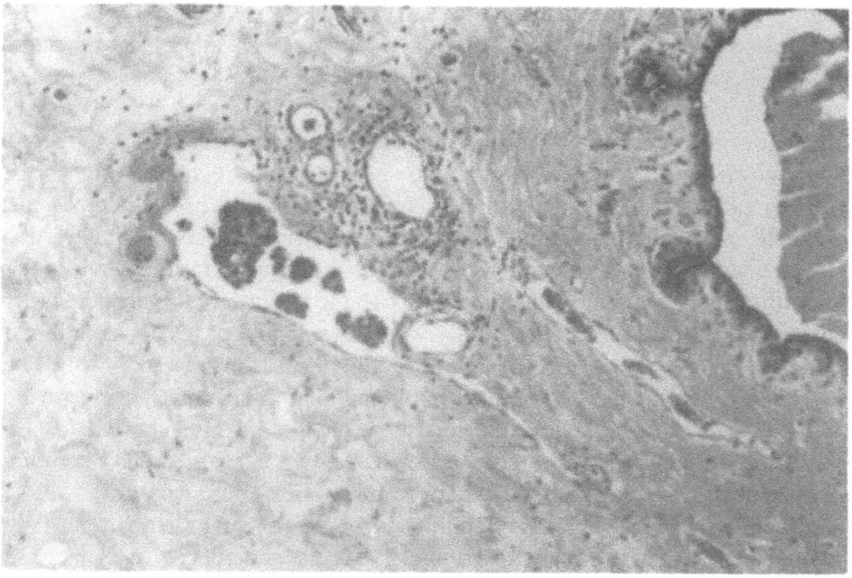

Figure 3. This photograph demonstrates intravascular tumor cell emboli in a breast carcinoma. A residual benign normal duct is present as well (upper right). Note that the small vascular channel which has been invaded exhibits fibrin deposition beneath the endothelium, possibly a consequent of intravasation.

erties to form metastatic colonies. Thus, circulatory microemboli may account for the presence of tumor cells in secondary organs.

Metastatic phenotype is considered a composite of many specific functional properties (Mart and Saini, 1992), including: (a) the ability to invade through vascular basal lamina from the outside, or intravasation; (b) avoidance of destruction while in circulation; (c) adherence to vascular endothelium at a secondary site; (d) invasion through vascular basal lamina from the inside, or extravasation; and (e) ability to migrate, proliferate, and establish support from host stromal cells at secondary sites.

Despite the clear analogies to invasive phenotype, the metastatic phenotype represents an extension of growth and differentiation derangements (Fidler et al., 1978). In particular, the growth of normal somatic cells is restricted to anatomically defined embryonic differentiation fields, whereas metastatic growth does not comply with this restriction. Thus, colonization of tissues outside of the home organ is a manifestation of profoundly inappropriate differentiation. Since vascular dissemination of epithelium has no physiologic basis outside of embryogenesis, it is perhaps more difficult to view metastasis as a disorder of somatic differentiation.

It is hardly a completely novel function since lymphocytes routinely circulate and emigrate through vascular walls into various tissues.

Vascular dissemination of neoplasms may occur via lymphatics, to lymph nodes, or via the blood circulation to virtually any tissue in the body. Lymphatic metastases involve regional lymph nodes; that is, the lymph nodes which receive direct lymphatic drainage from the primary organ. They often develop in carcinomas but, for unknown reasons, rarely in sarcomas. Metastasis to regional lymph nodes generally occurs in a sequential and thereby anatomically predictable manner. Lymphatic and hematogenous metastasis are not equivalent from either clinical or biological standpoints. Most carcinomas acquire the capacity for lymphatic metastasis before hematogenous metastasis. Although the reasons for this are not known, it may reflect easier access to the thinner-walled lymphatic spaces, fewer traumatic forces inside of lymphatics compared to blood vessels, or the lack of a need to extravasate when the cells reach the lymph node into which afferent lymphatics drain. Carcinomas having lymphatic metastases are more likely to have developed the capacity for hematogenous metastases. Hence, lymphatic metastasis is a clinically useful adverse prognostic feature in common forms of neoplasia.

Unlike metastasis to regional lymph nodes, the anatomical distribution of systemic metastasis is only partially explained by vascular anatomy. Thus, metastatic colonies tend to form in organs through which the venous drainage of the primary organ flows. This is the case, for example, in metastases of colorectal carcinomas, which nearly always occur in the liver due to portal blood drainage. It has long been observed, however, that the frequency of lung carcinoma metastases to the adrenal glands is far out of proportion to the amount of systemic blood flow to these organs. In fact, there are many examples of this so-called organ preference of metastasis. Most experts have invoked the "seed and soil" theory initially advanced by Paget to explain such findings (Morikawa et al., 1988; Rusciano and Burger, 1992). According to this theory, the colonization of heterologous tissues by neoplastic cells is determined, at least in part, by factors intrinsic to the neoplasm, the secondary tissue, or both. These factors are largely unknown but are presumed to reflect specific interactions between neoplastic and host-derived cells. For example, the molecular and cellular structure of stromal tissues in an organ may be particularly receptive to colonization by the cells of a given neoplasm. There is, after all, a great divergence in extracellular matrix composition and endothelial phenotype among various tissues in the body. Organ preference for metastasis may also reflect the phenotypic heterogeneity of different neoplasms from the same primary organ. In breast carcinomas, for example, neoplasms which express estrogen receptors preferentially metastasize to bones whereas estrogen receptor negative tumors preferentially colonize soft tissues.

The behavior of clinical human neoplasms suggests that evolution to the metastatic phenotype follows development of invasive phenotype. In general, metastatic capacity is a function of tumor volume which, in large part, reflects the number of population doublings which have evolved from the initial transformed cell, and therefore the number of opportunities for the acquisition of new phenotypic properties as a consequence of alterations in gene structure or regulation. Nevertheless, some invasive carcinomas acquire metastatic phenotype when they are very small and others only after reaching a substantially larger size. Unfortunately, it is often clinically impossible to establish whether a neoplasm has acquired the ability to colonize distant organs. Patients with no clinically detectable metastatic lesions when they present with, for example, colon or breast cancer, may in fact harbor microscopic, clinically inapparent hematogenous metastases. Such lesions, when they reach a detectable size, become disease recurrences. Inability to establish the presence of metastatic phenotype, apart from identifying nodal metastases, represents a major obstacle to optimizing therapy for cancer patients.

Many factors complicate experimental analysis of metastatic phenotype, not the least of which is the inherent inefficiency of the process. Even among clonal cell populations that have been highly selected for metastatic ability in animal tumor models, only 1–2% of injected cells give rise to colonies. Much of this inefficiency is thought to be accounted for by the relative hostility of the vascular lumen, where circulating neoplastic cells are exposed to trauma as well as elements of the host immune system. There is considerable evidence that metastasis is favored by interactions between circulating tumor cells and platelets. Although it is unclear how platelets facilitate metastasis, it may be speculated that they provide a protective barrier, or possibly augment extravasation by initiating thrombosis. Finally, platelet-derived growth factors (see later) may help initiate tumor-host interactions at the secondary site.

In addition to nonselective factors that contribute to metastatic inefficiency, presumably in a random way, the pathophysiology of metastasis is made even more complex by the existence within primary tumors of clonal neoplastic subpopulations, which differ among themselves in metastatic ability. In addition to genetic heterogeneity, there is some evidence to suggest that this phenotypic diversity may reflect modulation of metastatic phenotype through geographical, nutritional, or other (unknown) epigenetic influences on the metabolism of select cell populations. In this respect, the metastatic phenotype may not be expressed in a stable fashion. Functional heterogeneity of primary tumor subpopulations, interestingly, may also be an important factor in the generation of the metastatic phenotype through productive interactions between groups of cells having differing, but complimentary, phenotypic attributes (Heppner, 1989, 1991).

Clinical Aspects of Neoplastic Growth

Systemic (hematogenous) dissemination is currently the most common cause of death in patients with common malignancies. Metastatic disease, however, does not usually cause the presenting symptoms or signs of the neoplasm and by no means does it fully account for the medical problems caused by the growth and spread of neoplasms. The clinical manifestations and consequences of neoplastic growth are virtually limitless, but they can be divided into four broad categories as follows: (a) impairment of organ function at primary or metastatic sites; (b) nonphysiologic secretory activity by neoplastic populations; (c) secondary pathologic changes in neoplastic tissues; and (d) systemic metabolic derangements.

Neoplasms may impair organ function by replacing normal tissues with neoplastic tissues or by physically impeding the functions of residual normal tissues. An important example of impaired primary organ function due to replacement by neoplasm is the crowding out of normal hematopoietic precursors in bone marrow by leukemic cells, which results in peripheral pancytopenia that causes bleeding tendency and immunosuppression. Examples of dysfunction due to physical impingement by neoplasm are obstruction of a bronchus by lung carcinoma, causing recurrent pneumonias, or obstruction of the common bile duct in pancreatic carcinoma, causing jaundice.

Nonphysiologic secretory activity by the neoplastic population often results in recognizable clinical syndromes. An obvious example is inappropriate hormone secretion by endocrine neoplasms. Pheochromocytoma, a neoplasm of adrenal medulla, results in hypertension due to dysregulated elaboration and secretion of adrenergic substances. Similarly, pituitary adenoma may result in acromegaly due to growth hormone secretion. Secretion of ectopic hormones or hormonelike substances by nonendocrine neoplasias is called a paraneoplastic syndrome. Examples include hypercalcemia due to elaboration of a parathyroid hormonelike substance. This is not uncommon in lung carcinoma. A second paraneoplastic syndrome is caused by erythropoietin secretion from renal cell carcinoma, resulting in polycythemia. Clinical detection of nonfunctioning substances secreted by tumor cells is often useful for diagnosis/monitoring of cancer patients, since these may be assayed in serum. Examples of these so-called tumor markers include human chorionic gonadotropin (germ cell neoplasia), carcinoembryonic antigen (colon carcinoma), prostate specific antigen (prostate carcinoma), and α feto protein (germ cell neoplasia, hepatocellular carcinoma).

Secondary pathologic changes in neoplastic tissue such as ulceration, bleeding, or infarction frequently result in clinical symptoms or signs which implicate the organ system harboring the tumor. Examples include expectoration of blood (hemoptysis) in lung carcinoma or the passage of bloody urine (hematuria) in bladder/renal cell carcinoma. In most cases, such secondary pathologic changes are a reflection of the inability to maintain tissue integrity due to derangements of functional differentiation or interaction with host-derived supportive tissues.

Finally, patients with disseminated (widely metastatic) cancer often develop systemic abnormalities. The terminal stages of neoplasia, for example, is often characterized by a wasting syndrome, called cachexia.

Stage (Extent of Disease) and Grade (Aggressiveness of Disease)

Stage is a clinicopathologic concept, referring to the anatomical extent of neoplastic disease spread as determined by a combination of clinical, radiographic, and pathologic data. Criteria for Staging are highly specific for individual primary sites. They reflect anatomical considerations unique to various organ systems as well as the divergent biology of neoplasms which arise in different locations. Staging algorithms may thus be complex. In recent years, oncologists have developed an approach to Staging (the TNM system) which combines features of the primary tumor (T category) with status of regional lymph nodes (N category) and presence of hematogenous metastases (M category).

The T category describes features of local tumor extent such as size (diameter) of the mass or its depth of penetration into underlying host tissues. Size measurements are typically employed in solid organs, such as the breast, where neoplasms grow as a lump or an expansile mass. In contrast, depth assessments are a component of staging for neoplasms which arise from an epithelial surface and burrow into underlying structures. Depth assessments may be a linear measurement, as in malignant melanoma, or a histologic description of how deeply the tumor has penetrated. Invasion into the muscular coat, for example, is a staging parameter in colorectal carcinoma (see Figure 4). The T category also incorporates the presence of contiguous spread by neoplasm into neighboring structures. Examples of Stage-related local growth characteristics are the presence of pleural spread in lung cancers, the presence of skin involvement in breast cancers, or spread to the parametrial soft tissues by cervical cancers.

Design of N category staging algorithms for some malignancies reflects the clinical observation that prognosis is a function of the nodal tumor burden, not merely the presence of nodal disease. For example, Staging groups for breast carcinoma are defined by the number of axillary lymph nodes which harbor metastases. Patients with a solitary axillary node metastasis have a prognosis marginally worse than node negative patients, whereas those having 10 or more involved lymph nodes have an outcome approaching that of patients with hematogenous metastases. The N component of Stage refers specifically to regional node groups, or those within the normal lymphatic drainage of the organ. Thus, a breast carcinoma with supraclavicular node metastases is considered to have distant (as opposed to regional) metastases (i.e., since this lymph node does not receive direct lymphatic drainage from breast tissues).

Based on various TNM criteria particular to a given organ, the Stage is designated by Roman numerals which, in general, may be summarized as follows: I, small/superficial mass limited to organ; II, large tumor or some regional nodes positive; III, extension to adjacent structures/organs, or many regional nodes positive; and IV, hematogenous metastases.

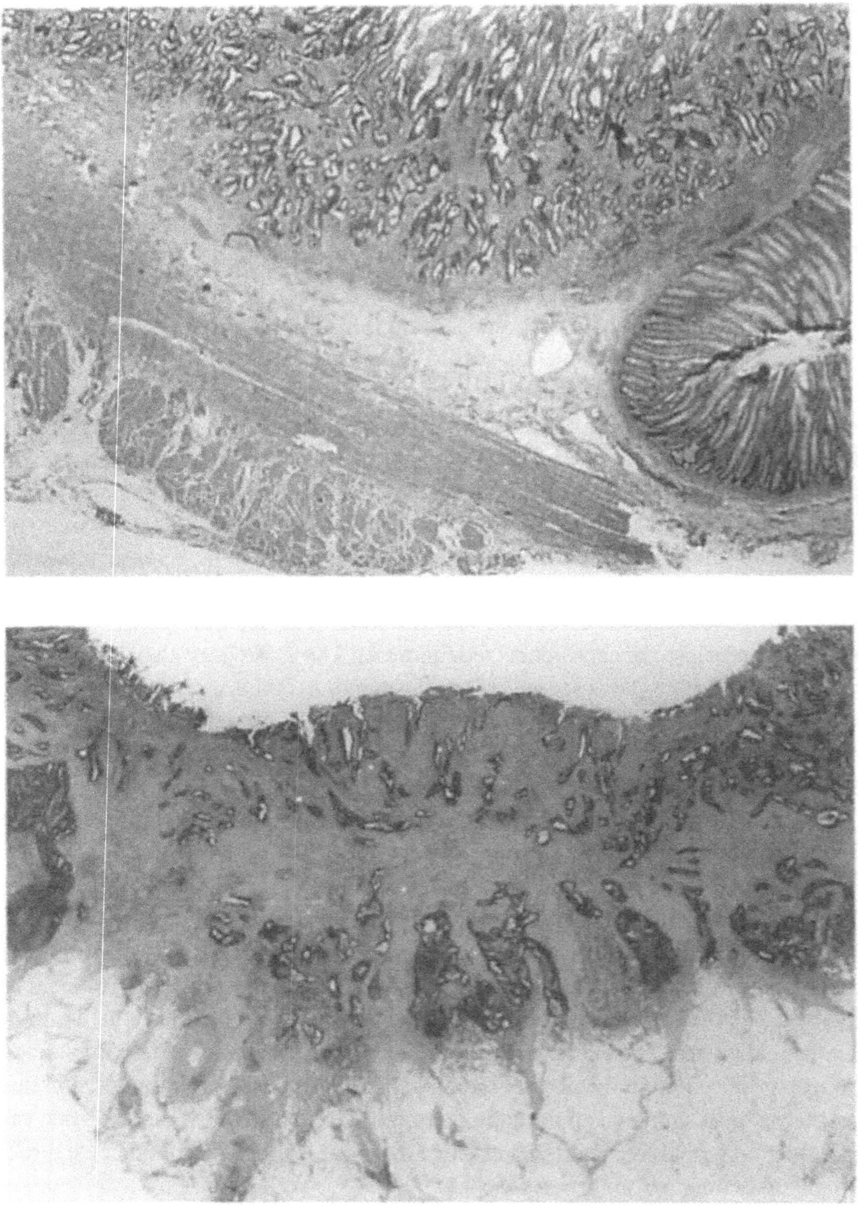

Figure 4. These low magnification photomicrographs exhibit different colorectal adenocarcinomas, at different stages. The neoplasm in the top panel illustrates invasion into the submucosal layer, but not into the muscularis. The neoplasm in the bottom panel, in contrast, shows complete obliteration of the muscularis propria by desmoplastic tumor stroma, with extension of neoplastic glands into pericolonic fatty tissue. Note the presence of surface ulceration.

Most human malignancies progress sequentially through these Stage groups. In other words, a colon carcinoma with liver metastases will usually be transmural and have pericolonic nodal metastases. But this is clearly not always true, since some node negative colorectal cancer patients relapse with distant (i.e., hematogenous) metastases.

Grading is an attempt to measure the aggressiveness of a neoplasm by histologic evaluation of differentiation, as previously described. As with Staging, morphologic criteria for Grading are specific to the particular type of neoplasm. Basically, however, the aggressiveness of a malignancy is determined by combination of: (a) mitotic index (which in large part determines growth rate); (b) neoplastic cell microarchitecture (i.e., presence of infiltrating single cells versus nests or glandlike structures) which reflects invasiveness; and (c) nuclear morphology (enlargement, chromatin density and pleomorphism) which is a measure of structural genetic pathology.

In a well-differentiated (i.e., low grade or less aggressive) carcinoma, a majority of the neoplastic cells will be forming glands or nests (i.e., differentiating). Accordingly, a relatively small proportion of neoplastic cells will be infiltrating host stroma as individual cells. In a poorly-differentiated (i.e., high grade, or very aggressive) carcinoma, most neoplastic cells will be invading. Consequently there will be fewer cohesive nests of neoplastic epithelium and more infiltrating single cells, as well as less evidence of cytoplasmic maturation (see Figure 5).

Mitotic rate and degree of invasiveness are strongly correlated since, according to the blocked ontogeny model, proliferation and maturation have a reciprocal relationship. Similarly, Stage and Grade are also correlated since invasiveness is one of the factors which contribute to metastatic phenotype. Poorly-differentiated carcinomas display angiolymphatic tumor cell emboli in tissue sections, and hence lymphatic metastases, more frequently than do well-differentiated carcinomas. Finally, growth rate and invasive capacity are correlated with degree of nuclear enlargement and chromatin staining alterations since features of abnormal differentiation are believed to be acquired through abnormalities of DNA structure which induce abnormal gene expression (see later).

The clinical relevance of Grade is variable, and depends on the tumor type. Grading is sometimes limited by subjectivity or intratumoral heterogeneity. It is most useful (in clinical treatment decisions) for sarcomas and lymphomas. Outcome of prostatic carcinomas, moreover, is exquisitely Grade-dependent. For most common tumors, the overall survival difference between well versus poorly differentiated neoplasms is 30–50%.

Natural History of Malignant Neoplasia

Most neoplasms have a reasonably constant, exponential growth rate throughout the majority of their life span. This is an important biological feature in neoplasia because it predicts that the length of time required for a single neoplastic cell to reach a mass of 1 cm in diameter, which is 30 population doublings, is three times

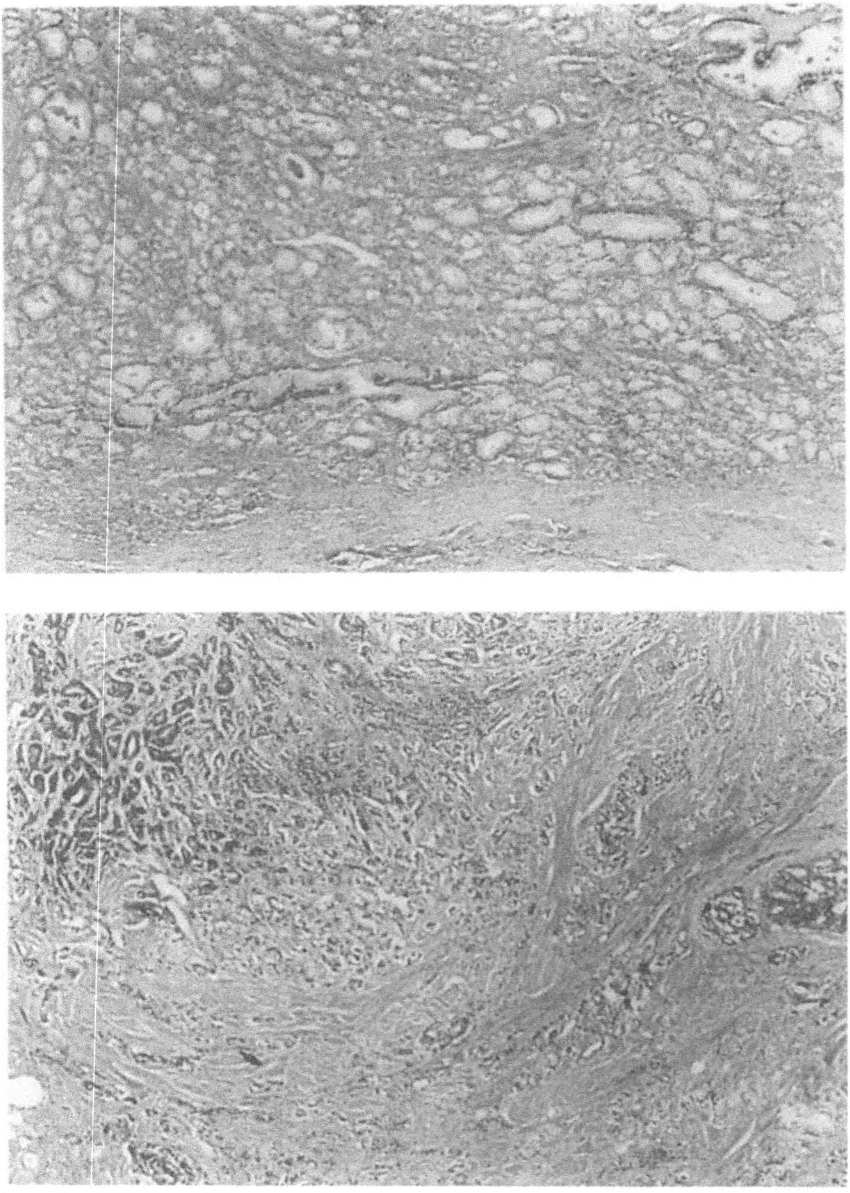

Figure 5. This figure illustrates the difference between a well-differentiated prostatic adenocarcinoma (top) and a moderate-to-poorly differentiated adenocarcinoma of the prostate (bottom). The neoplastic cells in the well-differentiated neoplasm demonstrate formation of abundant glandlike structures and the neoplasm has a relatively well-circumscribed (pushing) interface with surrounding host tissue. The more poorly differentiated neoplasm, in contrast, exhibits glandular differentiation only in some areas (top right) and has a more ragged, infiltrative growth pattern.

longer than the length of time required to grow from 1 cm to a near lethal 1 kg mass, which requires only an additional 10 population doublings. Thus, about 75% of the life span of (untreated) neoplasms occurs prior to development of symptoms, since small tumors are generally clinically silent. It is especially sobering to note that, on average, carcinomas acquire a metastatic phenotype between the eighth and 12th population doubling. The actual amount of time between the inception of a neoplasm and its detection depends on the doubling time (DT) of the neoplasm, which is a function of tumor type, primary location, and Grade (Table 4). For most common neoplasms, the interval between inception and clinical presentation is 2–5 years.

There is a striking correlation between the physiologic turnover rates of normal cells and the DTs of neoplasms derived from cell populations of that type. This supports the blocked ontogeny model of neoplasia since, to a great extent, cell turnover rate is tissue determined. An additional bit of evidence favoring the maturation arrest theory is that cell types which do not cycle in postembryonic tissue, such as cerebral neurons or cardiac myocytes, never give rise to neoplasms in adults. These cell types lack a postembryonic stem cell population.

Finally, although the natural history of a neoplasm may be viewed as a series of volume doublings akin to bacterial colony growth, we know it is far different from that. Recall that epithelial malignancies have a defined pre-invasive growth phase, an invasive (premetastatic) growth period and, often, a metastatic phase of growth. Thus, neoplasms evolve phenotypically, a phenomenon called tumor progression. As the blocked ontogeny model makes no explicit provision for phenotypic evolution, this theory may be difficult to reconcile with tumor progression. However, Pierce et al. have shown that the phenotype of the neoplastic population may evolve in response to environmental changes. Indeed, these authors have been able

Table 4. Neoplasms With Varying rates of Proliferation

Neoplasms with rapid proliferation (short DT)
acute leukemias/high grade lymphomas
childhood blastomas
germ cell malignancies
Neoplasms with moderate proliferation (intermediate DT)
colon carcinoma
lung carcinoma
breast carcinoma (highly variable, age dependent)
high grade sarcoma
Neoplasms with slow proliferation (long DT)
prostate carcinoma
thyroid carcinoma
salivary gland carcinomas
low grade lymphoma/chronic leukemia

to abrogate malignant phenotype of neoplastic cells by injecting them into developing blastocysts or primordial tissues. In other words, neoplasia represents an epigenetic process, at least in part and, under the proper environmental influences, may be reversible.

Others have emphasized genetic instability as the pathogenetic mechanism of tumor progression. According to this paradigm, structural lesions of cellular DNA acquired during growth of the neoplasm results in the repeated selection and expansion of clones which exhibit progressively more deranged properties of proliferation and differentiation. Deeper scrutiny of tumor progression, as well as furthering our understanding of abnormal differentiation and etiology of neoplasia take us into the realm of molecular genetics, which will be the focus of the next section of this chapter.

MOLECULAR AND GENETIC ASPECTS OF HUMAN NEOPLASIA

Molecular Pathogenesis of Invasion and Metastasis

Functionally, invasive and metastatic phenotypes involve aberrant expression of molecular families which mediate tissue growth and structure, including growth factors, cell adhesion molecules, and matrix proteases. The list of species in these molecular families is rapidly growing as investigators gain more detailed knowledge of normal tissue homeostasis and regeneration. The growth factors, adhesion molecules, and proteases implicated in neoplastic progression to date are the same species which mediate physiologic processes. This is in keeping with the notion that neoplasia represents a caricature of differentiation, as opposed to an acquisition of novel phenotypic traits.

As stated above, invasive epithelial neoplasia is histologically characterized by stromal cell proliferation and synthetic alterations of extracellular matrix. These alterations are induced largely via deranged elaboration and secretion of various growth factors which, under normal circumstances, are carefully regulated mediators of physiologic tissue cycling or tissue regeneration following inflammation (Ethier, 1995). There are at least six structurally distinct growth factor classes or families. These polypeptides interact with plasma membrane growth factor receptor molecules, through which the growth factor signal is received and transduced. A given growth factor receptor is generally capable of interacting with all of the growth factors in a particular family. The cellular effects of growth factors are pleiotropic and dependent not only on the differentiation status of the target cell but also on the presence of other growth factors. Growth factors, unlike hormones, are believed to mediate highly localized cellular interactions as inferred by their brief half lives, presence in extracellular matrices and, geographically limited range of activity (Pusztai et al., 1993).

In general, growth factor elaboration is a property inherent to the proliferative phenotype. Thus, growth factors are initially produced by the neoplastic populations. However, growth factor production is eventually induced in host cells as a secondary effect of stimulation by neoplasm-derived growth factors. Alteration of host support cell phenotype by neoplastic cell-derived growth factors is called a paracrine interaction. Neoplastic cells may also support their own growth by inappropriate elaboration of growth factors; this is called autocrine stimulation. It is important to note that the proliferation-differentiation status in normal tissues is driven by an equilibrium between completing stimulatory and inhibitory growth factor signals. In neoplasia, maturation arrest may be seen to shift the growth factor milieu to favor a dynamic, as opposed to static, condition of tissue activity.

Ingrowth of new vasculature (angiogenesis) is a prototypical host-tumor interaction. Tumors cannot sustain growth *in vivo* beyond a diameter of 1–2 mm without inducing neovascularization (Folkman, 1985). Angiogenesis is mediated by growth factors, as classically demonstrated by the ability of cell-free extracts from neoplastic cells to stimulate vascular sprouts in experimental bioassays (Blood and Zetter, 1990). It is possible, moreover, to modulate angiogenesis by manipulating or blocking the action of certain growth factors. Variability in the physiologic growth factor milieu between different organ systems may partially account for organ specific preference of metastasis, as noted earlier.

All normal cells, particularly epithelia, have intermolecular connections to one another or to the extracellular matrix. Intercellular attachments are essential for physiologic activities, such as maintaining the physical integrity, or barrier functioning of mucosal surfaces. Differential adhesion between various cell types in large part determines the organization of heterogeneous cell societies. Thus, cell adhesion selectivity is a general property of cells which is fundamental to morphogenesis and tissue structure. In human cells, there are at least two families of membrane-associated adhesion molecules, called integrins and cadherins, which are differentially expressed as a function of differentiation status (Albelda, 1993). Hormone-induced cyclic changes, for example, are associated with modulation in the level or type of integrin or cadherin molecule. Cell adhesion molecules are bound not only to other cells or extracellular matrix, but also to the cytoskeleton. They are intimately linked, therefore, with cell shape and metabolism (Daneker et al., 1989).

In order to invade, neoplastic epithelium must be able to detach from other neoplastic cells and form attachments with extracellular matrix components (Lester and McCarthy, 1992). Accordingly, membrane spanning adhesion molecules that mediate attachment to homologous cells are expressed abnormally in neoplasia. Several studies have demonstrated a causal relationship between invasive or metastatic phenotype and downregulated cadherin expression. Further, membrane receptors to extracellular matrix (ECM) constituents, such as laminin, may be increased (Ruoslahti et al., 1985). Synthetic ECM alterations are believed to facilitate the ability of neoplastic epithelial cells to both detach from homologous cells and establish a foothold during invasion. The stroma of malignant neoplasms often contains fetal glycoproteins, such as tenascin, or altered proteoglycans. Exit from the vasculature during metastasis (extravasation) also requires physical interactions between neoplastic cells and constituents of endothelium, vascular basal lamina, or tissues at the secondary site. This function also is believed to require pathologic expression of cell adhesion molecules.

Pathologists routinely exploit the abnormal adhesion properties of neoplastic cells in diagnostic testing procedures. In the uterine cervix, for example, the region of dysplasia is often quite small in area compared to the entire cervical mucosal surface. One reason why dysplastic cells routinely appear on Pap smears, however, is because they are less cohesive and readily dislodged or abraded by scraping. The same principle applies to cytologic detection of lung cancer cells in sputum or metastatic carcinoma cells in serous effusions. In general, adhesion correlates directly with differentiation. Poorly differentiated carcinomas are highly dyshesive and thereby tend to grow in small nests which fall apart readily. Well differentiated carcinomas, in contrast, have more cohesive populations which form larger masses of cells.

Basal lamina and fibrous extracellular matrix are highly insoluble, and thereby form a barrier which impedes passive physical crossing, particularly by epithelium. Extracellular matrices, as previously noted, actually are biochemically complex and diverse. They are comprised of a macromolecular complex which, depending on the organ, contains various combinations of specific collagen isotypes (there are at least five) for structural integrity, proteoglycans (there are several) for hydration, and regulatory functions and glycoproteins (fibronectin and laminin) to mediate cell adhesion. Basal lamina, as previously noted, is a specialized form of ECM which mediates epithelial structure and differentiation.

We have already cited inappropriate motility and cell adhesion as properties of the invasive phenotype. In order to invade, neoplasms are also believed to enzymatically degrade native ECM with proteolytic enzymes (Furcht et al., 1994). This was inferred decades ago from the blood clot or tissue lysing activity observed in cell-free extracts from malignant neoplasms. Proteases are physiologically important as mediators of tissue remodeling such as involution, trophoblast implantation, or wound repair. Neoplasms contain elevated levels of multiple proteolytic enzymes, including collagenases, cathepsins, elastases, and plasminogen activators. The levels of such proteases correlate with invasive or metastatic capacity in both experimental and clinical tumor systems.

Tumor associated proteases each catalyze highly site-selective molecular cleavages of specific ECM constituents. Due to the molecular complexity of ECM, tumor cell invasion is thus speculated to require the combined activity of multiple proteases. Some proteases are synthesized and secreted by the invading populations themselves. Fibroblasts and tumor infiltrating inflammatory cells are also induced to elaborate and secrete proteases within neoplasms, possibly via paracrine interactions. Thus, neoplastic cells likely cooperate intimately with host-derived populations in ECM degradation. Proteases are capable of activating one another *in vitro*. This has prompted some to speculate that protease activity of sufficient molecular diversity to alter ECM may require a proteolytic cascade mechanism analogous to the blood coagulation system.

ECM proteolysis has consequences apart from providing space for invading neoplastic populations. Molecular level alterations of various ECM constituents induce proliferation or phenotypic changes in various cells which are in contact with ECM via transmembrane linkages. Also, some growth factors are embedded in ECM and liberated by proteolysis. Further, cell migration is stimulated, or directed, by chemotactic signals from proteolytically altered ECM constituents or from circulation-derived substances which become proteolytically activated. These considerations do not apply exclusively to the neoplastic population per se. Proliferation, migration and differentiation of endothelium during angiogenesis are also mediated by proteolysis and ECM molecular architecture (Grant et al., 1989; Hlatky et al., 1994). Therefore, growth factor milieu, adhesion, and proteolysis are complimentary processes in neoplastic invasion and metastasis.

Cancer Etiology Determinant of Neoplastic Pathobiology

We have thus far described neoplasia in terms of altered cellular structure and function. We began, though, by asserting that neoplasia was caused by abnormalities of genetic structure and function. Thus, the etiology of neoplasia can really only be discussed in terms of genetic lesions and their consequences. These genetic lesions are very heterogeneous, thus accounting for the tremendous morphologic and clinical diversity of neoplasia. Some of this genetic heterogeneity reflects the potential variety of genes which may be structurally altered. As we, however, shall see, neoplasia results from lesions which involve a limited number of specific genes which control cell growth and differentiation. The heterogeneity of genetic pathology in neoplasia is also a consequence of the variety of structural aberrations to DNA which may occur (see Table 5).

A variety of agents or conditions may cause or predispose to the genetic lesions which characterize neoplasia. These include: (a) chemical carcinogens; (b) inheritance of specific gene mutations; (c) persistent, or chronic, inflammatory states; (d) radiation; and (e) infection with certain viruses. The reader should note that more than one of these factors, along with other unknown factors, likely act cooperatively to incite neoplasia.

Carcinogenic Agents

Many human malignancies are associated with accidental or self-inflicted exposure to carcinogenic compounds. Lung cancer (smoking) and mesothelioma (asbestos, smoking) are probably the best examples. Bladder carcinoma and leukemia have also been linked to environmental carcinogen exposure. Because chemical carcinogenesis is amenable to scientific analysis using laboratory animals, a good

Table 5. Structural DNA Pathology in Human Neoplasia

Cellular DNA content alterations
hypodiploidy
hyperdiploidy
hypertetrapolidy
Chromosomal alterations
numerical
gains
loss
structural
translocation
inversion
segmental deletion
Individual gene alterations
amplification
intrachromosomal (heterogeneously stained regions)
extrachromosomal (double minutes)
mutation
base pair transition
base pair transversion
base pair insertion/deletion

deal of knowledge has accumulated on this subject. It has thus become a paradigm of human tumorigenesis (Smith, 1991).

Chemical carcinogenesis is a multistep process. In the first step, initiation, the carcinogen induces irreversible DNA damage (i.e., mutation). Some initiators mutate DNA directly. Others do so after being partially metabolized by the target cell's cytochrome p450 oxygenase system, and are thereby indirect acting. The parent compound of indirect acting carcinogens is called the procarcinogen and the active metabolite is called the ultimate carcinogen. Initiating compounds are reactive electrophiles which mutate DNA by nonenzymatic oxidation. Not all carcinogen induced DNA damage leads to initiation since cells are capable of repairing damaged DNA (see later). In order to preserve the genetic damage, then, a round of cell division is required to copy the unrepaired mutation into a daughter cell genome, where it remains stable. Thus, mutations leading to neoplasia must involve precursor, as opposed to terminally differentiated, cells. In the second step of experimental chemical carcinogenesis, promotion, a non-mutational influence alters the initiated cells' metabolism in such a way as to complete malignant transformation. Transformation, is a term which, as will be recalled, is generally applied to cells grown in culture, and is defined by *in vitro* characteristics such as loss of contact growth inhibition, abnormal colonial morphology, and exogenous growth factor independence. As such, it is only indirectly related to the morphological, functional, and clinical features of human tumor cells *in vivo*. The effects of promoters are fully reversible and therefore must follow initiation. Induction of proliferation via alteration of differentiation status are cellular effects shared by promoters. Some promoters such as estrogen alter cell phenotype by inducing gene transcription in the target cell. Chemical promoters, such as TPA, activate cellular enzymes such as protein kinase C (PKC), which function to phosphorylate a wide variety of proteins associated with surface membrane receptors, ion channels, and regulation of proliferation (Blobe and Hannun, 1994). Note that cellular physiology is such that constitutive hyperfunction of key regulatory enzymes, such as PKC, may have profound effects. We will encounter this principle again when oncogenes are discussed. The reader should also note the importance of proliferation to the chemical carcinogenesis sequence. Induction of a proliferative phenotype within tissues harboring initiated cells is believed to facilitate transformation by leading to further genetic damage either by completing initiation or by causing additional genetic errors during DNA replication. Thus, abnormal proliferation, in a sense, drives a vicious cycle which leads to the phenomenon of tumor progression (Cohen and Ellwein, 1991).

Heredity

Heritability of neoplasia may occur due to Mendelian transmission of specific genetic lesions leading to a well-characterized syndrome or type of neoplasm. It may also become manifest as a familial predilection to development of neoplasia, sometimes of a particular type, but insufficiently well-defined genetically to establish an inheritance pattern. Several diseases are characterized by dominant Mendelian inheritance of cancer predisposition. Some of them—including familial retinoblastoma, familial polyposis coli, and neurofibromatosis—are caused by inherited (i.e., germline) mutations of tumor suppressor gene loci (see later) (Malkin and Friend, 1992). Other neoplasia syndromes are inherited in a recessive manner, and are caused by mutations causing functional defects of DNA repair enzymes. These include ataxia-telangiectasia, Bloom's syndrome, and Fanconi's anemia.

Despite their associations with heritable germline mutations, not all patients with retinoblastoma or colorectal carcinoma have familial syndromes. Further, the specific genes associated with these syndromes may become mutated, but somatically, as in spontaneous retinoblastoma, colorectal carcinoma, as well as other neoplasms from diverse primary sites. There is overlap, therefore, in the genetic basis of heritable and spontaneous neoplasia. This reflects the limited number of genes which regulate critical steps in cell proliferation and differentiation.

Both Mendelian and non-Mendelian genetic predisposition are well described in breast cancer. Approximately 5–10% of patients have a family history of the disease. Women having a first degree relative with breast cancer, overall, have an incidence of breast cancer 2–3 times that of the overall population. Lack of a consistent or demonstrable inheritance pattern for breast cancer in some families likely reflects the involvement of marginally penetrant predisposing genes, interaction with environmental factors, or both. Other families have a much stronger familial pattern of breast cancer. This has recently been found to reflect germline mutations of two specific genes, called BRCA-1 and BRCA-2.

Inflammatory Conditions

Some diseases which impose unremitting inflammation, thereby resulting in persistent tissue regeneration, result in elevated cancer risk for the involved organ. Specific examples include idiopathic inflammatory bowel disease and chronic gastritis. In each of these conditions, the malignancy evolves from a dysplasia, which may occur on a background of metaplasia. Inflammation, through its biochemical mediators, is believed to perform a promoterlike function in such forms of neoplasia by sustaining regenerative cell proliferation. As previously noted, uninterrupted cell proliferation is theorized to magnify the effects of carcinogens by increasing the likelihood that mutations will be transcribed into daughter cells. Cell division may also lead directly to DNA damage through errors in DNA replication or chromosomal segregation.

Radiation

Ultraviolet light damages cells in many ways, but its carcinogenicity results from mutagenic activity—the formation of pyrimidine dimers—by so-called UVB radiation (λ 280–320 nm). Recall, however, that cells are capable of repairing damaged DNA. To do this, a multienzymatic process termed excision-repair is employed in which an endonuclease opens the damaged segment, an exonuclease excises the damaged segment, a polymerase synthesizes a new strand using complimentary DNA, and a ligase splices the repaired segment. Patients with an autosomal recessive mutation of excision-repair endonuclease called xeroderma pigmentosum have a high incidence of skin cancer. Skin cancer is associated with ultraviolet radiation exposure in a dose related manner. There are three clinically distinct, common forms of skin cancer: basal cell carcinoma, squamous carcinoma, and malignant melanoma. All are particularly common in lightly pigmented Caucasians. Only melanoma is frequently associated with lethal clinical behavior.

Ionizing radiation (X-rays, gamma rays, subnuclear particles) is also mutational, and hence carcinogenic. Radiation acts indirectly by first producing free radicals from water molecules. These, in turn, damage DNA. In addition to dose, dose rate, and host factors, the type of radiation impacts carcinogenicity. Particle radiation, which has high linear energy transfer, is most potent. Leukemias and thyroid carcinomas are the most common neoplasms which follow ionizing radiation exposure.

Viruses

It has long been known that viruses, particularly retroviruses, are associated with development of neoplasia in animal species. Study of these animal tumors, and the viruses which induced them, has proven invaluable to our understanding of human neoplasms by leading to the concept of oncogenes. The transforming mechanisms of viruses are heterogeneous. First, viral transformation often requires nonproductive infection, which refers to an incomplete viral life cycle, without production and release of viral particles. This is logical since productive infection will destroy, rather than transform, the infected cell. Nonproductive infection is usually associated with integration of viral DNA into the host genome. This is believed to be an important event in virus associated neoplasia, since it may induce abnormal host gene expression either by insertion of viral DNA in or near critical gene regulatory sequences, or by causing structural damage to DNA during integration. Expression of virus-specific genes within infected cells also facilitates viral transformation. Certain viral gene products interfere with host cell proliferation and differentiation via binding interactions with cell cycle regulatory proteins (see later).

Viruses were once thought to be widely responsible for the genesis of human neoplasia, analogous to their key roles in many forms of neoplasia in animal species. However, detailed study has essentially limited the list of virus-associated neoplasms to those summarized in Table 6. This possibly reflects the limited number of retroviruses which infect human, as opposed to animal hosts. Viral oncogenesis is impacted by many factors intrinsic to the host or the host's environment. Obviously, not everybody with Epstein-Barr Virus (EBV) or human papillomavirus

Table 6. Viruses Associated With Human Neoplasia

Virus	*Linked Neoplasms*	*Associated Factors*	*Mechanism***
Human Papilloma (HPV)	Papillomas (skin), dysplasias, carcinomas (genital)	HPV genotype, immunosuppression	E6, E7 inactivate, p53, RB-1, respectively
Epstein-Barr (EBV)	Burkitt's lymphoma nasopharyngeal carcinoma	Other infections (malaria), geography (Asia, Africa), immune disorders+	(a) LMP-1 prevents apoptosis via bcl-2 interaction (b) EBNA-2 transactivation of host genes
Hepatitis B (HBV)	Hepatocellular carcinoma	Cirrhosis, toxins	HBx activates PkC
HTLV-1***	T-cell leukemia	Geography (Japan, Southern USA)	tax activates IL-2, GM-CSF, c-fos, c-sis

Notes: + X-linked immunodeficiency, HIV.

** Virus encoded gene product interferes with host oncogene/tumor suppressor gene product or interferes with host gene transcription.

*** HTLV-1, Human T cell leukemia virus, a retrovirus.

(HPV) infections develops cancer. In this regard, alterations of gene structure or expression induced by viral integration probably represent only one of multiple influences required to transform human cells *in vivo*. We shall return to viral oncogenesis during the discussion of oncogenes.

Genetic Basis of Neoplasia

In order to understand the relationship between genetic pathology and the basis of neoplasia, it is important to grasp: (a) the pathways of cellular growth signal transduction, (b) the regulation of cell division, and (c) the correlation between structural genetic pathology and its consequences at the cellular level.

Cellular Growth Signal Transduction

Earlier, we described how cellular phenotype is responsive to growth factors, which interact with plasma membrane spanning growth factor receptors. Thereby, somatic tissues are able to proliferate or differentiate according to prevailing environmental conditions. A given cell may express multiple receptors, each having specificity for certain growth factor ligands, some of which may be inhibitory to cell growth or division.

Ligand binding to its cognate receptor stimulates protein kinase enzyme activity from the intracytoplasmic moiety of the growth factor receptor by inducing a conformational shift (Ullrich and Schlessinger, 1990). This process is called signal transduction. Protein kinases phosphorylate their substrates, which may include other (second messenger) protein kinases, thereby modulating enzymatic activity or binding properties. Growth factor induced signal transduction results in a protein phosphorylation cascade, which induces pleiotropic effects by altering the functional status of diverse cytoplasmic proteins.

Either directly or through second messengers, protein kinase activity may modulate activity or structure of cellular proteins which regulate metabolic pathways, ion concentrations, cell shape, or cell adhesion, thus impacting phenotype. Protein kinase activity also results in changes at the nuclear level—either transcription of specific genes via promoter sequence interactions, or DNA replication (i.e., cell cycle progression) via activation of transcription factors such as *fos*, *jun*, or *myc*. In general, a cell is exposed to multiple, sometimes opposing signals to proceed with proliferation or differentiation, the effects of which are further modified by internal cellular factors. Thus, proliferation/differentiation are under a complex, multilevel regulation system which reflects an equilibrium between competing signals.

The Cell Cycle

The replication of dividing cells can be divided into a series of stages which proceed in a sequence as follows: (a) G_1 (for gap 1) is the period of active DNA transcription during which cellular constituents, such as organelles, are replicated; (b) S (for synthesis) is the period during which genomic DNA is replicated; (c) G_2 (for gap 2) is the interval during which mitotic apparatus is prepared; and (d) M (for mitosis).

In normal tissues, a majority of the cells do not cycle continuously. Cells may exit the cell cycle under three circumstances. First, there is a phase, called G_0, which corresponds to quiescent cells. G_0 cells are capable of re-entry into G_1, but only after growth factor stimulation. G_0 cells differ morphologically from G_1 cells. Since there is substantially less gene transcription, RNA content is low in G_0 cells and chromatin is condensed. A cell may also exit G_0, or any other phase of the cell cycle, by undergoing apoptosis (programmed cell death). Apoptosis, as we shall see, is becoming increasingly recognized as an important factor in the genesis and treatment of neoplasia. Finally, as previously noted, terminal differentiation constitutes a form of irreversible withdrawal from the cell cycle.

Progression through the cell cycle sequence is a carefully regulated process (Hartwell and Kastan, 1994). This ensures that cell proliferation does not exceed cell losses. Cycling cells must also complete DNA synthesis without error as well as initiate it from a normal (i.e., nonmutated) genome. For example, in order to consistently ensure that cell division will not precede under inappropriate conditions, entry into the S-phase typically requires stimulation by more than one growth factor. There are also molecular level restrictions, or checkpoints, on cell cycle progression, primarily at the G_0/G_1 and G_1/S interfaces. Progression through these checkpoints requires sequential synthesis of a series of enzymes, called cyclins, each of which is active during only one phase of the cell cycle. The cyclins, in turn, activate protein kinases (via binding) which phosphorylate the proteins required for DNA synthesis.

Several factors regulate cell cycle progression at the G_1-S interface. A critical element in the mechanism is the product of the retinoblastoma gene: Rb (Wiman, 1993). In its active, unphosphorylated form, Rb acts as a brake on cell division through binding interactions with two transcription factors (ERF and DP1) which are required for DNA synthesis. Deactivation of Rb via phosphorylation is mediated by cyclin-dependent kinase enzymes, the levels of which are unregulated in late G_1. The activity of G_1 cyclin dependent kinase is, in turn, indirectly modulated by another tumor suppressor gene product: p53. During mitosis, Rb activity is restored by a phosphatase enzyme with carefully regulated expression intervals. It should be noted parenthetically that p53 is induced under conditions of DNA damage, such as exposure to ultraviolet light. There are many other checkpoints on cell cycle progression. The products of the ataxia-telangiectasia gene, for example, act at the G_2-M transition. Thus, regulation of the cell cycle is accomplished by an interactive set of signal transduction systems which are responsive to a variety of internal and external influences.

Correlation Between Genetic Alterations and Phenotype

Many of the proteins which mediate checkpoints in control of apoptosis, cell proliferation, or signal transduction have been identified as products of oncogenes or tumor suppressor genes (Oren, 1992). Oncogenes are defined as genes which, when activated (i.e., overexpressed), facilitate neoplastic transformation. Virtually all of the oncogenes described to date are growth factors, growth factor receptors, second messenger proteins, or transcription factors. Thus, dysregulated levels of oncogene product or function promote cell proliferation and oppose terminal differentiation. Tumor suppressor genes are defined as genes which facilitate neoplastic transformation when deactivated (i.e., underexpressed). Fewer tumor suppressor genes than oncogenes have been described. However, tumor suppressor genes are just as relevant to human neoplasia as oncogenes (Levin, 1995). As previously described, two of the well-characterized tumor suppressor gene products (p53 and Rb) inhibit DNA transcription via binding interactions, thus opposing cell proliferation. Deactivation would thus favor a proliferative, and thus incompletely differentiated, phenotype.

The mechanisms by which oncogenes are activated, and tumor suppressor genes deactivated, are heterogeneous and not fully described. However, inappropriate function usually results from structural genetic pathology, as caused by mutagenic chemicals, radiation, or mitosis associated errors of DNA sequence or segregation. Neoplasms, particularly carcinomas, are associated with a wide variety of structural DNA pathology which includes complex karyotypic alterations, gene amplifications, and gene mutations. Oncogenes may be activated by chromosomal translocation, by point mutation, or by gene amplification (excess gene copy number). Tumor suppressor genes may be inactivated by gene sequence mutations which result in loss of protein function, or by loss of gene copies through partial or complete chromosomal deletions. Heterozygous tumor suppressor gene mutation may be inherited, as in familial retinoblastoma. Finally, viral gene products may interfere with either oncogene function or tumor suppressor gene function. By virtue of their fundamental influences on cell proliferation and differentiation, the oncogenes and tumor suppressor genes link structural genetic pathology to the classical functional and morphologic aspects of neoplasia.

It is important to note that although some oncogenes or tumor suppressor genes are consistently abnormal in tumors of a specific type, they may also be inappropriately expressed in neoplasms of other histologic types or primary sites. This is hardly surprising given their key roles in cell proliferation and differentiation.

It is not possible, within the limited scope of this chapter, to provide a comprehensive summary of all the known oncogenes and tumor suppressor genes in human neoplasia. We hope, however, to emphasize two important concepts by briefly presenting highlights of these genes. The first is the inextricable relationship between structural pathology involving certain genes and aberrations of differentiation or cell proliferation. The second is the manner in which alterations of specific genes produce complimentary, or synergistic effects on cellular phenotype (Adams and Cory, 1992).

Retinoblastoma Gene (Rb)

Rb is a tumor suppressor gene, located on 13q, which is always deactivated in a childhood neoplasm called retinoblastoma. The Rb gene product acts as a brake on cell cycling at the G_1/S interface. Retinoblastoma occurs in a familial form in which one mutant Rb allele is inherited, and the second is inactivated spontaneously in somatic tissue. The neoplasm also occurs in a sporadic form in which both alleles become deactivated in somatic tissue. Due to presence of germline mutation, patients with familial retinoblastoma are susceptible to other malignancies, particularly osteogenic sarcoma. Somatic deactivation of Rb has been observed in many spontaneous adult malignancies (e.g., breast, bladder carcinoma). Deactivation of the Rb gene in neoplasia may occur by any combination of 13q deletion and point mutation which alters Rb protein conformation. Deactivating Rb mutations alter the protein structure such that there is persistence of phosphorylated Rb or altered binding to myc (Goodrich and Lee, 1992). Nonmutational deactivation may also occur via binding by viral proteins. An example is the HPV-encoded E7, as occurs in squamous carcinoma of the cervix.

myc

The name of this oncogene, located on 8q24, is derived from the homologue in birds, which is altered by avian myelocytoma virus, a retrovirus known to cause B cell lymphomas in chickens. Humans have at least three myc homologs (L-myc, c-myc, and N-myc), the normal expression of which is tissue dependent. Presence of myc is required for G_1-to-S transition, as described earlier, via Rb interactions. The protein functions as nuclear transcription factor for DNA replication enzymes. Expression of myc has also been shown to inhibit terminal differentiation and regulate apoptosis, presumably via interactions with gene promoter sequences. Activation of myc in neoplasia has been shown to occur by at least two mechanisms. First, myc amplification commonly occurs in neuroblastoma and small cell lung carcinoma. Myc is also activated by a chromosomal translocation, t (8:14), which occurs in Burkitt's lymphoma. This rearrangement places the gene encoding sequence for myc next to promotor sequences for the immunoglobulin heavy-chain gene which, in B cells, is continually expressed, resulting in constitutive myc transcription.

p53

This tumor suppressor gene, located on 17p, is the most common target for acquired genetic pathology in human neoplasia (70% of colon cancers, 35% of breast cancers, and 50% of lung cancers) (Chang et al., 1993). A germline mutation of p53 results in Li-Fraumeni syndrome, a rare disease characterized by early age onset of breast carcinoma, leukemia, and sarcoma. In normal cells, wild type p53 represses cell proliferation by, among other things, indirectly regulating the activity expression of cyclin dependent kinases (CDK) via binding interaction with the promotor region of WAF-1, a gene which modulates CDK activity. It has also been shown that p53 induces apoptosis in cycling populations under conditions of DNA damage. Deactivation of p53 would therefore not only lead to unrestricted proliferation but would also result in genetic instability. Accordingly, loss of wild type 53 activity is correlated with presence of complex karyotypic alterations. Deactivation of p53 in neoplasia usually reflects a combination of chromosomal deletion and point mutation. As with Rb, wild type protein may also be sequestered by virus encoded proteins (i.e., HPV-encoded E6). Mutations of p53 occur in phylogenetically-conserved exons which encode binding region sequences.

bcl-2

The name of this oncogene, located on 18q 21, is an acronym for B cell lymphoma, a neoplasm often associated with activation of this gene. The function of the bcl-2 protein is to impede apoptosis, either by induction of antioxidant pathways or by stabilizing cell membrane constituents, preventing oxidative injury (Chiou et al., 1994). Thus, bcl-2 expression prolongs cell viability. Activation of bcl-2 in B cell lymphomas is effected by a translocation [t(14;18)], which is frequent in these tumors and places the bcl-2 encoding sequence next to the heavy-chain gene promotor region. Note the analogy to c-myc activation in Burkitt's lymphoma. Overexpression of bcl-2 may also occur secondary to interaction between a virus encoded protein (LMP-1 of EBV) and the bcl-2 promotor region.

ras

Ras is the most commonly mutated oncogene in human neoplasia (~ 30% of malignancies). Mutation is especially frequent in lung, pancreas, and colon carcinomas. The name is derived from retrovirus-induced rat sarcomas. The product of ras is a protein which functions as a second messenger present on the inner surface of plasma membrane (Lacal and Carnero, 1994). Second messengers such as ras transduce extracellular signals received by tyrosine kinase growth factor receptors to cellular effector molecules such as phospholipase and PKC. The sequence of plasma membrane signal transduction via ras is initiated when a growth factor binds to a tyrosine kinase growth factor receptor, such as PDGFR, EGFR, or ERBB-2. Receptor dimerization induced by growth factor binding results in formation of a bridging protein complex the bound receptor and ras. Ras is activated by exchanging GDP for GTP, mediated by guanine releasing factor (GRF) in the bridging protein complex. Active ras subsequently phosphorylates raf-1, resulting in a protein kinase cascade that, ultimately, activates nuclear transcription factors including myc. Other *in vitro* consequences of ras activity include altered cell shape and upregulation of protease synthesis or secretion. Ras is deactivated by GTP-ase activating proteins (GAPs) which catalyze GTP for GDP exchange. Activation of ras in human neoplasia is generally caused by point mutation, resulting in inability to hydrolyze GTP. This imposes a persistently activated state. Interestingly, one of the GAP species is neurofibromin (NF-1) which is mutated in neurofibromatosis, a dominantly inherited familial neoplastic syndrome.

EGFR and ERBB-2

These are oncogenes which function as transmembrane growth factor receptors with tyrokine kinase activity. EGFR is abnormally expressed in most squamous carcinomas, especially in lung, and ERBB-2 is activated in adenocarcinomas, especially from breast and stomach. As previously noted, ligand binding to growth factor receptors induces a conformational change in the extracellular domain which causes dimerization and leads to ras activation. Tyrosine kinase activity may alternatively activate the phosphatidylinositol pathway, leading to PKC activation and its diverse cellular consequences. Activation of ERBB-2 in neoplasia usually reflects gene amplification which is believed to cause spontaneous dimerization via increased concentration of oncoprotein.

Pathophysiology of Neoplasia

At this point it is possible to reconcile the presence of multiple genetic defects in neoplasia with the classical histologic sequence in epithelial neoplasia (i.e., dysplasia → invasion → metastasis) in a unified theory of cancer pathophysiology (see Figure 6). This theory, called multistep neoplastic progression, suggests that phenotypic traits associated with deregulated proliferation/differentiation, invasion, and metastasis are acquired in a stepwise fashion during the life span of the neoplasm through inherent genetic instability. In other words, structural genetic damage occurs in discrete steps, each of which in some manner alters gene expression such as oncogene activation or tumor suppressor deactivation. The sum total of these genetic defects leads to the malignant, or metastatic, phenotype. It is important to note that abnormal expression of several growth regulatory genes (probably six or seven) is required to induce a malignant phenotype. In addition to linking genetic pathology to morphology, the multistep tumor progression theory accounts for many other known aspects of neoplasia particularly clonal progression. Geneticists have long understood, using studies such as G6PD enzyme isotype homozygosity, that neoplasms are clonal in origin (i.e., they are derived from a single cell). Nevertheless, in order to conform to our theory, neoplasms must also display some evidence of heterogeneity. Indeed, at the cytogenetic level, malignant neoplasms display multiple karyotypes. Using sophisticated marker chromosome analysis, however, it can be demonstrated that various karyotypes are structurally related, and thus evolved from one another (not from different neoplastic progenitor cells). It has also been demonstrated that one clone (the most aggressive) may come

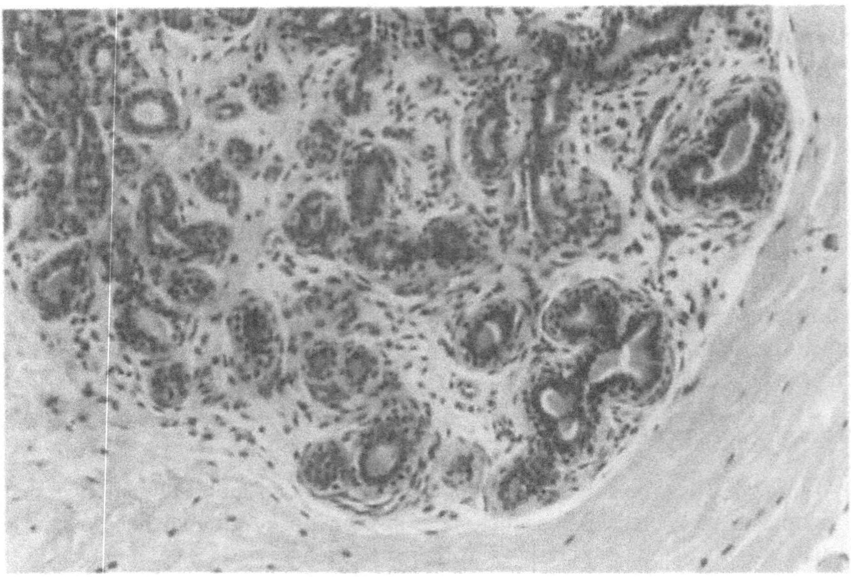

continued

Figure 6. This series of photographs illustrates the concept of tumor progression in neoplasia of the breast. The top panel (**A**) exhibits a benign lobule, with well-circumscribed rounded acini, each lined by only two layers of uniform benign cells. In hyperplasia (panel **B**), there is expansion of the acini and terminal ducts by a proliferation of relatively uniform, cytologically bland cells. In low grade forms of intraductal carcinoma (**C**) there is more pronounced distention and coalescence of acini by a population of cells exhibiting somewhat larger and more hyperchromatic nuclei. The reader will note that the distinction between florid forms of hyperplasia and low grade forms of intraductal carcinoma may be subtle. With high grade intraducted carcinoma (**D**), distinction from hyperplasia is more easily made, since there is considerable nuclear enlargement, hyperchromasia, and pleomorphism as well as foci of necrosis. Panel **E** is an immunostain for type IV collagen, a component of basal lamina. In the lower right there is a duct which is filled by malignant epithelial cells. Note presence of a defect in the basal lamina, through which neoplastic cells are growing into the surrounding host stroma. This process is reflected histologically by the presence of invading nests of neoplastic cells, as well as a desmoplastic host reaction, as is observed with the ductal structure illustrated in panel **F**.

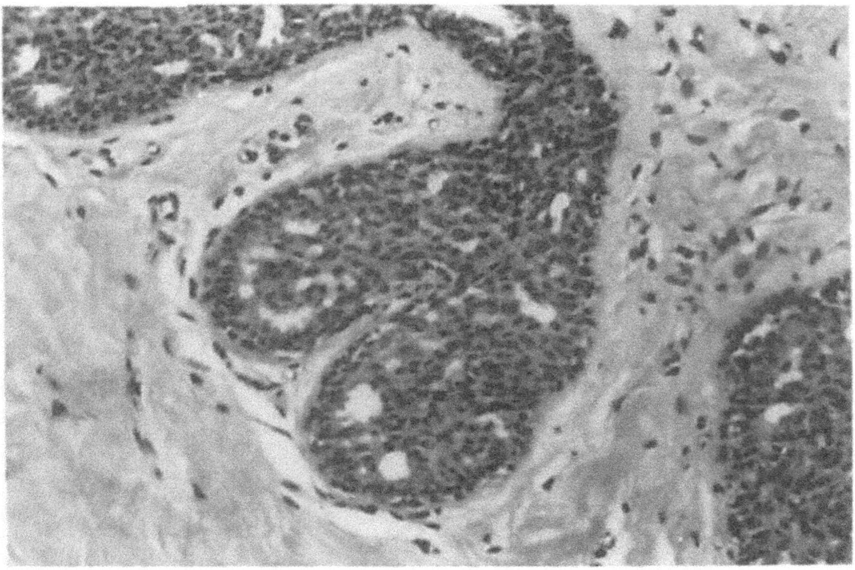

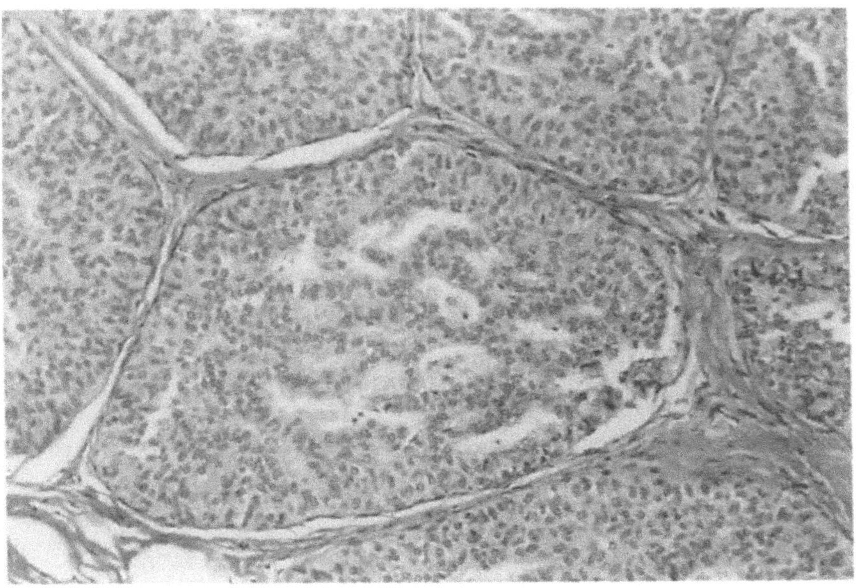

Figure 6. Continued

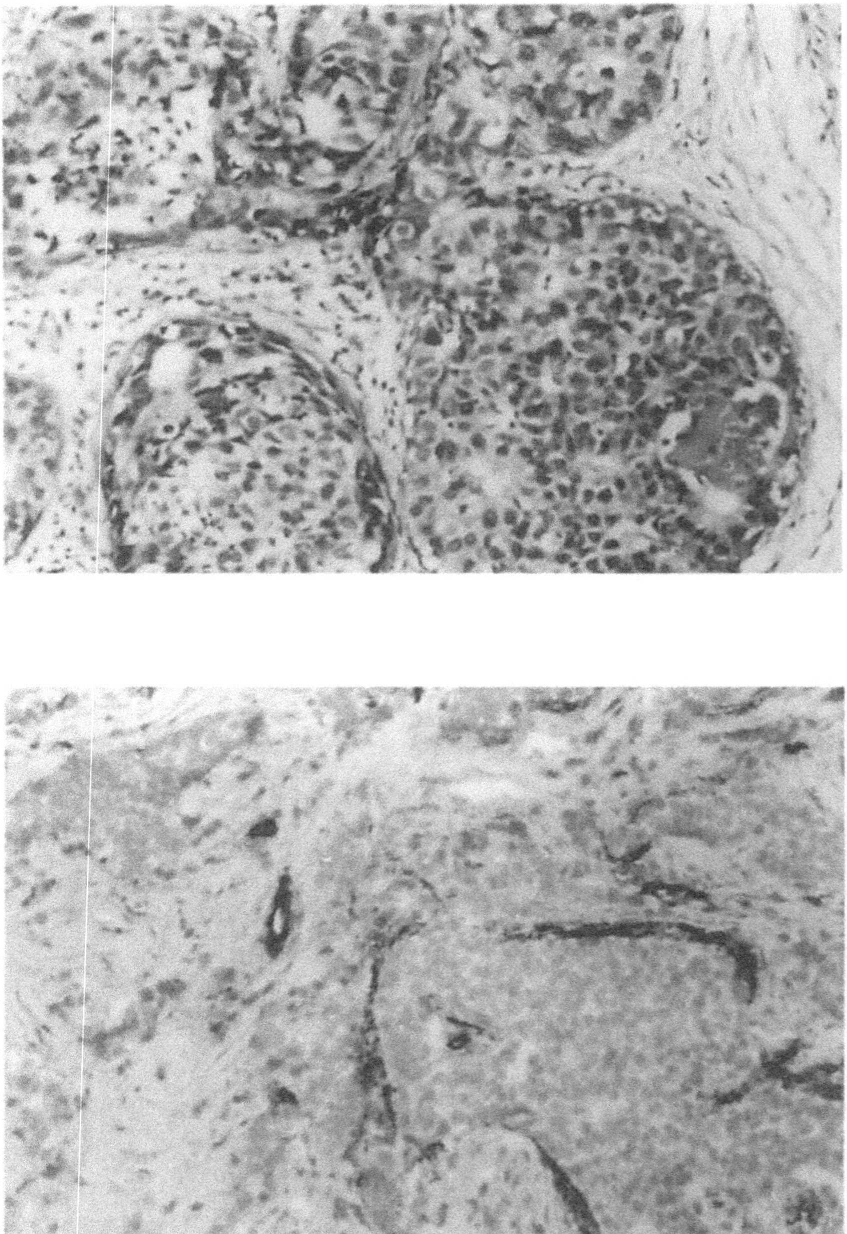

Figure 6. Continued

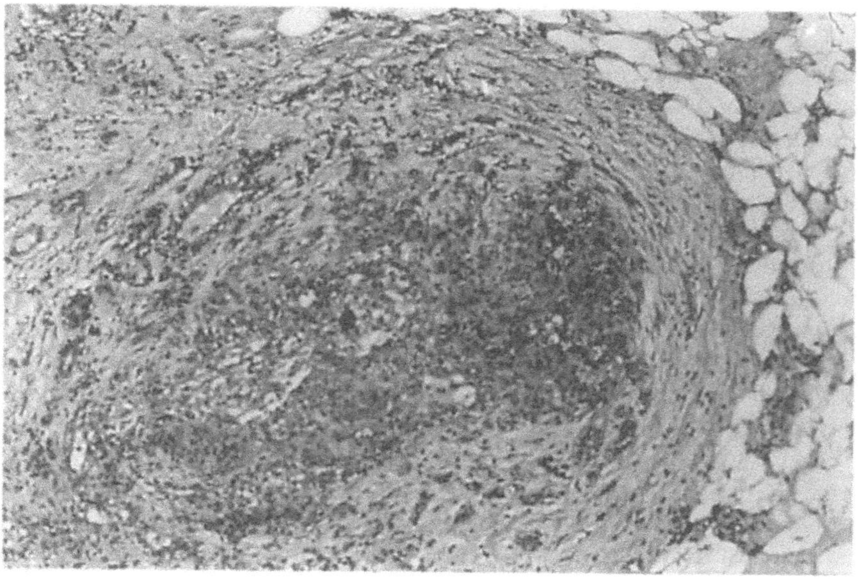

Figure 6. Continued

to numerically dominate a given neoplasm, presumably by nature of its growth advantage. However, such clonal dominance is not the rule, since interactions among different neoplastic subpopulations may result in a coordination of their proliferative activities, as well as their ability to metastasize and even respond to therapy. Therefore, individual neoplasms are also cellular societies, similar in concept at least to the cellular societies previously described in normal tissues. Once again we see how neoplasia can be viewed as a caricature of physiologic tissue growth and differentiation.

SUMMARY

We began our discussion of neoplasia with two unifying definitional concepts. First it was asserted that all forms of neoplasia can be seen as a failure of somatic tissues to differentiate appropriately. This failure was caused by arrested maturation of organ determined stem cell progeny which resulted in an immature or incompletely differentiated phenotype. From this fundamental concept it was possible to understand the histopathology of diverse forms of neoplasia, to derive an explanation for benign versus malignant phenotype, to theorize mechanisms which help explain invasion and metastasis, and even to account for the distinctive clinical features of neoplastic diseases.

The second unifying concept was that the primary, and likely causal, structural lesions inherent to neoplastic diseases involve the genome, either in the germline, in the somatic tissues, or in both. This genetic pathology was heterogenous but, not unexpectedly, it induced abnormal expression of specific genes which control tissue growth and differentiation. Some of these genes mediate responses to environmental stimuli whereas others regulate critical aspects of cell proliferation. The net effect of the genetic alterations, at any rate, is to impose a proliferative—and thereby incompletely differentiated—phenotype. Further, the extent of genetic pathology is directly related to degree of differentiation abnormality, and hence clinical behavior. Thus, the inextricable connection between morphologic pathology and genetic pathology in neoplasia is made whole. Finally, with the discovery of gene products, such as p53, which function to maintain genomic structure, we can theorize that genetic instability accounts for the process of tumor progression, in which phenotypic evolution within a neoplasm occurs over time.

The catalogue of genetic lesions involved in neoplasia is far from complete, as is the spectrum of consequences at the level of cell proliferation and differentiation. Moreover, we are a long way from understanding all of the mechanisms by which the genome becomes destabilized. Thus, our ability to predict whether neoplasia will arise, whether a neoplasm is likely to be lethal, or how it will respond to therapy, remains imperfect. Nevertheless, our understanding of the matrix of events that control the dynamics of normal differentiation and proliferation provides us with an organizational framework which can guide research efforts toward elucidating the detailed mechanisms which determine neoplastic development and behavior.

REFERENCES

Adams, J.M. & Cory, S. (1992). Oncogene co-operation in leukemogenesis. Cancer Surveys 15, 119–141.

Albelda, S.M. (1993). Role of integrins and other cell adhesion molecules in tumor progression and metastasis. Lab. Invest. 68, 4–17.

Blobe, G.C. & Hannun, Y.A. (1994). Regulation of protein kinase C and role in cancer biology. Cancer Met. Rev. 13, 411–431.

Blood, C.H. & Zetter B.R. (1990). Tumor interactions with the vasculature: Angiogenesis and tumor metastasis. Biochim. Biophy. Acta 1032, 89–118.

Chang, F., Syrjanen, S., Kurvinen, K., & Syrjanen K. (1993). The p53 tumor suppressor gene as a common cellular target in human carcinogenesis. Am. J. Gastroenter. 88, 174–186.

Chiou, S-K., Rao, L., & White, E. (1994). Bcl-2 blocks p53-dependent apoptosis. Molecular Cellular Biol. 14, 2556–2563.

Cohen, S.M. & Ellwein, L.B. (1991). Genetic errors, cell proliferation, and carcinogenesis. Cancer Res. 51, 6493–6505.

Daneker, G.W., Piazza, A.J., & Steele, G.D. (1989). Mercurio AM. Relationship between extracellular matrix interaction and degree of differentiation in human colon carcinoma cell lines. Cancer Res. 49, 681–686.

Ethier, S.P. (1995). Growth factor synthesis and human breast cancer progression. J. Natl. Cancer Inst. 7, 964–973.

Fidler, I.J., Gersten, D.M., & Hart I.R. (1978). The biology of cancer invasion and metastasis. Cancer Res. 28, 149–251.
Folkman, J. (1985). Tumor angiogenesis. Cancer Res. 43, 175–202.
Furcht, L.T., Skubitz, A.P.N., & Fields, G.B. (1994). Tumor cell invasion, matrix metalloproteinases, and the dogma. Lab. Invest. 70, 781–783.
Goodrich, D.W. & Lee, W.H. (1992). Abrogation by c-myc of G1 phase arrest induced by RB protein but not by p53. Nature 360, 177–179.
Grant, D.S., Tashiro, K.I., Segui-Real, B., Yamada, Y., Martin, G.R., & Kleinman, H.K. (1989). Two different laminin domains mediate the differentiation of human endothelial cells into capillary-like structures in vitro. Cell 58, 933–943.
Hart, I.R. & Saini, A. (1992). Biology of tumour metastasis. Lancet 339, 1453–1457.
Hartwell, L.H. & Kastan, M.B. (1994). Cell cycle control and cancer. Science 266, 1821–1828.
Heppner, G.H. (1989). Tumor cell societies. J. Natl. Cancer Inst. 81, 648–649.
Heppner, G.H. (1991). Cell-to-cell interaction in regulating diversity of neoplasms Scm. Cancer Biol. 2, 97–103.
Hlatky, L., Tsionou, C., Hahnfeldt, P., & Coleman, C.N. (1994). Mammary fibroblasts may influence breast tumor angiogenesis via hypoxia-induced vascular endothelial growth factor up-regulation and protein expression. Cancer Res. 54, 6083–6086.
Iozzo, R.V. (1995). Tumor stroma as a regulator of neoplastic behavior. Lab. Invest. 73, 157–160.
Lacal, J.C. & Carnero, A. (1994). Regulation of ras proteins and their involvement in signal transduction pathways (Review). Oncol. Reports 1, 677–693.
Lester, B.R. & McCarthy, J.B. (1992). Tumor cell adhesion to the extracellular matrix and signal transduction mechanisms implicated in tumor cell motility, invasion and metastasis. Cancer and Met. Rev. 11, 31–44.
Levin, A.J. (1995). Tumor suppressor genes. Sci. Amer. 28–37.
Malkin, D. & Friend, S.H. (1992). The role of tumour suppressor genes in familial cancer. Cancer Biol. 3, 121–130.
Morikawa, K., Walker, S.M., Nakajima, M., Pathak, S., Jessup, J.M., & Fidler, I.J. (1988). Influence of organ environment on the growth, selection, and metastasis of human colon carcinoma cells in nude mice. Cancer Res. 48, 6863–6871.
Oren, M. (1992). The involvement of oncogenes and tumor suppressor genes in the control of apoptosis. Cancer Met. Rev. 11, 141–148.
Pusztai L., Lewis, C.E., Lorenzen, J., & O'D. McGee, J. (1993). Growth factors: Regulation of normal and neoplastic growth. J. Pathol. 169, 191–201.
Ruoslahti, E., Hayman, E.G., & Pierschbacher, M.D. (1985). Extracellular matrices and cell adhesion. Arteriosclerosis 5, 581–594.
Rusciano, D. & Burger, M.M. (1992). Why do cancer cells metastasize into particular organs? BioEssays 14, 185–194.
Sell, S. & Pierce, G.B. (1994). Maturation arrest of stem cell differentiation is a common pathway for the cellular origin of teratocarcinomas and epithelial cancers. Lab. Invest. 70, 6–22.
Smith, P.J. (1991). Carcinogenesis: Molecular defenses against carcinogens. The Cancer Cell 47, 3–20.
Ullrich, A. & Schlessinger, J. (1990). Signal transduction by receptors with tyrosine kinase activity. Cell 61, 203–212.
Wiman, K.G. (1993). The retinoblastoma gene: Role in cell cycle control and cell differentiation. FASEB J. 7, 841–845.

Chapter 2

Cancer Induction by Ionizing Radiation

KENNETH L. MOSSMAN

INTRODUCTION

Ionizing radiation (X-rays and gamma rays) is an indispensable tool in the diagnosis and treatment of many human diseases. The National Council on Radiation Protection and Measurements (NCRP) estimates that approximately 190 million radiology examinations and nuclear medicine procedures are performed annually in the

Advances in Oncobiology
Volume 1, pages 43–59.

ISBN: 0-7623-0146-5

United States (NCRP, 1989). An additional 0.5 million cancer patients are treated by radiation each year (NCRP, 1987).

Although the benefits of radiation exposure in medical diagnosis and therapy are well known, the risks of exposure are not fully understood. Cancer is the chief somatic effect of radiation of principal concern at low doses. In spite of our extensive experience with radiation and the kinds of cancers known to be induced by radiation, many important questions about radiation as a carcinogen are yet to be fully answered. This chapter will emphasize radiation carcinogenesis in humans and will discuss the kinds of populations studied and the evidence for radiation-induced cancers in man, what we know about radiation carcinogenesis and what we do not know, and estimates of carcinogenic risks especially at low doses. Risks of radiogenic cancer will be discussed within the framework of medical uses of radioactive materials and radiation-producing equipment.

THE MAGNITUDE OF THE CANCER PROBLEM

Cancer is the second leading cause of death in the United States, accounting for approximately 1.2 million new cases (excluding skin cancer) and 525,000 deaths annually. One out of five deaths in the U.S. is from cancer (American Cancer Society, 1993). Cancer causation is not fully understood but it is known that various external factors (e.g., chemicals, viruses, and radiation) and internal or host factors (e.g., hormones, immune status, and genetic factors) play a decisive role. These internal and external factors may combine, or interact in a sequential fashion, to initiate and promote carcinogenesis and facilitate progression of tumor growth.

Ionizing radiation is probably the most thoroughly studied and well-known human carcinogen. We know more about the cancer causing effects of radiation than of most other known or suspected human carcinogens. Ionizing radiation, in comparison to other known human carcinogens, is weakly carcinogenic; it is responsible for only a small part of the cancer burden in the United States. As shown in Table 1, less than 5% of all cancer deaths may be attributable to ionizing radiation exposure. The primary factors contributing to cancer are dietary habits and tobacco consumption; together, they account for about two-thirds of all cancer deaths (Doll and Peto, 1981).

Table 2 provides a more detailed accounting of cancer deaths attributable to radiation exposure. Major sources of radiation exposure to the population are natural background, medical and dental radiodiagnostic and radiotherapy procedures, and exposures to various consumer products and other sources (e.g., building materials, smoke detectors, and occupational exposures). The number of cancer deaths attributable to radiation, as shown in Table 2, have been estimated from studies of human populations exposed to very high radiation doses. Cancer deaths attributable to low levels of ionizing radiation have never been actually observed. Published epidemiological studies have not clearly and consistently demonstrated that radiation exposure at low doses, such as used in diagnostic X-ray or nuclear

Table 1. Relative Importance of Various Factors Associated with Cancer

		Percent of all Cancer Deaths	
		Best Estimate	*Range*
1.	Diet (except food additives)	35	10–70
2.	Tobacco	30	25–40
3.	Infection (viruses, etc.)	10?	1–?
4.	Reproductive and sexual behavior	7	1–13
5.	Occupation (including ionizing radiation)[a]	4	2–8
6.	Alcohol	3	2–4
7.	Geophysical factors (including natural background radiation)[b]	3	2–4
8.	Pollution	2	<1–5
9.	Medicines and medical procedures (including ionizing radiation)[b]	1	0.5–3
10.	Food additives	< 1	–5[c]–2
11.	Industrial products	< 1	<1–2

Notes: [a] Ionizing radiation contributes approximately 0.3%.
[b] Ionizing radiation contributes approximately 50%.
[c] The negative estimate allows for the possibility of a protective effect of antioxidants and other preservatives.

Source: Doll and Peto, 1981.

Table 2. Cancer Deaths Attributable to Radiation Exposure

	No. of People Exposed per Year	*No. of Cancer Deaths per Year*	*% of Total Cancer Deaths*
All cases of cancer		525,000	100
Natural background radiation[a]	250,000,000	16,000[c]	3
Medical diagnosis and therapy	196,000,000[b]	9,000[c]	2
Consumer products and other sources	130,000,000	1,000[c]	<1
Total of radiation sources[a]	250,000,000	26,000[c]	5

Notes: [a] Excludes indoor radon which has been estimated by the U.S. Environmental Protection Agency to be responsible for 5,000–14,000 lung cancer deaths annually.
[b] Estimate based on number of radiologic examinations.
[c] Annual number of radiogenic cancer deaths is projected and not based on direct observations. Estimates are based on cancer risks extrapolated to low dose from epidemiological studies of human populations exposed at high dose.

Sources: NCRP, 1987; NCRP, 1989; NAS, 1990; American Cancer Society, 1993.

medicine, result in an increased incidence of cancer (Boice, 1992). The 2% of cancers attributable to medical and dental X-ray procedures cannot be identified in the U.S. population.

The number of cancer deaths attributable to radiation exposure shown in Table 2 does not include contributions of radon exposure. Radon has only recently been recognized as the major source of radiation exposure to the U.S. population, accounting for approximately 55% of the total population dose (NCRP, 1987). Based on epidemiologic studies of uranium miners exposed to very high levels of radon in occupational environments, the U.S. Environmental Protection Agency (EPA) has estimated that indoor radon may account for approximately 14,000 lung cancer deaths per year in the U.S. population. This number could range from 7,000 to 30,000 deaths per year (U.S. EPA, 1992). However, because no direct observations of cancer as a result of domestic radon exposure have been made, the lower limit of the EPAs estimate of lung cancer mortality should include zero. Table 2 also does not consider the contribution of ultraviolet light to skin cancer. Almost all of the more than 700,000 cases of basal cell and squamous cell skin cancer diagnosed each year in the U.S. are related to ultraviolet light exposure (American Cancer Society, 1993).

RADIATION AS A HUMAN CARCINOGEN: EPIDEMIOLOGICAL EVIDENCE

Evidence that radiation is a human carcinogen is derived from a number of epidemiological studies, many of which are shown in Tables 3 to 6. Exposures of human populations have resulted from military (Table 3), medical (Tables 4 and 5), and occupational (Table 6) uses of radiation. A brief description of some of the more important studies follows.

Military Exposures

The largest single source of data is the survivors of the atomic bombings of Hiroshima and Nagasaki in August, 1945 (Table 3). In the Life Span Study (one of several cohort-based epidemiological studies), conducted by the Radiation Effects Research Foundation, approximately 93,000 atomic bomb survivors and 27,000 nonexposed control persons are being studied with mortality and causes of death continuously updated. Individuals received doses ranging from less than 10 centigray (cGy) (1 cGy = 1 rad) to more than 500 cGy. The average dose to survivors was approximately 20 cGy. In the mortality survey from 1950–1985, a total of 6,000 cancer deaths have been observed; only about 350 excess cancers have been observed to date and could be attributable to radiation exposure (Shimizu et al., 1989; NAS, 1990; Schull and Weiss, 1992).

Various cancers have been observed in excess in the exposed population. The four major cancer types identified are: leukemia, lung cancer, female breast cancer,

Table 3. Cancer in Human Populations Exposed to Nuclear Weapons

Population	*Observed Excess Cancers*
Japanese atomic bomb survivors	leukemia, lung, breast, thyroid, esophagus, stomach, colon, ovary, bladder, myeloma
Marshall Islanders (fallout)	thyroid

Source: Boice, 1992.

Table 4. Cancer in Patients Exposed to Ionizing Radiation for Diagnosis or Therapy

Population	*Observed Excess Cancers*
Patients treated with radiation for cancer	
Cervical cancer	bladder, rectum, uterus, ovary, stomach, sarcomas, kidney; small leukemia risk
Childhood cancer	sarcomas, thyroid, brain, breast; no leukemia excess
Hodgkin's disease	possibly breast, stomach, thyroid, lung; no leukemia excess
Non-Hodgkin's lymphoma	leukemia
Breast	leukemia, uterus; small risk of second breast cancer
Patients treated with radiation for benign conditions	
Spondylitis	leukemia, lymphoma, lung, esophagus, stomach, central nervous system, bone
Tinea capitis	thyroid, brain, skin, leukemia
Thymus enlargement	thyroid, breast
Benign gynecological diseases	leukemia, colon, uterus, bladder
Benign breast disease	breast
Tuberculosis (fluoroscopy)	breast; no lung or leukemia
Patients exposed to diagnostic radiation	
Scoliosis (spinal X-rays)	breast
Children (prenatal)	leukemia, childhood cancer
Adults (general diagnostic procedures)	possibly multiple myeloma and CML only

Source: Boice, 1992.

Table 5. Cancer in Patients Exposed to Internal Radionuclides

Population (Radionuclide)	*Observed Excess Cancer*
German patients with bone disease (Ra-224)	bone; possibly liver and breast
Patients with thyroid cancer (I-131)	leukemia, bladder
Patients with polycythemia vera (P-32)	leukemia
Thorotrast patients (Th-232)	liver, leukemia, lung; possibly bone and others

Source: Boice, 1992.

and cancer of the thyroid. The first cancer reported was leukemia which began to appear in the exposed population two years after the bombing. However, not all leukemia types were equally affected. Acute leukemia and chronic granulocytic leukemia were substantially increased in the exposed populations but chronic lymphocytic leukemia incidence remained unchanged in survivors. Radiation-induced solid cancers became apparent 5 to 10 years (at a minimum) after leukemia induction. Only after 1974 did the cumulative excess of solid cancers since 1950 exceed the leukemia excess. The major solid tumors found in excess in the Hiroshima and Nagasaki populations have been neoplasms of the thyroid, female breast, and lung. In addition to these, increasing numbers of cancers of the esophagus, stomach, urinary tract, and lymphomas have been observed (NAS, 1990). The Hiroshima and Nagasaki experience has formed the basis for an extensive human data base which has been used in the development of radiation risk estimates and radiation protection standards. Supplementing the atomic bomb survivor data are a large number of smaller epidemiological studies involving medical uses of radiation.

Several populations have been studied for radiation-induced cancer as a result of fallout from nuclear weapons tests in the 1950s in the South Pacific and western United States. During one of these tests in 1954, over 200 inhabitants of the Marshall Islands (near the Bikini Island test site in the South Pacific) were accidently exposed to fallout radiation from the bomb. Whole body doses have been estimated to be 14–175 cGy and thyroid doses (resulting from incorporation of radioiodine from fallout) were as high as 700–1,400 cGy in children and 350 cGy in adults. Following 23 years of study, 244 subjects had thyroid nodules and seven thyroid cancers appeared. In a study of childhood malignancies, an increased incidence of leukemia was observed in children aged 0 to 14 who may have been exposed to fallout in Utah as a result of atmospheric nuclear weapons tests conducted in Nevada. In a preliminary study of over 3,000 military personnel present during atmospheric weapons testing, an increased incidence of leukemia has been reported. Interpretation of the results of the Utah and Nevada studies is complicated by the large uncertainty in the radiation doses received by individuals and other problems related to the nature of the study populations (Mossman, 1984).

Medical Exposures

Exposures to ionizing radiation for medical purposes have occurred in a variety of settings including radiotherapy for cancer, radiotherapy for benign disease, and radiodiagnostic studies (Tables 4 and 5). Medical radiation exposures, particularly as therapy for benign diseases and cancer, have been associated with excess cancer. Populations include individuals exposed internally from injected or ingested radionuclides or exposed externally from X-rays for ankylosing spondylitis, benign breast disease, ringworm of the scalp, hyperplastic disorders of the lymphoid tissues in the head and neck, benign gynecological disorders, cancers of the cervix and breast, and childhood cancers. In many situations, patients received doses of hundreds of centigray to treated areas (NAS, 1990; Boice, 1992; Schull and Weiss, 1992).

An international study of cancer incidence following radiotherapy for cervical cancer was conducted on more than 200,000 women in 15 countries. A small but significant leukemia risk was detected. After a minimum latent period of about 10 years, risk of second tumors following radiotherapy increased with time and reached nearly twofold in long-term survivors (Boice, 1992).

The mortality experience of over 14,000 persons treated between 1935 and 1954 in 87 British radiotherapy clinics for ankylosing spondylitis (a rheumatoid condition of the spine) indicate a significant excess of leukemia. Radiation doses averaged approximately 380 cGy to the active bone marrow of these patients. Leukemia risk peaked 3–5 years after treatment. No excess in chronic lymphocytic leukemia was observed in these patients, a finding consistent with the studies of the Japanese atomic bomb survivors. Lung cancer, stomach cancer, and other solid tumors were also seen in excess in the spondylitic patients (Boice, 1992; Schull and Weiss, 1992).

Excess breast cancer has been observed in women treated for postpartum mastitis and in women who received multiple chest fluoroscopies to manage pneumothorax treatment for tuberculosis. In a study of over 600 women treated for postpartum mastitis, a significant excess in breast cancer was observed in which the average cumulative breast dose was 350 cGy. X-ray fluoroscopic examinations were frequently used to monitor lung collapse treatments for tuberculosis during 1935–1954. The most recent study included over 2,500 women in Massachusetts who received an average of 88 fluoroscopies for an average total dose of 79 cGy. Excess breast cancer did not appear until 10–15 years after exposure and was observed to be proportional to radiation dose. Exposures during the adolescent and teenage years was associated with the highest risk (NAS, 1990; Boice 1992; Schull and Weiss, 1992).

Irradiation of children for ringworm of the scalp (tinea capitis) has been associated with excess incidence of thyroid, brain, and skin cancers. In one study of over 10,000 children excess thyroid cancer was observed even though the average dose to the thyroid was only 9 cGy. In another study of 2,200 tinea capitis patients, no excess thyroid cancers were found, but brain and skin cancers were observed in excess (Mossman, 1984). Thyroid cancer excess has also been observed in children given radiotherapy for tonsil and nasopharyngeal conditions. In one study involving exposure of over 3,600 individuals for whom vital statistics were known, 8% had thyroid cancer (Boice, 1992).

Excess liver cancer and leukemia have been observed in patients given Thorotrast (a colloidal preparation of thorium-232 dioxide) for radiographic imaging primarily for the diagnosis of suspected brain diseases. Use of Thorotrast began in 1928 but was stopped in 1935 after it was learned that it caused liver cancer. The injected thorium was deposited in body tissues and resulted in continuous alpha particle exposure at low dose rates (about 25 cGy/year liver dose rate and 16 cGy/year endosteal dose rate following a 25 ml injection of Thorotrast). Surveys of over 3,000 patients in Denmark, Portugal, and Germany showed excessive liver cancers (Mossman, 1984; Boice, 1992; Schull and Weiss, 1992).

Thyroid cancer treated by high dose I-131 has been associated with excess leukemia and breast cancer. However, because of the small number of patients involved, the relation between I-131 therapy and incurred cancer risk has to be regarded as questionable in these thyroid cancer patients (Mossman, 1984).

Studies linking diagnostic X-ray exposure and cancer have, in general, failed to show a correlation. Studies linking diagnostic radiation with adult leukemia are inconsistent. In one report from California involving 136 cases and a like number of controls, an association between diagnostic X-ray, particularly low back X-rays, and chronic myelogenous leukemia was found. However, in another study involving 1,000 patients with leukemia and lymphoma, little evidence was found that diagnostic X-ray procedures were causally associated with the cancers (Boice, 1992). In another study to evaluate the risk of radiation-induced cancer mortality following cardiac catheterization (25–30 cGy per examination), no increase in cancer mortality was observed in a group of over 4,800 children with congenital heart disease (Mossman, 1984).

Several studies have suggested a correlation between *in utero* X-ray exposure and an increased risk of leukemia in childhood. Most reports seem consistent with a 40% increased risk of childhood leukemia associated with low dose intrauterine exposure to diagnostic radiation between 1–10 cGy just before birth. It remains to be established whether adult cancer is increased by fetal irradiation, although there has been recent evidence that this may be the case. In analyzing the childhood leukemia studies, it has been suggested that selection factors, related to the medical reasons why women received prenatal X-rays, were responsible for at least a portion of the increased leukemia risk. The absence of any childhood leukemia in atomic bomb survivors exposed *in utero* supports this selection theory. While it is now generally accepted that prenatal X-ray is associated with an increased risk of childhood cancer, the causal nature of this association and the level of risk remain uncertain (NAS, 1990; Boice, 1992; Schull and Weiss, 1992).

Occupational Exposures

Studies of individuals occupationally exposed to ionizing radiation also provide evidence of association between cancer and radiation (Table 6). The first neoplasm linking radiation exposure and cancer was a skin cancer in a radiologist reported in 1902. In a survey of over 6,000 radiologists practicing between 1920 and 1969, excess cancers of the skin, leukemia, and aplastic anemia were noted particularly in those radiologists practicing in the 1920s and 1930s. These early radiologists were exposed to large doses of radiation because of the primitive X-ray equipment used and poor radiation practices employed. Today, with the use of high technology equipment and films and knowledge of radiation safety practices, modern day radiologists apparently no longer have excess risks of leukemia and skin cancer (Mossman, 1984).

Excess bone cancers have been observed in luminous dial painters. In a study of nearly 1,500 women in the U.S. radium watch dial industry, excess bone sarcomas were observed as a result of the ingestion of large quantities of bone seeking isotopes of radium. Workers ingested the radium by licking paint brushes in order to make fine tips. Head cancers, particularly of the paranasal sinuses, were also found in excess in these workers, presumably the result of radon gas (a product of radium decay) accumulation in the sinuses (Mossman, 1984).

In a study of 4,000 uranium miners, excess cancer mortality as a result of lung cancer has been observed. Lung irradiation in miners (mean dose 400–500 cGy) was due to inhalation of radon gas. The magnitude of the radiation risk is not certain, however, because of the significant smoking histories of many of the miners. Some studies suggest that smoking and radiation interact additively in elevating the risk for lung cancer; other studies suggests a greater than additive or multiplicative interaction (NAS, 1988). Although lung cancer in uranium miners as a consequence of high exposures to radon gas is well recognized, the public health effects of radon in the domestic environment (generally orders of magnitude lower in concentration than that found in mines) remains highly questionable. Several epidemiological studies in China, Sweden, and the United States have failed to provide compelling evidence that indoor radon increases lung cancer risk. Studies of indoor radon have to be interpreted with caution because of the inherent difficulties in accurately

Table 6. Cancer in Populations Occupationally Exposed to Radiation

Population	*Observed Excess Cancer*
Radiologists	leukemia, skin; possibly breast, thyroid, myeloma; no lung
Underground miners	lung
Luminous dial painters	bone; possibly myeloma

Source: Boice, 1992.

estimating past exposures and in adequately adjusting for the effects of smoking (Boice, 1992; Schull and Weiss, 1992).

Cancer mortality in nuclear workers has been the subject of several investigations but results have been generally inconsistent (NAS, 1990). The mortality experience of about 31,500 male and 12,600 female workers employed between 1944 and 1978 at the Hanford nuclear installation in Richland, Washington revealed no evidence for increasing risk with increasing film badge exposure for any cancer except for multiple myeloma. The most recent analyses of Hanford workers revealed a strong healthy worker effect and a significant deficit in cancer mortality, including leukemia (Boice, 1992).

Recently, radiation risks were reported to be an order of magnitude higher than currently accepted (NAS, 1990) in a report of Oak Ridge National Laboratory workers. An excess of leukemia was reported but the risk was observed to decrease with increasing levels of exposure. Problems with the study design and data analysis suggests that considerable caution must be exercised in accepting these unusual results. Analysis of the data failed to take into account the influence of various socioeconomic factors. Furthermore, chronic lymphocytic leukemia, which is known not to be increased by radiation, was grouped with other cancer types. Compared to the general population, the Oak Ridge workers were at a 21% significantly reduced risk of dying of cancer (Boice, 1992).

A proportional mortality study of naval shipyard workers suggested an increased risk of cancer and leukemia among nuclear workers in Portsmouth, New Hampshire. However, results of this study have not been borne out by other studies. Interpretation of these results is complicated because the use film badge exposures is only an approximation of actual organ doses, and the contributions of natural background over the working lifetime (i.e., several decades) is often greater than the occupational dose. Furthermore, confounding factors such as the contribution of other occupational and nonoccupational carcinogens are usually not considered (Mossman, 1984; NAS, 1990; Boice, 1992).

PRINCIPLES OF RADIATION CARCINOGENESIS

Based on a large body of laboratory animal and human radiation exposure studies (some of which have been discussed above), a number of generalizations can be made about radiation carcinogenesis.

Ionizing radiation is a weak carcinogen. Ionizing radiation from the natural background and medical sources accounts for a very small percentage of the cancer burden in the United States (Tables 1 and 2). Cigarette smoking, dietary factors, and ultraviolet light exposure are major human carcinogens which cause the vast majority of cancers. Table 7 compares the efficiency of ionizing radiation and other mutagenic and carcinogenic agents in producing DNA damage. The amount of DNA damage produced by ionizing radiation is orders of magnitude lower than for most other agents for equal cell killing efficiency.

Table 7. Yields of DNA Damage for Various Agents at Equal Cell Killing Efficiency

Agent	*DNA Lesion*	*Number of Lesions per Cell per D_{37}* [a]
Ionizing radiation	ssB[b]	1,000
	dsB[c]	40
	DPC[d]	150
Bleomycin A2	ssB[b]	150
	dsB[c]	30
UV light	thymine dimer	400,000
	ssB[b]	100
Hydrogen peroxide	ssB[b]	<2,600,000
Benzo[a]pyrene 4,5-oxide	adduct	100,000
Aflatoxin	adduct	10,000
1-Nitropyrene	adduct	400,000
Methylnitrosourea	7-methylguanine	800,000
	O^6-methylguanine	130,000
	3-methyladenine	30,000
2-(N-Acetoxy-N-acetyl)amino-fluorine	adduct	700,000

Notes: [a]D_{37}, dose of agent required to reduce survival of cells to 37% of the number exposed.
[b]ssB, single strand breaks.
[c]dsB, double strand breaks.
[d]DPC, DNA-protein cross-links.

Source: Ward et al., 1988.

Radiation-induced cancers are indistinguishable from naturally occurring cancers. Because radiation-induced cancers are indistinguishable from naturally occurring cancers, it is impossible to show, in individual cases, that a particular cancer was radiation-induced. Radiation induction of cancer can only be inferred by studying large populations of exposed individuals and comparing the specific cancer incidence in the exposed population with that in an appropriate control population (Mossman 1984). Although there is ample evidence that radiation causes cancer in man and laboratory animals, the mechanisms of radiation carcinogenesis remain speculative. The origin of cancer is hypothesized to be the result of mutational events to critical genetic loci that lead to the development of cancer. This "somatic mutation hypothesis" leads to the following outcomes: (a) spontaneous mutations should produce an irreducible background level of cancer, (b) agents which change the host genome by mutation or by the addition or deletion of genetic mutation (e.g., ionizing radiation) should increase the background cancer incidence, (c) abnormalities in the genome that favor mutations should increase the risk of cancer, and (d) inheritance of an initiating mutation should strongly predispose the host to cancer (Knudson, 1985).

The important and rapidly developing field of molecular biology is having a significant impact on our knowledge and understanding of the carcinogenesis process. There are at least six known tumor-suppressor genes which if deleted or inactivated can increase the risk of cancer, including the RB or retinoblastoma gene on chromosome 13 and the WTG-1 gene on chromosome 11, which has been implicated in Wilm's tumor. Errors in the p53 gene on chromosome 17 are among the most fascinating genetic defects associated with human cancer. Defects in the p53 gene have been implicated in cancers of the colon, lung, esophagus, breast, liver, brain, and tissues of the reticuloendothelial and hematopoietic systems (Fabrikant, 1992).

Radiation-induced cancers appear in almost every tissue of the body. Almost all of the human cancers can be induced by ionizing radiation. However, chronic lymphocytic leukemia, Hodgkin's disease, cancer of the prostate, cancer of the uterus and cervix, cancer of the testis, and cancer of the mesothelium and mesentery appear to be exceptions. It is unclear why these particular cancers are not induced by radiation exposure; yet other cancers, such as female breast, thyroid, and lung, can be induced by radiation (Mossman, 1984).

Sensitivity to radiation carcinogenesis varies considerably for various tissues and organs. Among the numerous tissues and organs known to be susceptible to radiation-induced carcinogenesis, there is a wide range of sensitivities. The types of cancers with the highest sensitivity to radiation induction are female breast, thyroid, and leukemia (especially myeloid leukemia); cancers with moderate sensitivity include lymphomas (excluding Hodgkin's disease), lung (bronchus), stomach, colon, liver and biliary tract, pharynx, and pancreas; cancers with low sensitivity include kidney, bone, skin, brain, and salivary glands. Why organs and tissues vary in sensitivity is not known. Apparently, the sensitivity to radiation induction is unrelated to the spontaneous incidence for most cancers. For instance, thyroid cancer has a low spontaneous incidence but a very high radiation sensitivity; skin cancer has a high spontaneous incidence but a low radiation sensitivity; prostate cancer has a very high spontaneous incidence and apparently cannot be induced by radiation. Female breast cancer is an example of a human cancer with a high spontaneous incidence and a high radiation sensitivity (Mossman, 1984).

Cancer induction by radiation has a long latent period. A feature of radiation-induced cancer is a long period of time between the radiation exposure and the appearance of the cancer. This time period, which may be decades for some tumors, is referred to as the latent period. It is unclear whether the latent period represents an induction period in which the transformed cells prepare to divide and form a tumor, or represents a development period in which the tumor assumes characteristics of a malignancy involving growth and spread, or a combination of these (Mossman, 1984). Latent periods are different for human tumors. Leukemia has the shortest latency (5–10 years). For solid tumors, the mean latent period appears to be 20–30 years. The mean latency and the distribution of latent periods in an irradiated population for a given tumor type may depend on several factors including the age distribution of the population and radiation dose. The characteristic long latent period complicates the study of cancer induction by radiation. During the long latent period, irradiated individuals may be exposed to other carcinogens and other disease processes which can confound the interpretation of the epidemiological data (Mossman, 1984).

Age, sex, and other host factors may modify cancer risk. Various host factors may contribute to cancer induction by radiation. Breast cancer and thyroid cancer have a significant dependency on sex. Radiogenic breast cancer and thyroid cancer are higher in females and consequently, the total cancer risk is greater for females than for males. Breast cancer occurs almost exclusively in women; the natural incidence of thyroid cancer is almost three times greater in females than in males (American Cancer Society, 1993). Radiogenic risks for cancer of the lung and esophagus also appear to be higher in females (Schull and Weiss, 1992). Age at the time of exposure is probably the most important determinant of risk for cancer. Individuals exposed before age 20 have higher risks than those exposed at later ages. Breast cancer risk is highest when exposure occurs in the second decade of life (Mossman, 1984; Schull and Weiss, 1992).

Several other host related factors may contribute to radiation risk. Hormonal status may influence radiation risk for breast and thyroid cancer. However, in general, sensitivity to radiation may not be necessarily related to hormones. Although the female breast and thyroid have high radiation sensitivity, the prostate and uterus (also under hormonal influence) are not known to be susceptible to radiation carcinogenesis. Certain specific groups in the general population may be at higher risk for cancer induction by radiation. Risk for radiation-induced thyroid cancer is apparently higher in Jewish children than in other ethnic groups. Individuals with diseases characterized by genetic-based deficiency in DNA repair capacity, including ataxia telangiectasia and xeroderma pigmentosum, are at higher risk for cancer. Cigarette smoking increases the radiation risk for lung cancer in uranium miners although the exact nature of the interaction remains unknown (Mossman, 1984).

Dose-response relationships are different for different cancers. The relationship between radiation dose and excess cancer incidence is different for the various cancer types known to be induced by radiation. Leukemia data can be fit by a curvilinear dose response model. This model predicts that radiation effects per unit dose are lower at low dose than at high dose. Other neoplasms, such as bone cancer, can be fit by a threshold dose response function. The threshold for bone cancer appears to be about 500 cGy. Breast cancer data may be fit by a linear no-threshold model. Unlike the curvilinear model, this model predicts that the radiation effects per unit dose are equal at low doses and high doses (Mossman, 1984).

Risk estimates of radiation-induced cancers are based primarily on high dose data. Evidence that ionizing radiation is a human carcinogen is derived almost exclusively from a number of epidemiological studies of populations exposed to high doses. Although no epidemiological studies have conclusively shown that radiation at low doses causes cancer, it is nevertheless assumed that radiation is carcinogenic at low doses. A popular view within the scientific community has been that the risk of developing cancer is directly proportional to dose, even at very low doses, and that any dose, no matter how small, potentially may be damaging (the linear no-threshold model). Other more complex dose-response models, such as the quadratic and linear-quadratic, have also been used to predict the carcinogenic effects of low doses of radiation. Dose-response models are important in estimating risks of cancer induction following low doses of radiation since the statistical inaccuracies of direct observations at low doses are formidable.

Dose extrapolation must be viewed with caution. First, insufficient data exist in the low dose region to distinguish one extrapolation model from another. Thus, estimates of health effects at low doses are associated with large uncertainties. Second, the extrapolation process, especially for the well-known linear, no-threshold model, assumes that carcinogenic mechanisms operative at high dose are also relevant at low dose. This assumption is probably not valid because significant cell killing occurs in the high dose range (> 100 cGy) but to a much smaller degree, if at all, at low dose. Cell killing may play an important role in carcinogenesis by promoting cell division in surviving cells. Thus, use of simple extrapolation models may overestimate risks of carcinogenesis at low doses.

RADIATION RISKS

Table 8 summarizes current risk estimates for radiation-induced cancers. Risk is expressed in terms of fatal cancers per centigray. Thus, risk estimates for cancer incidence will differ from the risks shown because of different mortality rates for cancer sites. For instance, the risk estimate for thyroid cancer incidence is approximately 10 times higher than the value in the table because of the low mortality rate for thyroid cancer (approximately 10%). Lung cancer, on the other hand, has a high mortality rate (about 90%) and, consequently, the lung cancer incidence risk will be approximately the same as the mortality risk listed in the table.

These risk estimates have been based primarily on studies of the Japanese atomic bomb survivors. Accordingly, risk estimates reflect the natural incidence of cancer in the Japanese population. For instance, the high radiological risk for stomach cancer (Table 8) reflects the high natural incidence of stomach cancer in the Japanese rather than a high radiation sensitivity to carcinogenesis. Because the natural incidence of stomach cancer is lower in Americans (and in Japanese who adopt the American diet), radiation risks for stomach cancer in Americans would be expected to be lower. The risk estimates listed in Table 8 must be used with caution when applied to the U.S. or other non-Japanese populations.

The total cancer mortality risk due to radiation is small compared to the baseline lifetime cancer mortality rate. Approximately 20% of the population may be expected to die as a result of cancer (Table 8). This risk is increased by 0.25% for each centigray of radiation dose (to the whole body).

Table 8. Lifetime Risk Estimates for Fatal Cancer at Low Doses

Cancer	*Risk per Centigray*[a]
Leukemia	5×10^{-5}
Bone surface	0.5×10^{-5}
Breast (females only)	4×10^{-5}
Lung	8.5×10^{-5}
Thyroid	0.8×10^{-5}
Colon	8.5×10^{-5}
Esophagus	3×10^{-5}
Skin	0.2×10^{-5}
Stomach	11×10^{-5}
Liver	1.5×10^{-5}
Bladder	3×10^{-5}
Ovary	2×10^{-5}
Other (combined remaining tissues)	5×10^{-5}
Sum of fatal cancers	50×10^{-5} (1 in 2,000) per centigray
Baseline cancer mortality	0.15 (1 in 6.7) to 0.25 (1 in 4)

Note: [a] 1 centigray = 1 rad.

Diagnostic X-ray examinations pose little risk of radiogenic cancer. Doses to various tissues and organs (e.g., bone marrow, breast, lung, and thyroid) range from cGy to about 0.50 cGy, depending on the nature of the study (e.g., chest, barium enema, barium swallow, CT), number of views, and fluoroscopy (Laws and Rosenstein, 1978). Assuming a whole body dose of 0.2 cGy for a radiodiagnostic study, the total lifetime cancer mortality risk due to radiation is approximately 1/10,000. This risk represents a 0.05% increase in the baseline cancer mortality risk.

Radiation therapy exposes small volumes of tissue to large radiation doses. Relatively little is known about the risk for radiogenic secondary tumors in individuals who have been treated with radiation for a primary malignancy. Some evidence has been derived from studies of radiotherapy patients treated for cervical, ovarian, and breast cancers, for Hodgkin's disease and non-Hodgkin's lymphoma, and in certain instances of tumors in childhood (Fabrikant, 1992). In an analysis of the overall carcinogenic risk to radiotherapy patients, a lifetime leukemia risk of 0.7% and solid tumor risk of 0.3% has been estimated (Cumberlin et al., 1989). Risk appears to be higher in children because they have more years of life, compared to elderly patients, during which they are subject to risk.

SUMMARY

Ionizing radiation at high doses is known human carcinogen. Numerous population studies involving military, medical, and occupational uses of radiation clearly show that leukemia and a variety of solid tumors may be induced by radiation. However, at low doses of radiation, typically encountered in diagnostic radiology and nuclear medicine, the evidence for cancer causation is much less compelling. Most epidemiological studies are inconsistent. Only a few studies suggest a significant association between radiation and cancer. Even in these investigations, the causal nature of such associations and the levels of risk remain uncertain.

Radiation risks for cancer have been based primarily on studies of Japanese survivors of the atomic bombings. Extrapolation of data derived from these studies, using a linear no-threshold model, has been the basis for predicting cancer risk at low doses. The derived risk estimates suggest that radiodiagnostic studies, including nuclear medicine and X-ray examinations, are associated with small risks for cancer. Radiotherapy for cancer, which involves substantially higher doses, are associated with higher risks, but these risks are still very small considering the benefits derived from the treatments.

REFERENCES

American Cancer Society (1993). Cancer Facts and Figures–1993. Atlanta, GA: American Cancer Society, Inc.

Boice, J.D. (1992). In: The Biological Basis of Radiation Protection Practice (Mossman, K.L. & Mills, W.A., eds.), pp. 89–120, Williams and Wilkins, Baltimore.

Cumberlin, R., Dritschilo, A., & Mossman, K. (1989). Carcinogenic effects of scattered dose associated with radiation therapy. Intl. J. Radiat. Oncol. Biol. Phys. 17, 623–629.

Doll, R. & Peto, R. (1981). The causes of cancer: Quantitative estimates of avoidable risks of cancer in the United States today. J. Natl. Cancer Inst. 66, 1191–1308.

Fabrikant, J.I. (1992). In: Radiation Protection in Medicine (Poznanski, A.K., ed.), pp. 111–121, Proceedings of the Twenty-eighth Annual Meeting of the National Council on Radiation Protection and Measurements (NCRP), NCRP, Bethesda, MD.

Knudson, A.G. (1985). Hereditary cancer, oncogenes, and anti-oncogenes. Cancer Res. 45, 1437–1443.

Laws, P. & Rosenstein, M. (1978). A somatic dose index for diagnostic radiology. Health Physics 35, 629–642.

Mossman, K. (1984). Ionizing radiation and cancer. Cancer Investigation 2, 301–310.

National Academy of Sciences (NAS) (1988). Health Risks of Radon and Other Internally Deposited Alpha Emitters (BEIR IV). National Academy Press, Washington, DC.

National Academy of Sciences (NAS) (1990). Health Effects of Exposure to Low Levels of Ionizing Radiation (BEIR V). National Academy Press, Washington, DC.

International Commission on Radiological Protection (ICRP) (1990). 1990 Recommendations of the International Commission on Radiological Protection. ICRP Publication 60. International Commission on Radiological Protection, Oxford, UK.

National Council on Radiation Protection and Measurements (NCRP) (1987). Ionizing Radiation Exposure of the Population of the United States. NCRP Report No. 93. National Council on Radiation Protection and Measurements, Bethesda, MD.

National Council on Radiation Protection and Measurements (NCRP) (1989). Exposure of the U.S. Population from Diagnostic Medical Radiation. NCRP Report No. 100. National Council on Radiation Protection and Measurements, Bethesda, MD.

Schull, W. and Weiss, K. (1992). Radiation carcinogenesis in humans. Adv. Radiation Biology 16, 215–258.

Shimizu, Y., Kato, H., Schull, W., & Preston, D. (1989). Studies of the mortality of A-bomb survivors 9. Mortality, 1950–1985: Part 1. Comparison of risk coefficients for site-specific cancer mortality based on the DS86 and T65DR shielded kerma and organ doses. Radiation Res. 118, 502–524.

U.S. Environmental Protection Agency (1992). A Citizen's Guide to Radon (second edn.). U.S. Government Printing Office 1992-329-782, Washington, DC.

Ward, J.F., Limoli, C.L., Calabro-Jones, P., & Evans, J.W. (1988). In: Anticarcinogenesis and Radiation Protection (Cerutti, P., Nygaard, O.F., & Simic, M., Eds.), pp. 321–327, Plenum, New York.

Chapter 3

Environmental Causes of Cancer

J.S. MALPAS

Advances in Oncobiology
Volume 1, pages 61–78.

ISBN: 0-7623-0146-5

INTRODUCTION

Cancer affects the whole of the animal kingdom, and seems to be an inevitable consequence of the eukaryotic state. Its causation in man has been the subject of intense study, which has as its ultimate goal the eradication of the agents responsible from the environment.

It is evident that the risk of developing cancer in man increases with age, which is such an important factor that it has to be taken into account in all data related to incidence. The age-specific incidences for such common cancers as those of the breast and colon rise sharply at about 50 years of age. These facts could suggest that the body is programmed to develop cancer as the tissues age. The wide variation in types of common cancer throughout the world is against this notion: it is hardly likely that Glaswegians are programmed for lung cancer, while populations in Kazakhstan, Transkei and Bulawayo are programmed for esophageal cancer. It is much more likely that increasing incidence with age is due to cumulative exposure to an environmental hazard.

A second important argument for environmental influence arises out of a study of incidence. It has been noted that when populations migrate, the incidence of cancers from which they suffer alters, and they show a gradual acquisition of the cancer pattern of the population they have joined. This, it must be emphasized, is a gradual progress, and will take several generations before it is complete. A good example is the incidence of stomach cancer, which gradually falls in Japanese living on the West Coast of North America.

For certain cancers there is now overwhelming evidence for an environmental cause. In the case of certain industrial exposures, this is now so convincing that legislation has been introduced to protect the community. Certain social activities, such as smoking tobacco or drinking alcohol, occupational exposures to chemicals such as 2-naphthylamine or asbestos, or the administration of certain pharmaceutical agents, can be shown to be associated with cancer and, more importantly, the degree of exposure correlates with frequency. Final proof is provided by the fact that withdrawal of the noxious agent reduces the incidence.

It must be admitted that cancer occurs in children, but it is rare, and there is increasing evidence of a strong genetic component in many childhood cancers. It is also true that a number of common cancers such as breast and colon have no easily demonstrable environmental cause. Nevertheless, the evidence points overwhelmingly in the direction of an environmental hazard. The number of substances known to cause cancer is now so large, and the substances themselves so widespread, that as Morris Goldblatt once remarked, it is a lucky man that succeeds in getting out of this world alive!

MEASURING THE RISK

Cancer epidemiology is the study of the pattern of cancer in population. An excellent review of the nature of epidemiological evidence and the contribution this can make to the prevention of disease is the Rock Carling Fellowship research published by Richard Doll in 1967. Appropriate measurements are the first step in identifying environmental causes, and some description of the advantages and disadvantages of the statistical methods used is required. In many countries cancer registries collect accurate data on diagnoses. These are used to describe the incidence of various cancers, the difference in incidence between males and females, the incidence at various ages and in various socio-economic classes, occupations, different localities, and even over the course of time in different cohorts of population. Even ethnic group or religion may provide important clues as to etiology; for example, the low rates of tobacco and alcohol-related cancers in the Mormon population in the United States. Collection of these data will allow the calculation of the proportional risk for a particular group affected by a type of cancer. The term relative risk is frequently used, and is defined as the ratio of the incidence rate in the particular population under consideration for, say, exposure to a particular agent, compared to the rate in a population that is not exposed. If a cancer is very rare, the absolute risk is a better measure of the effect; for example, the number of patients with mesothelioma in a population exposed to asbestos.

INCIDENCE

Incidence is the number of new cases that occur in a population, and the incidence rate is the number of new cases over a period of time. As cancer is relatively uncommon, large populations need to be observed, and a convenient figure is usually that of the number of new cases occurring in 100,000 of the population in one year. It has been mentioned previously that age is a most important factor, so that for comparison of data age-specific rates are used, as these alone have meaning when different populations are to be compared. Clearly, comparison of an aging population in Western Europe with that of a developing country would be of little value.

Incidence rates are most frequently used, but it may also be convenient to express the risk of a particular group being affected by cancer in terms of another population. Thus, the standardized mortality ratio (SMR) for, say, deaths from acute leukemia in an area over a given time can be calculated by applying national age- and sex-specific rates to the number of people in the area in each age/sex group. The rate observed (O) to the expected (E) frequencies gives the SMR, which can be expressed as:

$$SMR = 100 \times \frac{O}{E}$$

MORTALITY

Mortality rates are calculated and interpreted in exactly the same way as incidence rates. Since they record deaths rather than diagnosis, and because modern treatment may be curative, diagnosis does not always equate with death. Examples of this are the successful surgical treatment of colon or breast cancer. Although this is a disadvantage of mortality statistics, mortality rates are more available, and while they may underestimate the true incidence of a particular cancer and, since most data are used for comparisons so long as the death rates are approximately equal in the two populations, mortality rates may be valid for such a purpose. They are, however, unhelpful where there has been a steady improvement in the curability of a particular cancer, when they are useless as a guide to the changing risk for that disease.

SOURCES

There are many useful publications which are invaluable sources for epidemiologists. These include the International Classification of Disease (ICD), a publication by the World Health Organization (WHO); the World Health Statistics Annual, also collected and published by WHO; and Cancer Incidence in Five Continents (now edited by Muir et al., 1987), a compilation of cancer statistics gathered from cancer registries around the world.

DESCRIPTIVE EPIDEMIOLOGY AND THE EVIDENCE FOR AN ENVIRONMENTAL CAUSE

The collection of accurate data on incidence or mortality may be helpful in detecting possible environmental causes of cancer.

Geographical Evidence

The geographical distribution of different types of cancer and comparison of markedly high or low rates of incidence inevitably gives rise to questions which are difficult to answer. Data for these studies are usually collected from administrative areas as by-products of naturally-gathered statistics and great variation may be seen internationally and nationally. Table 1 gives some examples of the wide range of incidence of individual cancers. As will be seen from this table, the incidence rates for different cancers vary widely from one part of the world to another: cancers of the lip, esophagus, nasopharynx, and skin (melanoma) being particularly notable. This suggests that environmental factors are more likely than genetic causes for such variations.

These variations imply that diet, social habits, or other aspects of the environment are important. An obvious example is the exposure to strong sunlight in Queensland, Australia as a cause for melanoma of the skin. When incidence data are collected within a country, the differences may not be so marked; they nevertheless provide convincing evidence for an environmental, rather than a genetic, factor because the genetic background would be likely to be the same for the whole nation. For example, very high local incidence of nasopharyngeal cancer has been shown to occur in China. These tumors, found in the population of Southern China (the so-called "Canton tumors") are now known to be associated with Epstein-Barr virus, and possibly with a promoting factor from the Tung oil tree (*Aleurites fordii*), which produces a carcinogen, a diterpine ester, also found in *euphorbiacae*.

Once a factor is suspected of being closely associated with the occurrence of cancer (for example, cigarettes with lung cancer), convincing evidence to support

Table 1. Worldwide Variation in Incidence of Cancer at Various Sites: Cases Per 100,000 Population

Site	*ICD No.*	*Males*			*Females*		
		Highest	*Lowest*	*Ratio*	*Highest*	*Lowest*	*Ratio*
Lip	140	Canada (N'fdland.)	Japan		Australia	U.K.	
		15.1	0.1	151.0	1.6	0.1	16
Nasopharynx	147	Hong Kong	U.K.		Hong Kong	U.S. (Iowa)	
		30.0	0.3	100	12.9	0.1	129
Oesophagus	150	France	Rumania		India	Czech Rep.	
		29.9	1.2	24.9	12.4	0.3	41.3
Melanoma	172	Australia (Q'land)	Japan		New Zealand	India	
		30.9	0.2	154.5	68.1	1.2	56.8

Note: Data based on Muir et al. 1987.

such a thesis is provided by a dose-response relationship. Although the consumption of cigarettes per capita may not be entirely accurate, since there will be variations in the degree of smoke inhalation, nevertheless there is a striking correlation with the incidence of lung cancer and the number of cigarettes smoked (Table 2).

Occupation

This will be discussed in more detail later, but it is important in a general discussion to point out how valuable occupation has been in identifying carcinogenic hazards. Mortality data have been most valuable because registry information on occupation has, until now (on the whole) been deficient in describing the person's place of work, the various environmental factors (chemical or physical), and whether the noted occupation was the only one for the person registered. It is not uncommon for a person to have a number of jobs, and perhaps the most relevant will be those of adolescence or early adulthood (which are of course the ones most easily forgotten). A number of significant relationships have been demonstrated between the cause of death and occupation. In England and Wales, it has been shown that electrical workers are at a high risk for leukemia, while in the U.S. workers in the food processing, printing, metal fabrication, and machinery industries are more liable to develop this condition. In England and Wales, butchers run a higher risk of lung cancer, and woodworkers are more likely to develop nasopharyngeal cancer. All these observations have given rise to more intensive studies to discover an etiological agent.

Because occupation is such a broad term, and little information may exist as to specific carcinogenic agents associated with a particular occupation, job exposure matrices, which define different chemical and physical factors for each occupation, are now being constructed. These will be of great help in calculating the risk from working in a particular occupation.

Study of Migrants

Observation of the variation in cancer incidence in migrant populations has been of great importance because it provides a naturally-conducted "experiment" which

Table 2. Consumption of Cigarettes and Mortality From Lung Cancer

Country	*Consumption Cigarettes/Adult/Year*	*Mortality Carcinoma of Lung in 1970s*
Norway	500	30
New Zealand	1,500	60
United Kingdom	2,000	80
United States	3,000	100

Note: Adapted from Doll and Peto, 1981.

can help determine the relevance of the genetic element in cancer causation. One of the best studies in this field has been the work on gastric cancer in Japanese migrants to the West Coast of the U.S. Japanese migrants who had been born in Japan showed a lower rate of incidence of gastric cancer than fellow Japanese who remained in Japan, but a higher rate than that of the indigenous white population. Japanese born in the U.S. had an even lower rate of incidence, though it was noted that the incidence of other gastrointestinal cancers such as carcinoma of the colon gradually increased towards the level of incidence of the indigenous population. After several generations, gastric cancer rates in the descendants of Japanese migrants approximate the rates in the U.S. This can be interpreted as the gradual withdrawal of an exogenous factor causing gastric cancer, while there is a gradual increasing exposure to either an exogenous factor causing colon cancer, or the withdrawal of a protective factor for the other gastrointestinal cancers.

The reason for the gradual changes seen is that most migrant populations tend to retain the customs and habits of their homeland for a period of time. Only as the population is assimilated do the subsequent generations acquire the local lifestyle. Migrant population studies have been criticized on the basis that migrants are highly selected—they are young, highly-motivated, and may have been particularly stressed—but it is hardly likely that this would have affected their genetic makeup.

There are a number of other important studies in migrants that have confirmed the gastric cancer story. These include a study on breast cancer and melanoma in migrant populations in Israel. Here, the individual duration of a sojourn in Israel was taken into account, and in each malignancy a trend to develop the indigenous rates for both cancers was seen with increasing duration of residence; this is very suggestive of a local environmental factor.

EVIDENCE FOR ENVIRONMENTAL FACTORS AS A CAUSE OF CANCER FROM STUDIES AT SPECIFIC SITES

Careful study of specific cancers at specific sites has shed light on the etiology of those cancers. As cancer causation is multifactorial, with some contribution from genetic factors, it is often difficult to identify a specific cause. It would be beyond the scope of this chapter to consider in detail every site-specific cancer, but examples have been given which illustrate the most convincing findings from these studies. Extensive reviews are available in Alderson (1986).

Oral Cancer

There is firm evidence that smoking or chewing tobacco, use of oral snuff, and drinking alcohol are causal in this cancer. Additional promoting factors are betel nut and lime. People working in manufacturing or sale of alcohol have increased rates of lip cancer. Important confirmatory evidence comes from a study of migrant populations. In migrants who continue to chew tobacco and betel nut (for example, as in Singapore) the incidence is as high as in the original population. When the same migrants go to Fiji, on the other hand, they stop this habit, and the incidence of the cancer falls to that of the indigenous population.

Esophagus

This raises a different problem because, while alcohol is a definite cause (especially the drinking of Calvados (distilled apple spirit) in Normandy), there is also a high incidence among the abstemious Moslem population in the region of Iran, Turkey, and South Russia. It is possible that either a genetic susceptibility or perhaps a nutritional factor such as iron deficiency may be involved in the Iranian population.

Stomach Cancer

This is of great interest, as there has been a universal decrease in cancer of the stomach worldwide, averaging 1 to 2% per annum. The reason for this is not clear. Genetic factors such as associations with blood group A or gastric atrophy have a weak influence, but the evidence that environmental agents such as nitrosamine or tobacco products are involved is also very weak. There has been a suggestion that protective foods containing vitamin C are more widely consumed and may provide some protection against an unknown noxious agent, although this has recently been questioned. The more frequent screening of populations by endoscopy has also been advanced as a possible reason for the fall in incidence, but this is unlikely, since the greatest fall has occurred in women (who are much less likely to be screened).

Liver Cancer

Primary cancer of the liver only will be considered. The large numbers of patients with primary liver cancers in sub-Saharal Africa and southeast Asia (totaling some 250,000 cases a year) makes this a very important cancer, far exceeding the numbers seen as a result of the rare, genetically-determined condition hemochromatosis. Chronic hepatitis B is endemic in the areas of maximal incidence and there is a causal relationship. There must be an important promoting factor, either another agent or perhaps malnutrition, as other populations with a high incidence of hepatitis B (such as Greenland Eskimos) do not suffer from primary liver cancer.

Bladder Cancer

A number of studies have indicated an increased relative risk for bladder cancer in smokers. The most convincing case for a chemical etiological factor is provided by a study of workers in the rubber industry or in chemical industries producing the aromatic amines, naphthylamine, benzidine, and auramine. The recognition of these as causal agents did much to prevent the deleterious effects of 4-amino biphenyl and alerted industry to its probable carcinogenic activity, leading to its early prohibition.

The examples of environmental agents causing cancer outlined above, and discussed by specific site, are certainly not comprehensive. Lung cancer and its relationship to tobacco will be discussed later, but these examples do show how, for many cancers, a cause has been identified and then removed, with a subsequent fall in incidence. In some cases the effect has even been pre-empted. It also illustrates how, once a causative agent is found (e.g., hepatitis B virus), it may be eliminated by immunization or possibly by the use of interferon in infected areas. Many cases such as breast cancer, ovarian cancer, colon cancer, sarcomas, etc. have no easily identifiable cause, and remain unresponsive to environmental manipulation.

CAUSATION

No consideration of the environment and cancer would be complete without a discussion of the concept of causation. Ethical constraints, of course, place a total embargo on any trial of suspected carcinogenic agents in man. A few "natural" experiments, such as atomic bomb explosions or chemical factory incidents, have yielded valuable information, but these events are, fortunately, rare. As the previous section indicates, it is possible to identify factors, either simple (i.e., chemicals, such as 2-naphthylamine) or complex (i.e., tobacco), which are in some way associated with cancer.

The word association implies a degree of caution. All considerations are by inference rather than absolute proof, and cause is a very strong word. The concept of bacteria as a cause of disease was argued extensively at the end of the nineteenth century. The Henle-Koch postulates or criteria became the basis for the acceptance of a particular organism as a cause of disease. To gain acceptance, the organism has to be found only in the diseased patient, it should be absent from tissues of persons not suffering from the disease, and the organism should be able to induce a similar disease in experimental animals.

There are, of course, difficulties in applying analogous criteria to cancer causation. These include the multifactorial nature of cancer causation, the long latent period between the exposure to an inducing factor and the occurrence of cancer, the different susceptibilities of individuals or races to a particular cancer under study, and the ethical restrictions on human experimentation. Nevertheless, Sir Austin Bradford Hill in 1965 was able to list a series of criteria which, if obeyed, indicated a causal relationship. These are given in Table 3.

Strength of association. Hill considered this one of the most important factors in deciding on causation. He quoted both scrotal cancer and cigarette smoking, with a 200× and 30× relative mortality, respectively, in chimney sweeps and smokers, as examples of strong causational data; he also pointed out that ratios of observed to expected numbers are much more important than the actual incidence.

Table 3. The Bradford Hill Criteria

1.	Strength of association
2.	Consistency
3.	Specificity
4.	Temporality
5.	Dose-response relationship
6.	Plausibility
7.	Coherence
8.	Experimental evidence
9.	Analogy

Consistency was next on the list, and he meant how frequently the effect was seen by other observers in different places at different times. As an example, the Surgeon General of the U.S. Public Health Service had found 20 retrospective and 7 prospective inquiries into the association of smoking and cancer of the lung, which supported the observations of Doll and Bradford Hill in the United Kingdom.

Specificity. Hill was careful not to overemphasize this aspect, which he defined as an association between a specific environmental exposure and a particular site and type of cancer. For example, nickel refiners in South Wales, working in the early part of the 20th century, had a high incidence of nasal cancer. This is quite clearly a causal relationship, but there is always the possibility that there was another cause which was not appreciated at the time.

Temporality. This in essence is a discussion of the "chicken and egg" problem. For example, do workers who develop a particular cancer in an occupation enter that occupation for reasons which are associated with a particular susceptibility to that cancer?

Dose-response relationship. Further confirmation of a causal relationship is provided when increased exposure to the factor considered responsible results in a greater incidence. Two examples are (a) increased concentrations of dust in an industrial process are accompanied by an increased incidence of respiratory disease, or (b) increased numbers of cigarettes smoked are accompanied by an increased incidence of lung cancer.

Plausibility. Hill considered this the least important criterion, largely because knowledge of the biological process involved might not be available at the time.

Coherence. The cause and effect interpretation of the data should not seriously conflict with the known factors of the natural history and biology of the disease.

Experimental evidence. Occasionally it is possible to conduct a large-scale experiment in man which is ethically correct. For example, reducing the dust produced in an industrial process, changing a carcinogenic agent (such as a particular oil) for a non-carcinogenic agent, stopping or reducing smoking, may

Table 4. Industrial Processes Entailing Exposures That Are Carcinogenic to Humans

Aluminum production
Auramine manufacture
Boot and shoe manufacture and repair
Coal gasification
Coke production
Furniture and cabinet making
Hematite mining (radon exposure)
Iron and steel founding
Isopropyl alcohol manufacture
Magenta manufacture
Painting and decorating
Rubber industry

Note: After Tomatis et al., 1990.

give strong support for a causal hypothesis. More recently, tests for mutability and carcinogenicity in the laboratory are being considered increasingly important.

Hill also considered that argument by **analogy** might be of additional help. The fact that an agent causes a cancer enhances the possibility that a similar agent might also do so.

These nine postulates served as the basis for a great, number of successful epidemiological studies conducted in the late 20th century, and have developed into such powerful tools as cohort studies and case control investigations. It is possible, however, that the era of identification of major environmental carcinogens by traditional epidemiological methods is now coming to an end. Successes have included tobacco smoking, the identification of a number of hazardous chemicals used in industry, and more recently, radiation hazards.

It would be impossible to detail all the steps by which a particular environmental hazard has been identified. A list of such hazards is given in Table 4. An outline of the work on the identification of naphthylamine among the aromatic amines recorded by Alderson (1986) is illustrative of the approaches that may be used.

THE CARCINOGENICITY OF AROMATIC AMINES: A CLASSIC STUDY

This work started with the observation by Rehn (1895), who observed three patients with bladder tumors in a group of 45 men manufacturing fuchsin. The three men had been exposed for 15, 20, and 29 years, respectively. In the period before World War I he identified a further 20 cases, and proposed a causal relationship. Germany at that time was a world center of chemical production, but during World War I

Table 5. Occurrence of Bladder Cancer in Dye Workers

Agent	*Observed (O)*	*Expected (E)*	*O/E*	*95% Confidence Limits*
1-Naphthylamine	6	0.70	8.57	3.1–18.7
Benzidine	10	0.72	13.89	6.6–25.5
2-Naphthylamine	26	0.30	86.67	56.6–127.0
Mixed	81	1.48	54.73	43.5–68.0

Note: After Alderson, 1986.

chemicals had to be produced by other countries including Great Britain, and it was soon noticed that bladder cancer was occurring in workers in other countries as well as Germany.

In 1954 Case (Case and Mosker, 1954) assembled a nominal roll of 2,466 males employed by 21 firms who had been exposed to four dyes for at least six months between 1921 and 1950. This was checked against a file of death certificates mentioning bladder cancer. A comparison of expected against observed figures, allowing for age, sex, and calendar period, is given in Table 5. An extension of this study to 1,555 men who had a similar length of exposure to the aniline dyes magenta and auramine was carried out by Case and Pearson. Their results are shown in Table 6.

While these studies were in progress, a large excess of bladder cancer was reported from an English county borough in which there was a rubber processing plant. It was pointed out that naphthylamines were used as antioxidants, and a survey of bladder cancer was therefore carried out in the rubber industry and associated industries such as the manufacture of rubber insulating cables. An increasing number of deaths from bladder cancer during the period 1921 to 1951 was noted in both these industries. The International Association for Research in Cancer (IARC) concluded that benzidine and 2-naphthylamine were proven causal agents in bladder cancer. The 1-naphthylamine compound (noted in Table 5) is probably carcinogenic by virtue of its contamination with 2-naphthylamine. These substances are now prohibited. This account has been given in some detail, as the aromatic amines provide an illustration of a successful epidemiological study which has led to an important public health measure.

THE LABORATORY CONTRIBUTION TO THE IDENTIFICATION OF ENVIRONMENTAL CAUSES OF CANCER

With the advancement of knowledge of cancer causation at the cellular and molecular levels, laboratory studies of mutagenesis and carcinogenesis have become increasingly important. Tests for these properties have proved sufficiently reliable to be included in the mandatory requirements for the introduction of new drugs through the assessment by such bodies as the Committee on Safety of Medicines and the Federal Drug Administration (FDA).

Perturbation of the structure of DNA by (for example) a point mutation may affect the activity of an oncogene or a tumor suppressor gene. By eliminating, for example, the p53 gene (a tumor suppressor gene) by a mutation, unsuppressed growth or cancer will develop. Mutagenic agents therefore may, under certain circumstances, become carcinogenic. A considerable effort has gone into developing short-term reliable tests for mutagenicity, of which the Ames test (or one of its many derivatives) is one of the most frequently used. In this test, the endpoint is the induction of mutation in the gene that governs synthesis of histidine in highly mutable strains of the bacterium *Salmonella typhimurium*. A number of carcinogenic chemicals require metabolic activation before they can produce an effect on DNA, so that liver slices are included in the assay system. An example of this is the conversion of inert cyclophosphamide to the active form, 4-peroxyhydroxycyclophosphamide.

A number of surveys have shown that correlation between carcinogenicity and positivity in the Ames test is quite good, ranging from 60–90% in humans. In animal systems, 80% of rat carcinogens give a positive response to the Ames test. For an excellent critique of testing, the IARCs "Cancer: Causes, Occurrences and Control" should be consulted.

DETAILED ACCOUNT OF ENVIRONMENTAL FACTORS CAUSING CANCER

These can be described under several headings, and IARC data available from 1987–1990 have been subdivided to detail industrial processes, chemicals (certainly or probably carcinogenic), drugs and other agents (either certainly or probably carcinogenic to humans) (Tables 6, 7, and 8). For chemicals and groups of chemicals and drugs probably carcinogenic to man, see the IARCs "Cancer: Causes, Occurrence and Control" (1990).

Table 6. Occurrence of Bladder Cancer With Other Dyes

Agent	*Observed (O)*	*Expected (E)*	*O/E*	*95% Confidence Limits*
Aniline	1	0.83	1.20	0.0–6.7
Auramine	3	0.13	23.08	4.6–67.4
Magenta	6	0.45	13.33	4.9–29.0
Auramine + magenta	0	0.20	0	0

Table 7. Chemicals or Mixtures of Chemicals That Are Carcinogenic to Humans

Aflatoxins	Erionite
4-Aminobiphenyl	Nitrogen mustard
Arsenic	2-Naphthylamine
Asbestos	Nickel
Benzine	Radon
Benzidine	Talc with asbestos
Bis-chloromethylether	Vinyl chloride
Chromium	

Note: After Tomatis et al., 1990.

Table 8. Drugs That Are Carcinogenic to Humans

Azathioprine	Estrogen
Chlorambucil	Oral contraceptives
Cyclophosphamide	Phenacetin mixtures
Cyclosporin	Procarbazine
Diethyl stilbestrol	Thiotepa
Melphalan	VP-16 (Etoposide)
Myeleran	

Note: After Tomatis et al., 1990 with additions.

AGENTS OTHER THAN CHEMICALS OR DRUGS THAT ARE ENVIRONMENTAL CAUSES OF CANCER

Tobacco

The recognition that tobacco smoking was responsible for carcinoma of the lung occurred in the 1950s, with the brilliant work of Doll and Hill in the U.K., and Wynder and Graham in the U.S. All forms of tobacco ingestion were causative, and there was a definite dose relationship. Duration of exposure was also a feature. Tobacco alone produced squamous carcinomas, although the incidence of small cell cancer and adenocarcinoma was also increased. Other cancers are also common in smokers: carcinoma of the lip, nasopharynx, larynx, and esophagus (as might be expected) are often seen. It is surprising that carcinoma of the pancreas is also more frequent. Bladder and urological cancers appear more commonly in smokers. So far, women are less affected than men, which is probably the result of different smoking habits in the past, rather than to any inherent resistance in women to the adverse effects of tobacco. High tar content of cigarettes or cigars appears to be an important factor in causation, and other exposures (for example, to asbestos, alcohol and possibly ionizing radiation) concurrent with smoking, increase the risk.

The influence of environmental tobacco smoke is now a major issue. The U.S. National Research Council produced a summary of ten case control studies and three cohort studies, and confirmed an increased relative risk of cancer in those who were exposed to environmental tobacco smoke. This has important implications for areas such as public transport, public buildings, restaurants, shops, etc. Tobacco is now a very important public health issue. The fact that it is a cause of a common cancer, and that its removal from the public domain would save millions of lives throughout the world, ensures that it takes pride of place in national health care programs.

Alcohol

There is consistent strong evidence that the intake of alcohol increases the risk of cancers of the oral cavity, pharynx, larynx, esophagus, and liver. Details of the carcinogenic risks are given in IARC (1988). Much evidence comes from the study of populations abstaining from alcohol, such as Mormons in the U.S., and from comparing findings with those from populations such as brewery workers, who are exposed to a higher than average intake of alcohol. Further confirmation comes from a definite dose relationship. Relative risks in males and females who drink varying daily amounts of alcohol in northern France are shown in Table 9.

Viruses

The relationship between viruses and cancer is complex. For example, although certain viruses such as herpes simplex can induce cancer (for instance carcinoma of the cervix), the tumor at a later stage does not appear to harbor the virus. Herpes simplex virus Type 2 (HSV2) is known to produce mutations in the cellular genome, and these must be the first step in cancer formation. In the case of viruses such as Epstein-Barr virus, there is good evidence of causation in Burkitt's lymphoma, but

Table 9. Relative Risks for Esophageal Cancer in Relation to Average Daily Alcohol Consumption in Brittany, France

Alcohol Consumption (g Ethanol/Day)	*Males*	*Females*
0.40	1.0	1.0
41–80	3.8	5.6
81–120	10.1	11.0
>121	101.0	—

Note: Risks are related to those persons who drank 0–40 g ethanol/day and smoked 0.9 g of tobacco daily.

Table 10. Viruses Associated With Cancer

Epstein-Barr virus (EBV)
Herpes simplex virus type 2 (HSV2)
Human immunodeficiency virus (HTLV- 1, HTLV-3)
Human papillomavirus
Hepatitis B

the evidence is less secure in nasopharyngeal cancer. Viruses which are closely associated with cancer are given in Table 10.

Epstein-Barr Virus (EBV)

EBV was isolated from African Burkitt's lymphoma 30 years ago. It is widespread, affecting populations in all parts of the world. It is implicated in infectious mononucleosis (a benign condition), in Burkitt's lymphoma, and in nasopharyngeal carcinoma. It is found in lymphomas in patients with immunodeficiency as a result of immunosuppressive therapy following transplantation, or in patients with acquired immunodeficiency syndrome (AIDS).

Human T Lymphotropic Virus Type 1 (HTLV-1)

Adult acute T cell leukemia was first recognized in Japan and later in the Caribbean islands. It is a rapidly fatal condition associated with hypercalcemia and is resistant to therapy for leukemia. Antibodies to HTLV-1 are found in 80% of patients. Transmission of the virus occurs through breastfeeding and probably through sexual intercourse. There is a long latent period between exposure to HTLV-1 and the onset of leukemia.

Human Immunodeficiency Virus (HIV)

HIV, leading to immunosuppression, high-grade non-Hodgkin's lymphoma and Kaposi's sarcoma was first reported in homosexual males in the U.S. in the mid-1980s. It is also possible that Hodgkin's disease, oral cancer, colon, and even pancreatic cancer are more common in AIDS patients, but this needs confirmation.

Hepatitis B

Epidemiological and laboratory studies have indicated a close association between hepatitis B and hepatocellular carcinoma. The association is restricted to chronic infection with hepatitis B, and the presence of HBsAg, the hepatitis B surface antigen. HBsAg positivity correlates with the incidence of hepatocellular carcinoma, which is 100 times more common in HBsAg-positive patients than in normal controls. Ultimate proof of the causality will be provided by a study following mass vaccination programs now being undertaken in China and Gambia, which will, it is hoped, reduce the incidence of hepatitis B and (eventually) of hepatocellular carcinoma.

Pollutants

Air pollution, in the form of smog heavily contaminated with either sulphur compounds or carbon monoxide and other hydrocarbons, is a very obvious and unpleasant environmental hazard of city life. There is evidence that lung cancer is more frequent in urban than rural areas, but it is difficult to convict air pollutants, as any arguments are confounded by occupation, smoking habits, and other exposures. In general it is considered that although air pollution may cause lung cancer, its contribution is very minor compared with that of cigarette smoking. There is no doubt that exposure to high local concentrations of hydrocarbons such as benzine, or to asbestos dust or nickel compounds, is closely associated with lung cancer (in the case of hydrocarbons, with leukemia).

Diet

A great deal of attention has been focused on diet as a possible cause of cancer. Dietary studies present great difficulties in that they are only able to cover a fairly brief period; diets are normally very complex and difficult to quantitate, and there is a possibility that different diets may contain varying amounts of protective agents such as vitamin A. High fat content, or diets containing excessive amounts of meat, have been implicated in colon cancer. However, case control studies have produced equivocal answers. This has also been true for a proven carcinogen such as nitrosamine, or foods containing nitrites which can be converted to nitrosamines. Some studies have shown an excess of stomach cancer for high nitrate diets, but others have shown a protective effect of nitrates with a lower incidence of gastric cancer. It is probable that the relationship is far too complex to be dissected out easily by the relatively crude tools presently available.

CONCLUSION

There can be no better conclusion than that of Sir Richard Doll, ending the Rock Carling Fellowship, where he states:

> "To sum up, we now know how to avoid the occurrence of a number, still relatively small, of the various types of cancer. Setting aside the benefits of earlier diagnosis, we might be able to prevent about 40 per cent of the cancer deaths that occur annually in men in Britain, and a somewhat smaller proportion—about 10 per cent—in women. In addition there is good reason to believe that a large proportion of the remaining types is, in principle, preventable, and with continued research we may learn how to prevent them within the next two or three decades."

That must continue to be our aim.

REFERENCES

Alderson, M. (1986). Occupational Cancer. Butterworths, London.

Case, R.A.M. & Hosker, M.E. (1954). Brit. J. Prev. Soc. Med. 8, 39–50.

Doll, R. (1967). The Rock Carling Fellowship—Prevention of Cancer: Pointers from Epidemiology. The Nuffield Provincial Hospitals Trust.

Doll, R. & Peto, R. (1981). The Causes of Cancer: Quantitative Estimates of Avoidable Risks of Cancer in the United States Today. Oxford University Press, Oxford.

Hill, A.B. (1965). The environment and disease: Association or causation. Proc. Roy. Soc. Med. 58, 295–300.

IARC (1988). ARC Monographs on the Evaluation of Carcinogenic Risks to Humans, Vol. 44. Alcohol Drinking. Intl. Agency for Research on Cancer (IARC), Lyon.

Muir, C., Waterhouse, J., Mach, T., Powell, J., & Whelan, S. (eds.) (1987). Cancer Incidence in Five Continents, Vol V. IARC Publications, Vol 88. Intl. Agency for Research on Cancer (IARC), Lyon.

Rehn, L. (1895). Archiv. für Klinische Chirugie 50, 588–600.

Tomatis, I., Actio, A., Day, N.E., Heseltine, E., Kaldor, J., Miller, A.B., Parkin, D.M., & Riboli, E. (eds.) (1990). Cancer: Causes, Occurrence and Control. Intl. Agency for Research on Cancer (IARC), Publication no. 100, Lyon.

Chapter 4

Progestin Regulation of Cellular Proliferation

ELIZABETH A. MUSGROVE and

ROBERT L. SUTHERLAND

Advances in Oncobiology
Volume 1, pages 79–98.

ISBN: 0-7623-0146-5

INTRODUCTION

The control of normal cellular proliferation involves a delicate balance between different influences within the extracellular environment, including those from hormones, growth factors, cytokines, and cell-cell interactions, which provide signals to stimulate or inhibit cellular proliferation. Among the diverse molecules that control rates of cell proliferation, steroid hormones and in particular the sex steroid hormones (androgens, estrogens, and progestins), regulate some of the largest physiological changes (both positive and negative) in the rates of growth and cell proliferation observed in normal tissues, e.g., in the development of the normal mammary gland at puberty and the cyclic changes in the uterus during the menstrual cycle. Although the characterization of steroid hormone receptors and definition of the molecular mechanisms by which they regulate specific gene expression has become increasingly well understood (Carson-Jurica et al., 1990), the molecular basis of proliferation control by steroid hormones remains undefined. Within this area studies on estrogen action are the most developed (Dickson and Lippman, 1987; Musgrove and Sutherland, 1991).

The regulation of growth and development of most female sex organs involves a balance between the actions of the two major female sex steroid hormones, estradiol and progesterone. While estrogen, acting in concert with other hormones and growth factors, appears to be the main drive to proliferation in these tissues, progesterone has two principal functions in normal mammalian physiology. First, progesterone is involved in preparing the uterus for implantation of the fertilized ovum and making nutrients available for its subsequent development. Second, progesterone causes the glandular elements of the mammary gland to grow and develop into secretory epithelium with the ultimate effect of acting in concert with other hormones, particularly prolactin, to facilitate milk production. In simplistic terms, progesterone might be seen as the "differentiating" female sex steroid which inhibits the "proliferative" effects of estrogen and directs the tissue towards its normal differentiated function. However, progesterone is not always "antiproliferative" and in some tissues induces proliferative responses of its own. In the case of the induction of stromal proliferation in the uterus this represents a corollary of its primary function in facilitating implantation; in the case of its stimulation of lobuloalveolar development in the mammary gland, such an action is a requirement for subsequent lactation, the ultimate differentiated function of this organ.

This chapter focuses on the effects of progesterone and synthetic progestins on cell proliferation, an area of cell biology that has not been widely studied from a mechanistic viewpoint. Emphasis is placed on the two most important progestin target tissues, the uterus and the mammary gland, since the increasing pharmacological use of progestins in oral contraceptives and hormone replacement therapies raises important questions of potential side effects including the possibility that increased cell proliferation may increase the risk of developing breast cancer. These potentially adverse effects should be balanced against the beneficial effects of progestins in decreasing the incidence of endometrial and ovarian cancer and the established role of synthetic progestins as first- and second-line therapy for the treatment of established endometrial and breast carcinoma. Several recent reviews address in more detail the biology of progestins and their antagonists (Clarke and Sutherland, 1990; Horwitz, 1992; Staffa et al., 1992).

PROGESTIN EFFECTS ON CELLULAR PROLIFERATION *IN VIVO*

The effects of progestins on cell proliferation *in vivo* have been investigated in a number of progestin target tissues from a range of species, with the majority of studies concentrating on the uterus and mammary gland. Two principal approaches have been employed: (a) progestins were administered to immature, ovariectomized, or intact animals and the resultant effects on cellular proliferation and differentiation assessed; or (b) correlative studies have related changes in cell proliferation to changes in serum progesterone levels.

Effects of Progestins in the Uterus

The uterus of the ovariectomized mouse has been the most widely utilized experimental system for studies on estrogen and progesterone control of cellular proliferation *in vivo*. In this system estradiol alone causes a major mitogenic response in epithelial cells but not in the connective tissue stroma, while progesterone administration significantly alters this proliferative response to estradiol. Pretreatment with progesterone completely inhibits estrogen-induced epithelial cell proliferation while sensitizing the stromal cells to respond to estradiol with increased mitosis. The progestin-induced switch in proliferation from epithelium to stroma is an essential prerequisite for implantation and decidualization in the mouse and rat, emphasizing the differentiation-inducing role of progesterone. It is also of major interest that in the mouse model, while progesterone inhibits estrogen-induced mitosis in the epithelium, it acts synergistically with estrogen to stimulate stromal proliferation. These data highlight the cell specificity of progesterone action and its ability to both stimulate and inhibit cell proliferation in different cell types.

These early studies, summarized in more detail in Clarke and Sutherland (1990), also provided insight into the cell kinetic basis of progestin antagonism of estrogen-induced epithelial cell proliferation. While prior administration of progesterone or simultaneous administration of estrogen and progesterone completely blocked the estrogenic response, no inhibition was seen when progesterone was given as little as 2.5 hours after estradiol. Since estrogen induces synchronous progression of mouse uterine epithelial cells through the cell cycle, it seems likely that progesterone inhibits mitosis by an action early in G_1 phase and is without effect on progression through late G_1, S, and G_2 phases (Figure 1), a conclusion

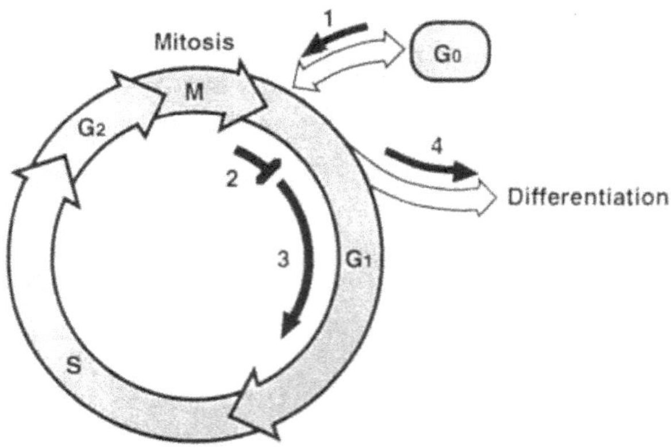

Figure 1. Cell cycle phase specific effects of progestins. The four phases of the cell cycle: G_1, DNA synthesis or S phase, G_2, and mitosis are illustrated. Cells can leave the cell cycle and enter a resting or G_0 phase which allows re-entry to the cell cycle. Alternatively, cells can leave the cell cycle to enter an irreversible program of cell differentiation. Examples of the effects of progestins are shown, i.e., 1. progesterone induction of resting G_0 cells into the cell cycle in mouse uterine stroma; 2. inhibition of cell cycle progression in early G_1 phase in uterine and mammary epithelial cells; 3. accelerated progression of cells through G_1 phase in mammary carcinoma cells; and 4. terminal differentiation in the uterus and mammary gland.

subsequently confirmed by more mechanistic studies on breast cancer cells in culture (see later). The inhibition of epithelial cell proliferation by progesterone is associated with morphological changes characteristic of epithelial cell differentiation.

In the progesterone-treated stroma, a single injection of estradiol results in synchronous entry of 30–40% of stromal cells into S phase while a second injection of estradiol produces no further effect. It thus appears that progesterone stimulates resting stromal cells to enter the cell cycle where estrogen accelerates their passage through a single round of replication by shortening G_1 phase. This replication is thought to be a prerequisite for the differentiation of stromal cells into decidual cells and their withdrawal from the cell cycle.

These observations in the mouse uterus identified critical issues for consideration in defining the molecular basis of progestin effects on cell proliferation. In particular, the complexity of the response was illustrated by observations on cell type specificity of progestin responsiveness, i.e., between epithelium and stroma, and the dependence of the proliferative response on the temporal relationship between administration of estradiol and progesterone. In addition, documentation of the inhibitory effects of progestins on epithelial cell proliferation provided evidence to support the pharmacological use of progestins in the treatment of endometrial carcinoma. Perhaps more importantly from a mechanistic viewpoint, the data on

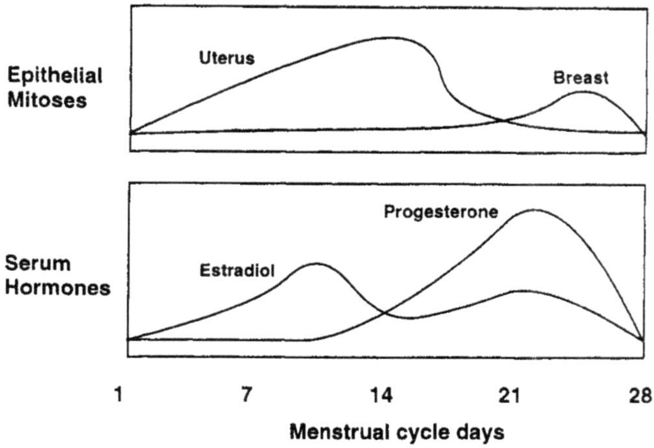

Figure 2. Changes in uterine and breast epithelial mitoses and serum estradiol and progesterone levels during the menstrual cycle. Redrawn with permission from Clarke and Sutherland, 1990.

cell cycle kinetics illustrated the inhibition of cell cycle progression in early G_1 phase in epithelial cells and activation of resting or G_0 stromal cells, and identified the fundamental paradox of progestin control of replication, i.e., these compounds have both stimulatory and inhibitory effects on target cell proliferation. Thus, any unifying model of progestin action must accommodate this apparent paradox, a goal that has yet to be attained.

Several of the principles of progesterone action identified in experimental animal models were subsequently confirmed in the human. The cyclical histological changes in the endometrium during the human menstrual cycle are well correlated with changes in circulating concentrations of estrogen and progesterone (Figure 2). Cyclical changes in proliferation occur, with DNA synthesis being maximal around the time of ovulation then decreasing to low levels until the end of the cycle. DNA synthesis in both the glandular epithelium and stromal elements increases as the serum estrogen concentration increases, while the rise in serum progesterone in the postovulatory phase results in the disappearance of mitoses in both epithelial and stromal cells indicating that progesterone is able to inhibit estrogen-induced proliferation in these cells. The hypothesis that progesterone inhibited estrogen-mediated DNA synthesis was confirmed by progestin inhibition *in vivo* of estrogenized postmenopausal endometrial proliferation. The inhibition of mitosis was accompanied by induction of glandular secretory activity and demonstrated the ability of progestins to promote differentiated function in postmenopausal women on estrogen therapy. Progestins are also known to inhibit the growth of endometrial carcinoma tissue, confirming their predominant inhibitory effect on epithelial cell proliferation in the mammalian uterus.

Effects of Progestins in the Mammary Gland

Although mammary epithelial development involves complex interactions between a number of different hormones which vary between species and at different stages of development, the principal role of progesterone is in promoting lobuloalveolar development in the adult gland. Progesterone does not appear to be necessary for ductal development, which occurs at adolescence mainly under the influence of estrogen in combination with pituitary factors, probably prolactin and growth hormone (Borellini and Oka, 1989).

The immature and ovariectomized adult mouse have provided the most extensively utilized experimental models for studies on the hormonal control of mammary gland development. Progesterone is not involved in the first two phases of mouse mammary gland development, i.e., prenatally or at sexual maturation. Rather, progesterone is involved in the last major phase of development which occurs at pregnancy with the development of lobuloalveolar structures which fill the interductal spaces. Lobuloalveolar growth in the ovariectomized, hypophysectomized, adrenalectomized adult mouse requires estrogen, either growth hormone or prolactin, and progesterone. Thus the major difference between stimulation of ductal and lobuloalveolar cell proliferation during development is the additional requirement for progesterone in the latter cell type (Clarke and Sutherland, 1990).

In the ductal epithelium of the fully mature gland, progesterone and estradiol alone both stimulate proliferation, though progesterone is considerably more effective (Haslam, 1988). Administration of the two steroids together resulted in a marked synergistic, although transient, effect which was attributed in part to the ability of estrogen to increase progesterone receptor levels (PR) and progestin responsiveness. These studies were interpreted as evidence that in the mature mouse progesterone, rather than estrogen, has a major role in promoting epithelial cell proliferation (Clarke and Sutherland, 1990). This adds another degree of complexity to progestin action by providing evidence for developmental differences in the hormonal responsiveness of mammary epithelial cells, i.e., progestins are not required for ductal epithelial cell proliferation during development but have a major influence on these cells in the mature gland.

The human breast responds to the fluctuations in serum hormone levels during the menstrual cycle with cyclical changes in breast volume and cellular morphology. Contributing to these effects are cyclical changes in the mitotic activity of the epithelium; the highest proliferative activity was noted in the intralobular terminal ducts. Studies in which mitoses were evaluated histologically showed that both mitosis and cell loss through apoptosis varied in a cyclical manner during the menstrual cycle, with mitoses being maximal on days 23–26 (Figure 2), although there was variation in the ability to detect mitoses in the early follicular phase. The stimulus for this wave of epithelial cell proliferation in the late secretory phase of the cycle is presently unknown. The effect coincides with, or immediately follows, a rise in the serum concentrations of both estrogen and progesterone (Figure 2).

The inference can thus be drawn that these hormones, alone or in combination, directly or indirectly, may be responsible, perhaps in an analogous manner to the synergism seen in the mature mouse mammary gland.

The roles of estrogen and/or progesterone in mediating breast epithelial DNA synthesis have yet to be investigated directly in humans, but the issue has been examined using *in vitro* techniques such as organ culture, transplantation of normal human breast tissue into nude mice, and primary cell culture. The consensus to emerge from these studies is that estrogen is capable of stimulating breast epithelial growth *in vitro*. Responses to progesterone, however, were variable and no clear consensus on a stimulatory role for progesterone or an ability to inhibit the estrogen-mediated effect has yet emerged, although progestin treatment inhibits the proliferation of steroid-responsive breast epithelial cells in short-term culture. Breast tissue is composed of ductal and lobular-alveolar epithelial elements, encased by myoepithelial cells, and surrounded by stroma and it is likely that these different cellular elements will respond differently to progestins and other growth regulators. Further experimentation in this area is critical to understanding the role of progesterone in the control of cell proliferation in the human breast. However, the fact that synthetic progestins are effective agents in the treatment of metastatic breast cancer as well as endometrial carcinoma argues for a major antiproliferative role for progestins in breast epithelial cells. Clinical data relating objective responses following progestin treatment to the presence of PR support a direct receptor-mediated effect on breast carcinoma cells, a conclusion supported by studies *in vitro*.

PROGESTINS/ANTIPROGESTINS AND CELL PROLIFERATION *IN VITRO*

The effects of progestins on cell proliferation *in vitro* have been investigated in a number of experimental systems, including organ culture of uterine or mammary tissues and cell cultures of normal and neoplastic uterine and mammary epithelium. The majority of studies have used tumor cell lines where the predominant effect is growth inhibition. Breast cancer cell lines have been studied in most detail, but the limited data available from other tumor types are generally in agreement with these findings. Progestin treatment of breast or endometrial cancer cell lines, growing at optimal rates in culture media containing fetal calf serum, results in a concentration-dependent decrease in the rate of cell proliferation (Clarke and Sutherland, 1990; Murphy et al., 1992). It was initially suggested that progestins only inhibited estrogen-induced cell proliferation but it is now clear that progestins can inhibit responsive cells stimulated to proliferate by a range of hormonal and growth factor mitogens (Clarke and Sutherland, 1990). These effects are mediated via the PR since cell lines lacking these receptors are unresponsive and the potency of a range of progestins is related to their affinity for PR. The presence of PR, however, does not guarantee a growth inhibitory response to progestins since some PR-containing breast cancer cell lines are unresponsive (Sutherland et al., 1988). Possible explanations for this anomaly include aberrations in receptor structure and function and/or post-receptor events involved in normal growth control. Thus, in most cell lines that are responsive to progestins *in vitro* progestin treatment results in decreased cell proliferation which is mediated via the PR. Such experimental models have yielded the most detailed data currently available on the mechanisms by which progestins control cell proliferation (Sutherland et al., 1988; Musgrove et al., 1991).

Treatment of exponentially growing breast cancer cells with progesterone or synthetic progestins resulted in accumulation of cells in G_1 phase and a corresponding decline in S phase cells. The marked decrease in the proportion of S phase cells following progestin treatment was not immediately apparent, but took place only after 12 hours or more of exposure to progestin (Figure 3A). Although the delay exceeded the length of G_1 phase in the cell line studied (13.5 hours), the point of action is within G_1 phase since progression through S phase and G_2 + M was unaltered and growth curves showed insignificant differences between control and treated cell numbers during the first cell cycle after administration. Thus there was a delay between the time of administration and the mean point of action within G_1 (Sutherland et al., 1988). This study, therefore, demonstrated that cells in very early G_1, just following mitosis, are sensitive to growth arrest by progestins, while cells at other stages of the cell cycle are insensitive, a conclusion identical to that reached following studies of uterine epithelium *in vivo* and confirming a predominant growth inhibitory effect of progestins on both uterine and mammary epithelial cells.

In breast cancer cells the growth arrest following treatment with even the most effective concentrations of progestin is profound but transient, and despite the continuous presence of the drug, cells initially arrested in G_1 phase resume proliferation. The time and rate of re-initiation of proliferation are concentration-dependent, so that overall effects on proliferation remain although growth arrest is transient. Similarly, in endometrial carcinoma cells the growth inhibition appears to be more profound in the first 96 hours of exposure, suggesting that transient arrest is not just a feature of the response of breast cancer cells. These effects may be accounted for by a loss of sensitivity to progestins as a result of the well-documented decrease in PR levels following progestin treatment (Clarke and Sutherland, 1990; Horwitz, 1992). Such data also emphasize the cyclical nature of progestin effects, a pattern common to their normal physiological function *in vivo*.

The first cell cycle kinetic studies in breast cancer cells described above were performed under conditions designed to maximize the rate of cell proliferation. Such conditions are ideal for examining growth inhibition, but may preclude detection of any stimulatory effects since the cells are growing at near maximal rates. Thus, further studies were undertaken using different experimental conditions (insulin-supplemented growth in serum-free medium) in which the growth rate of control cells was significantly reduced. In these experiments the previously described inhibitory effect of progestin was still apparent, but was preceded by a

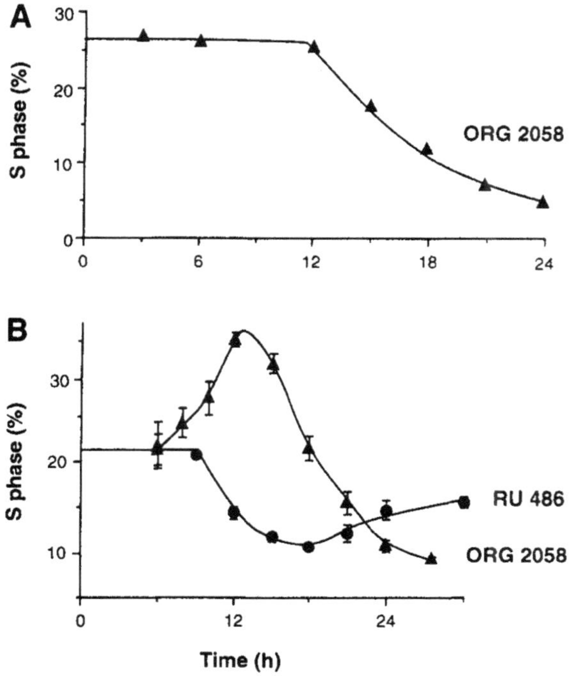

Figure 3. Comparison of temporal changes in % S phase after progestin or antiprogestin treatment. T-47D human breast cancer cells proliferating at maximal (**A**) or sub-maximal rates (**B**) were treated with either synthetic progestin (ORG 2058) or antiprogestin (RU 486) and the proportion of cells in S phase determined at intervals. The horizontal scales have been adjusted for the differing growth rates in these conditions. Data have been redrawn from Sutherland et al., 1988 and Musgrove et al., 1991.

transient stimulation of cell cycle progression (Musgrove et al., 1991). A cohort of stimulated cells moved through G_1 phase and reached S phase approximately eight hours after the commencement of progestin treatment, increasing the proportion of cells synthesizing DNA (Figure 3B). These cells replicated their DNA at the normal rate, but arrested in G_1 following cell division, such that after prolonged exposure the inhibitory effect predominated and the proportion of cells in S phase was reduced. The proportion of the total cell population which was stimulated depended on the growth rate, suggesting that rather than stimulating growth-arrested cells to re-enter the cell cycle as progestins do in uterine stromal cells, progestins accelerated breast cancer cells which were already in "cycle." In turn, this implies that a direct or indirect target of progestin action is a rate-limiting step in cell cycle progression through G_1 phase (see later). Finally, these studies on breast cancer cells *in vitro* provided evidence for two distinct effects of progestins on cell cycle progression within the one cell type: an initial stimulatory effect which accelerated the rate of progression of cells in G_1 phase, allowing a cohort of cells to transverse the cell cycle at a greater than normal rate and a later inhibitory effect occurring in early G_1 phase and resulting in delayed progression through G_1 phase (Sutherland et al., 1988; Musgrove et al., 1991). To date, such data are only available for a restricted series of breast cancer cell lines and urgently require expansion into other progestin responsive cell systems.

Further information on progestin control of cell proliferation would be expected from studies on progestin antagonists, the synthetic antiprogestins. Like progestins, the antiprogestin RU 486 inhibits breast cancer cell proliferation *in vitro* resulting in G1 accumulation and a decrease in the proportion of cells synthesizing DNA (Thomas and Monet, 1992; Musgrove and Sutherland, 1993). This effect is concentration-dependent and mediated via the PR. Additionally, most progestin-induced responses, particularly those on cell proliferation, can be inhibited by antiprogestins.

Since both progestins and antiprogestins inhibit cell proliferation, progestin antagonists are often described as having progestinlike actions on proliferation. This would be consistent with the properties of other steroid antagonists, e.g., nonsteroidal antiestrogens like tamoxifen, which act as agonists under some circumstances. However, comparison of the cell cycle kinetic changes accompanying progestin and antiprogestin treatment of breast cancer cells indicates that growth inhibition by antiprogestins is mediated by mechanisms which are different from those of progestins. The changes in cell cycle phase distribution after antiprogestin and progestin treatment differ in both the absence of initial stimulation of cell cycle progression after antiprogestin treatment and the timing for the decrease in the proportion of cells entering S phase; the %S phase declines 9–12 hours after antiprogestin treatment of T-47D cells but 15–18 hours or later after progestin treatment (Figure 3B). Thus, growth inhibition by RU 486 is unlikely to result from progestin agonist activity. However, acceleration of cell cycle progression by progestins leads to changes in S phase fraction just before inhibition by antiprogestins is apparent, at the time expected if this part of the response was mediated via the same targets as antiprogestin inhibition (Musgrove and Sutherland, 1993). In addition, antiprogestins can antagonize the stimulatory effects following treatment with progestin. Therefore, a more attractive interpretation of currently available data is that antiprogestin inhibition of cell cycle progression and progestin acceleration of cell cycle progression occur by opposite effects on a single pathway.

MECHANISMS OF PROLIFERATION CONTROL

Since steroid hormone receptors, including PR, are ligand-activated transcriptional regulators (Carson-Jurica et al., 1990), regulation of the expression of specific target genes is central to the mechanisms proposed to account for steroid regulation of cell cycle progression. The demonstration of steroidal control of cell cycle progression at defined points within G_1 phase (Musgrove and Sutherland, 1991) indicates that steroids have effects on key cell cycle regulatory genes, the products of which determine rates of G_1 progression. Such an interpretation is compatible with current concepts of mammalian cell cycle control where environmental signals act within G_1 phase to regulate rates of cell proliferation. Significant advances in our understanding of the molecular basis of cell cycle control in mammalian cells have emerged in recent years due to the discovery and functional analysis of nuclear proto-oncogenes, tumor suppressor genes, and the cell cycle regulatory cyclins and cyclin dependent kinases (Hunter, 1991; Motokura and Arnold, 1993). A number of these genes are potential targets of steroid hormone action and are discussed in detail below. However, there has been considerable debate over the issue of whether steroids directly control the expression of cell cycle regulatory genes, or whether the expression of cell cycle regulatory genes is modulated subsequent to steroidal regulation of growth factor or growth factor receptor gene expression (Dickson and Lippman, 1987; Musgrove and Sutherland, 1991).

Regulation of Growth Factor and Growth Factor Receptor Genes

The autocrine hypothesis for loss of normal growth control in the progression of normal cells to neoplasia proposed that transformed cells express the growth factors necessary to stimulate their own proliferation. Following the development of this hypothesis, Lippman and colleagues postulated that the effects of steroids and their antagonists on breast cancer cell proliferation were mediated predominantly via effects on growth factor gene expression (Dickson and Lippman, 1987). Thus the stimulatory effects of estrogen were proposed to be mediated by the increased production of autocrine transforming growth factor α (TGFα) by breast carcinoma cells, and paracrine insulin-like growth factor I (IGF-I) by adjacent stromal cells while antiestrogens had the opposite effect on these growth stimulatory molecules and activated TGFβ, a potent growth inhibitor of epithelial cells. Increased growth factor action therefore provides one possible mechanism underlying stimulation of breast cancer cell cycle progression by progestins, since the effects of progestins on these cells include increased expression of mRNA for the mitogens, epidermal growth factor (EGF) and TGFα and for the EGF receptor, through which both these growth factors exert their effects. Progestin induction of EGF and TGFα mRNA occurs before increases in the rate of entry into S phase are detected, a result compatible with a growth factor-mediated response. However, although treatment with these growth factors causes a qualitatively similar response to that observed after progestin treatment, in that some T-47D cells are stimulated to move through G_1 phase into S phase, fewer cells are affected than after progestin treatment and the affected cells do not reach S phase until six hours later than cells stimulated by progestins. These data argue against regulation of growth factor production as the principal mechanism underlying the acute progestin-induced stimulation of cell cycle progression observed in breast cancer cells (Musgrove et al., 1991).

The levels of EGF and TGFα mRNA remain elevated for more than 24 hours after progestin treatment, although progestin inhibition of proliferation is apparent within that time (Musgrove et al., 1991). Furthermore, there is discordance between the relative expression of these growth factor genes and sensitivity to progestin treatment in breast cancer cell lines. Thus decreased production of EGF or TGFα is unlikely to account for progestin inhibition of breast cancer cell growth. In one endometrial carcinoma cell line, progestin treatment decreased the expression of TGFα mRNA and the addition of TGFα partially reversed the growth inhibitory effects of progestins. However, in another endometrial cancer cell line, growth inhibition by progestins was not accompanied by modulation of TGFα expression, suggesting that although decreased expression of TGFα may contribute to growth inhibition in some circumstances, it is not a generally applicable mechanism (Murphy et al., 1992).

The effects of progestins on TGFβ have also been examined as a possible mechanism for mediating their effects on cell growth. TGFβ generally has growth inhibitory effects on epithelial cells and therefore increased expression or activation of TGFβ might contribute to growth inhibition (Gong et al., 1991). However, progestins cause a decrease rather than an increase in TGFβ mRNA and some cell lines which are sensitive to progestins are not sensitive to the growth inhibitory effects of TGFβ (Murphy et al., 1992). Thus, overall there are scant data which support growth factor regulation as a mechanism for acute changes in the rate of cell proliferation following progestin treatment.

Regulation of Proto-oncogene Expression

A group of genes known as the immediate-early growth response genes are rapidly induced upon mitogenic stimulation of quiescent cells, without the need for new protein synthesis. A subset of these genes, which include members of the *fos* and *myc* families, are required for cells to progress through G_1 phase: inhibition of their expression or function by antisense techniques or antibody microinjection prevents entry into DNA synthesis after mitogen stimulation of quiescent cells (Weinberg, 1989; Bravo, 1990). The Fos and Jun proteins are components of the transcription factor AP-1. Similarly, Myc forms heterodimers with transcriptional activity (Meichle et al., 1992). Induction of c-*fos*, c-*jun*, and c-*myc* will thus result in the regulation of other genes, which may themselves be involved in the regulation of cell cycle progression. In support of a key role for these genes in growth control, deregulated expression of *fos*, *jun*, and *myc* can induce malignant transformation, i.e., these genes are proto-oncogenes.

The role of induction of the components of AP-1 in growth factor-induced mitogenesis prompted investigation of their role in steroidal regulation of proliferation. In breast cancer cells, progestins have been reported to either transiently increase or decrease c-*fos* mRNA levels. If the enhanced expression of this gene is involved in the accelerated cell cycle progression induced by progestins, antiprogestin antagonism of this response would be predicted as part of the mechanism by which the progestin stimulation of proliferation is prevented. However, the effects of antiprogestin treatment on progestin induction of c-*fos* were equivocal, and other studies have noted discordance between c-*fos* induction and mitogenesis in breast cancer cells, raising doubts over the significance of progestin induction of c-*fos* (Musgrove et al., 1991). Several members of the *jun* family are also regulated: c-*jun* mRNA levels increase within three hours of progestin treatment while *junB* mRNA decreases within one hour. These changes are accompanied by decreased AP-1 transcriptional activity, which might potentially contribute to growth inhibition (Alkhalaf and Murphy, 1992).

In one endometrial cancer cell line, progestin treatment reduced the levels of both c-*jun* mRNA and protein but in another only transiently reduced the level of c-*jun* mRNA. The sustained reduction in c-*jun* expression was accompanied by a concentration-dependent decrease in AP-1 transcriptional activity. This effect was partially reversed by cotransfection of c-*jun*. Furthermore, transient transfection of c-*jun* but not *junB* reduced the effects of progestin treatment on cell proliferation (Alkhalaf et al., 1993). These experiments are difficult to interpret since in the transfection experiments increased expression of the c-Jun protein would occur over a limited portion of the duration of the experiment and only in a minority of the cell population. However, the possibility that altered abundance and/or activity of AP-1 complexes may modulate progestin responsiveness remains.

Ectopic expression of c-*myc* or microinjection of Myc protein reduces some of the growth factor requirements for progression from quiescence to proliferation. Activation of c-*myc* in quiescent fibroblasts is sufficient to both trigger DNA synthesis and permit sustained proliferation (Meichle et al., 1992). Thus there is clear evidence that c-*myc* has a regulatory role in G_1 phase, although this has yet to be formally demonstrated in steroid-responsive cells. In progestin-treated T-47D breast cancer cells, c-*myc* is rapidly but transiently induced, reaching a maximum after 1–2 hours treatment (Figure 4), similar to the effects of stimulation by mitogenic growth factors. Consistent with the hypothesis that induction of this gene is involved in progestin stimulation of cell cycle progression, simultaneous antiprogestin treatment effectively eliminated the progestin induction of c-*myc* and cell cycle progression. Treatment with antiprogestin alone resulted in a rapid, profound decrease in c-*myc* mRNA. This was apparent within 30 minutes of treatment, and within two hours c-*myc* mRNA levels had decreased by >90%. These data are consistent with an association between progestin regulation of c-*myc* expression in breast cancer cells and subsequent cell cycle progression. It has not yet been demonstrated that c-*myc* regulation is causal for the changes in cell cycle progression, but this is being actively investigated. However, since the regulation of c-*myc* substantially precedes any detectable changes in cell cycle progression, it is not merely a consequence of these changes.

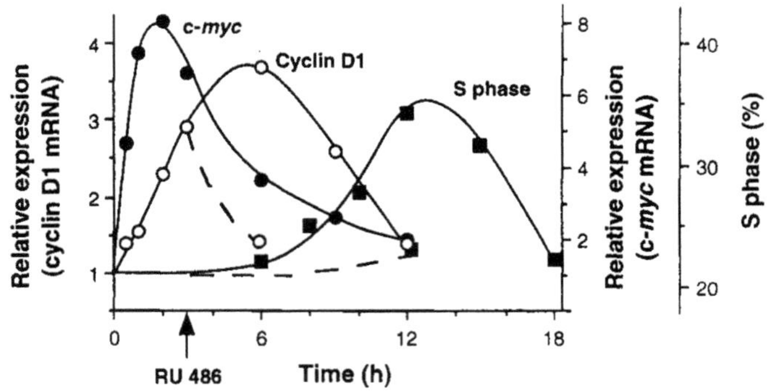

Figure 4. Effect of progestin stimulation on c-*myc* and cyclin D1 gene expression. T-47D cells proliferating in submaximal rates in insulin-supplemented serum-free medium were treated with ORG 2058 and harvested for Northern blot analysis at intervals thereafter. Parallel flasks were treated with RU 486 which was added three hours after progestin. The consequent effects on cyclin D1 and % S phase are shown in dashed lines. Redrawn from Musgrove et al., 1993.

Regulation of Cyclin Gene Expression

The discovery and functional analysis of the cell cycle regulatory cyclins, cyclin-dependent kinases (CDKs) and inhibitors of these kinases (Motokura and Arnold, 1993), has suggested additional targets for steroid action in recent years. Cyclins and CDKs are the regulatory and catalytic subunits, respectively, of cell cycle-regulated kinases. Mammalian cells contain multiple cyclins and CDKs. The members of each family share sequence homology within specific motifs which are thought to have functional significance. Some cyclins are particularly closely related, e.g., cyclins D1, D2, and D3, and thus form subgroups within the cyclin family. There is now clear evidence for tissue-specific expression of the D-type cyclins and distinct roles in the control of differentiation as well as proliferation, indicating that these genes are not redundant but may have complementary functions (Motokura and Arnold, 1993). Some cyclins are capable of binding to multiple CDKs, e.g., cyclin D1 can activate CDK4 and -6, presumably to mediate multiple functions; differential expression of the cyclins allows further scope for cell- and tissue-specific roles for particular cyclin/CDK complexes.

The sequential transcriptional activation of cyclin genes and consequent transient accumulation and activation of different cyclin/CDK complexes is the central mechanism for a series of control points in the mammalian cell cycle (Motokura and Arnold, 1993). In growth factor-stimulated cells, for example breast cancer cells, cyclins D1, D3, E, and A are sequentially induced during G_1 phase progression and entry into S phase (Figure 5). Microinjection studies with anticyclin D1 antibodies or antisense oligonucleotides have shown cyclin D1 to be necessary for entry into S phase. The effects of alterations in cyclin D1 expression in breast cancer

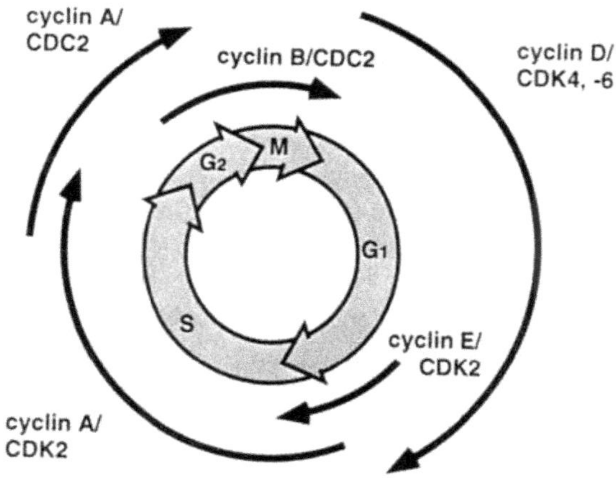

Figure 5. Sequence of cyclin-CDK complex formation throughout the cell cycle. Times of maximum activity of cyclin/CDK complexes are shown schematically. After mitogen stimulation of breast cancer cells, for example, progress through the cell cycle from early G_1 phase is accompanied by induction of cyclin D1, then cyclin D3, followed by induction of cyclin E at the G_1/S phase boundary and induction of cyclin A upon entry into S phase. The abundance of the CDKs is not regulated by progress through the cell cycle, and thus the formation and subsequent activation of cyclin/CDK complexes is driven by the transiently increased abundance of the cyclin subunit.

cells were examined by generating T-47D cells expressing human cyclin D1 under the control of a metal-inducible metallothionein promoter (Musgrove et al., 1994). In cycling cells, induction of cyclin D1 following zinc treatment resulted in an increase in the number of cells progressing through G_1 and in the rate of transition from G_1 to S phase, indicating that cyclin D1 is rate-limiting for progress through G_1 phase. Similar data obtained using rodent fibroblasts indicate that this function is likely to be universal in cells which express cyclin D1. In addition, in T-47D breast cancer cells arrested in early G_1 phase after growth factor deprivation, zinc induction of cyclin D1 was sufficient for completion of the cell cycle, a process requiring growth factor stimulation in control cells. Together these observations provide evidence for a central role for cyclin D1 in breast cancer cell proliferation, since the level of cyclin D1 regulates both the initiation and rate of cell cycle progression.

Cyclin D1 is induced by progestin treatment indicating that this gene is a likely mediator of progestin-induced increases in G_1 transit. The induction occurs later than the induction of either of the immediate-early genes c-*fos* or c-*myc*, but is apparent within 1–2 hours (Figure 4). In addition, progestin induction of cyclin D1 and stimulation of cell cycle progression are both prevented by antiprogestin, added either simultaneously or after a three hour delay. Induction of c-*myc* expression by progestins is transient, and by three hours c-*myc* mRNA levels are beginning to decline. It might be expected, therefore, that addition of antiprogestin three hours after progestin treatment would be too late to prevent the consequences of c-*myc* induction. Since antiprogestin addition three hours after progestin treatment prevents stimulation of cell cycle progression, regulation of cyclin D1 appears to be more closely connected to progestin stimulation of proliferation than does regulation of c-*myc*. Thus, a major potential target for progestin-induced increases in cell cycle progression through G_1 phase is cyclin D1, a gene with a recently well-documented function in controlling G_1 transit in a number of different cell types.

The molecular basis of cell cycle arrest in early G_1 following progestin treatment is less well defined but is also likely to involve effects on the expression and/or function of G_1 cyclins, particularly the D-type cyclins and their kinase partners. Cyclin D-dependent actions are regulated at many levels other than regulation of cyclin D1 abundance alone, suggesting that regulation of the function of associated kinases, particularly CDK4, might also play a critical role in the control of cell cycle progression in breast cancer cells. Recently a number of endogenous inhibitors of CDK activity have been identified. These include: $p16^{INK4}$ and $p15^{INK4B}$, which specifically inhibit the catalytic activity of cyclin D/CDK complexes; $p27^{KIP1}$, which inhibits both cyclin D/CDK4 and cyclin E/CDK2 complexes and appears to provide a link between the functions of these kinases; and $p21^{WAF1/CIP1}$, a general inhibitor of cyclin/CDK action. The potential downstream targets for mediating environmental effects on rates of cell proliferation, including the effects of progestins and their antagonists, thus include cyclins, CDKs, and CDK inhibitors.

CONCLUSIONS AND FUTURE DIRECTIONS

Studies in a diverse range of progestin-target tissues provide evidence for both stimulatory and inhibitory effects of progesterone and its synthetic analogues, the progestins, on cell proliferation. The fact that many of the effects observed *in vivo* are also apparent *in vitro* provides strong evidence that progestins can act directly on PR-expressing target cells to mediate these changes in cell proliferation. However, indirect effects mediated by secondary changes in the extracellular environment through progestin-induced alterations in the hormonal milieu or cell-cell interactions may also contribute to some of the changes observed *in vivo*.

The documentation and mechanistic analysis of both a stimulatory and inhibitory effect within a single cell type, i.e., human mammary carcinoma cells, has allowed further insight into these processes. An important outstanding question is whether or not these two events are separate or components of a single biological process. Preliminary attempts to dissect the stimulatory from the inhibitory effects employing different concentrations and timing of administration of progestins and antiprogestins in breast cancer cells failed to separate the two effects, consistent with the hypothesis that the two are linked. A possibility consistent with the effects of progestins *in vivo* is that the initial transient acceleration of breast cancer cell cycle progression by progestins is a manifestation of a necessity for proliferation before differentiation, the subsequent arrest in early G_1 reflecting cells leaving the cell cycle to enter a program of terminal differentiation. Support for such a conclusion is available from studies of both uterine and breast tissue, outlined above. Progestin treatment of breast cancer cells in tissue culture leads to increases in several markers of differentiated function but likely mechanisms for induction of differentiation have not been examined. However, both proto-oncogene and cyclin gene expression are regulated during differentiation as well as proliferation, suggesting that further investigation of the expression and function of these molecules may give insight into mechanisms of progestin-induced differentiation.

If this hypothesis were correct one would not expect to see prolonged stimulation of cell proliferation in the absence of growth arrest and differentiation. There are, however, reports in the literature that progestins induce sustained cell proliferation but these need to be documented in more detail. If this is indeed the case further refinement of the current, undoubtedly oversimplistic, model of progestin action on proliferation is required. An alternative scenario is that the stimulatory and inhibitory processes are separate and regulated by different gene networks. Under these circumstances it is simple to envisage expression of the stimulatory pathway in the absence of the inhibitory pathway in a cell type-specific manner. Another potential stimulatory mechanism for which evidence currently exists is the situation in the mouse stroma where progesterone stimulates resting G_0 cells to enter the cell cycle. Given that progesterone secretion and action is cyclic and transient, such cells upon entering the cell cycle may no longer be exposed to progesterone in the immediate environment, but could be stimulated to continue proliferation under the influence of other hormones and growth factors. Again, under these circumstances the net effect would be sustained stimulation of cell proliferation. However, at this time such models are purely speculative and would require good *in vitro* systems of progestin-induced cell proliferation to define more fully.

The molecular basis of control of cell cycle progression by progestins in mammary epithelial cells is becoming increasingly well understood due to the availability of well-characterized experimental models and recent major advances in the understanding of mechanisms of cell cycle control. Although significant progress has been made in dissecting the molecular basis of the stimulatory effect with involvement of c-*myc* and cyclin D1, less is known about the inhibitory effect. This is an area of current active investigation and is also likely to involve regulation of expression and function of the same genes and other interrelated molecules such as the tumor suppressor gene RB. However, full characterization of progestin-induced growth inhibition is likely to require a deeper understanding of the effects of these steroids on cellular differentiation.

Net effects on the growth of cell populations are the sum of effects on cell birth rates, i.e., cell proliferation and cell death rates. The cyclic changes in the mammary gland and uterus are known to involve programmed cell death, apoptosis. To date, the effects of progestins on apoptosis are poorly defined but may well contribute to important physiological regulation in progestin-target tissues and therefore urgently require further investigation.

Priorities for future research in this area include the development and characterization of a broader spectrum of *in vitro* model systems which display varying degrees of stimulation and inhibition of cell proliferation in response to progestins. Such *in vitro* models are often more useful for dissecting molecular mechanisms than studies on tissues *in vivo* but it is critical to a more advanced understanding of progestin pathophysiology that processes identified *in vitro* are validated *in vivo*. Given the availability of suitable models and the major recent advances in understanding the control of cell proliferation in mammalian cells, it seems only a matter of time before a clear understanding of progestin effects on cell proliferation will be forthcoming. Such information will undoubtedly aid in the interpretation of currently available and future epidemiological data on the relationship, if any, between cancer incidence and exposure to progestins through the use of oral contraceptives and hormone replacement therapy. It may also assist in the optimal clinical usage, and development of new progestin agonists and antagonists for a number of indicators, including contraception, infertility, menopausal symptoms, and the treatment of several common carcinomas.

SUMMARY

Progestins have diverse effects on cell proliferation in progestin target tissues which are developmentally regulated as well as being species- and cell type-specific. Both stimulatory and inhibitory effects are well documented but it is still unclear whether the stimulatory effects are sustained, i.e., whether they result in continued rounds of cell replication or result in transient effects on a given cell population, inducing a single round of replication before cells undergo terminal differentiation. The inhibitory effects have been more clearly defined, particularly for endometrial epithelium *in vivo* and for both endometrial and breast carcinoma cells *in vitro*. Progestin-induced inhibition of cell proliferation results from a block in cell cycle progression in early G_1 phase immediately after mitosis indicating that progestins inhibit the production and/or functions of genes critical to progression through early G_1 phase. These genes have not been identified but candidates include the proto-oncogene c-*myc* and the G_1 cyclins, particularly cyclin D1 and its kinase partners.

Recent studies employing human mammary carcinoma cells as models have documented a progestin-induced stimulation of cell cycle progression in G_1 phase. This effect appears to be confined to cells that are already in G_1 phase and leads to accelerated progression through this phase of the cell cycle. Since other phases appear unaffected, the resultant effect is a significantly reduced cell generation time, i.e., an increased rate of cell proliferation. Whether or not this is a good model for the stimulatory effects of progestins *in vivo* remains to be established. Studies of changes in gene expression accompanying increased G_1 transit have demonstrated that this effect is preceded by induction of c-*myc* and cyclin D1. Since delayed administration of a progestin antagonist inhibited cell cycle progression and cyclin D1 induction and the effects of progestins were mimicked by inducible expression of a transfected cyclin D1 construct, there is strong evidence that transcriptional activation of cyclin D1 is causal for progestin stimulation of cell proliferation.

These data, generated from well-characterized experimental models *in vitro*, have facilitated progress in our understanding of mechanisms of progestin action. The hypotheses developed from these systems urgently need to be validated in a broader spectrum of progestin target tissues including female reproductive organs *in vivo*. Such data are essential to an understanding of the further physiology of progestin action and the optimal use of progestins in the management of human disease.

REFERENCES

Alkhalaf, M. & Murphy, L.C. (1992). Regulation of c-jun and jun-B by progestins in T-47D human breast cancer cells. Mol. Endocrinol. 92, 1625–1633.

Alkhalaf, M., Murphy, L.J., & Murphy, L.C. (1993). Enhanced c-jun activity alters responsiveness to medroxyprogesterone acetate in Ishikawa human endometrial carcinoma cells. Mol. Endocrinol. 93, 1634–1641.

Borellini, F. & Oka, T. (1989). Growth control and differentiation in mammary epithelial cells. Environ. Health Perspect. 80, 85–99.

Bravo, R. (1990). Growth factor-responsive genes in fibroblasts. Cell Growth Differ. 1, 305–309.

Carson-Jurica, M.A., Schrader, W.T., & O'Malley, B.W. (1990). Steroid receptor family: Structure and functions. Endocr. Rev. 11, 201–220.

Clarke, C.L. & Sutherland, R.L. (1990). Progestin regulation of cellular proliferation. Endocr. Rev. 11, 266–302.

Dickson, R.B. & Lippman, M.E. (1987). Estrogenic regulation of growth and polypeptide growth factor secretion in human breast carcinoma. Endocr. Rev. 8, 29–43.

Gong, Y., Anzai, Y., Murphy, L.C., Ballejo, G., Holinka, C.F., Gurpide, E., & Murphy, L.J. (1991). Transforming growth factor gene expression in human endometrial adenocarcinoma cells: Regulation by progestins. Cancer Res. 51, 5476–5481.

Haslam, S.Z. (1988). Progesterone effects on deoxyribonucleic acid synthesis in normal mouse mammary glands. Endocrinology 122, 464–470.

Horwitz, K.B. (1992). The molecular biology of RU 486. Is there a role for antiprogestins in the treatment of breast cancer? Endocr. Rev. 13, 146–163.

Hunter, T. (1991). Cooperation between oncogenes. Cell 64, 249–270.

Meichle, A., Philipp, A., & Eilers, M. (1992). The functions of myc proteins. Biochim. Biophys. Acta 1114, 129–146.

Motokura, T. & Arnold, A. (1993). Cyclins and oncogenesis. Biochim. Biophys. Acta 1155, 63–78.

Murphy, L.C., Dotzlaw, H., Alkhalaf, M., Coutts, A., Miller, T., Wong, M.S.J., Gong, Y., & Murphy, L.J. (1992). Mechanisms of growth inhibition by antiestrogens and progestins in human breast and endometrial cancer cells. J. Steroid Biochem. Molec. Biol. 43, 117–121.

Musgrove, E.A., Hamilton, J.A., Lee, C.S.L., Sweeney, K.J.E., Watts, C.K.W., & Sutherland, R.L. (1993). Growth factor, steroid and steroid antagonist regulation of cyclin gene expression associated with changes in T-47D human breast cancer cell cycle progression. Mol. Cell. Biol. 13, 3577–3587.

Musgrove, E.A., Lee, C.S.L., Buckley, M.F., & Sutherland, R.L. (1994). Cyclin D1 induction in breast cancer cells shortens G_1 and is sufficient for cells arrested in G_1 to complete the cell cycle. Proc. Natl. Acad. Sci. USA 91, 8022–8026.

Musgrove, E.A., Lee, C.S.L., & Sutherland, R.L. (1991). Progestins both stimulate and inhibit breast cancer cell cycle progression while increasing expression of transforming growth factor α, epidermal growth factor receptor, c-*fos* and c-*myc* genes. Mol. Cell. Biol. 11, 5032–5043.

Musgrove, E.A. & Sutherland, R.L. (1991). Steroids, growth factors and cell cycle controls in breast cancer. In: Regulatory Mechanisms in Breast Cancer. (Lippman, M.E. & Dickson, R.B., eds), pp. 305–331, Kluwer Academic Publishers, Boston.

Musgrove, E.A. & Sutherland, R.L. (1993). Effects of the progestin antagonist RU 486 on T-47D cell cycle kinetics and cell cycle regulatory genes. Biochem. Biophys. Res. Commun. 195, 1184–1190.

Staffa, J.A., Newschaffer, C.J., Jones, J.K., & Miller, V. (1992). Progestins and breast cancer: An epidemiologic review. Fertil. Steril. 57, 473–491.

Sutherland, R.L., Hall, R.E., Pang, G.Y.N., Musgrove, E.A., & Clarke, C.L. (1988). Effect of medroxyprogesterone acetate on proliferation and cell cycle kinetics of human mammary carcinoma cells. Cancer Res. 48, 5084–5091.

Thomas, M. & Monet, J.-D. (1992). Combined effects of RU 486 and tamoxifen on the growth and cell cycle phases of the MCF-7 cell line. J. Clin. Endocrinol. Metab. 75, 865–870.

Weinberg, R.A., ed. (1989). Oncogenes and the Molecular Origins of Cancer. Cold Spring Harbor Laboratory Press, Cold Spring Harbor.

Chapter 5

Tumor Angiogenesis and its Control by Tumor Suppressor Genes

PETER J. POLVERINI

Advances in Oncobiology
Volume 1, pages 99–117.

ISBN: 0-7623-0146-5

INTRODUCTION

The formation of new capillary blood vessels, a process termed angiogenesis, is one of the most pervasive and fundamentally essential biological processes encountered in mammalian organisms. Angiogenesis is an important event in a number of physiological settings where it is tightly regulated in both time and space by divers families of proangiogenic cytokines and naturally occurring inhibitors. Angiogenesis is also central to the etiology and pathogenesis of a number of disease processes with solid tumor formation being the most thoroughly investigated example of an "angiogenesis-dependent" disorder. Studies conducted over the past 25 years have unequivocally established that the growth, progression, and metastasis of solid tumors is strictly dependent on their ability to induce the sustained growth of new capillaries. Although there is a wealth of data on the role of proangiogenic mediators in the pathogenesis of tumor neovascularization, only recently has attention focused on the contribution of naturally occurring inhibitors to this process. Based on recent work from several laboratories it is now abundantly clear that most if not all angiogenesis-dependent disease processes are a consequence of not only the overproduction of normal or aberrant forms of proangiogenic mediators, but also the result of a relative deficiency in angiogenesis inhibitors. In this chapter I will describe these multifunctional mediator systems, discuss how they function in normal and pathological angiogenesis, and how loss or inactivation of tumor suppressor genes can lead to disregulated angiogenesis. The implications of these findings in the development of novel strategies for the treatment of solid tumors and other angiogenesis-dependent disorders will also be discussed.

ANGIOGENESIS IS A FEATURE OF PHYSIOLOGICAL AND PATHOLOGICAL PROCESSES

Tissue regeneration and the repair of wounds, the cyclical proliferation of the nutrient-rich endometrium in preparation for implantation of the fertilized egg, and the development of the embryo and its supporting tissues are biological events that are strictly dependent on the rapid yet temporary ingrowth of new capillary blood vessels, a process termed angiogenesis (Folkman and Cotran, 1976; Folkman, 1985; Folkman and Klagsbrun, 1987; Polverini, 1995). In adult organisms capillary endothelial cells divide relatively infrequently. Turnover rates for endothelial cells are typically on the order of several months or years (Engerman et al., 1967; Tannock and Hayashi, 1972). Yet when called upon, as in response to hormonal signals during menses, following the release of proangiogenic cytokines from inflammatory cells, or as a consequence of the activity of proteolytic enzymes that release angiogenic mediators sequestered in the extracellular matrix, endothelial cells lining venules will systematically degrade their basement membrane and proximal extracellular matrix, migrate directionally, divide, and organize into new functioning capillaries all within a matter of days (Figures 1 and 2). This dramatic

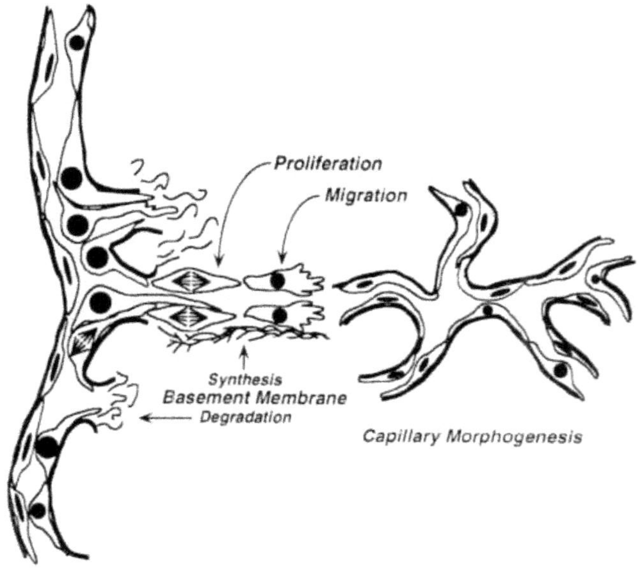

Figure 1. Diagram of the sequence of events involved in the process of angiogenesis.

amplification of the microvasculature is nevertheless temporary, for as rapidly as the new capillaries are formed, they virtually disappear within a matter of days or weeks, returning the tissue microvasculature to its status quo. It is this feature of transient growth and regression of capillaries that primarily distinguishes physiological from pathological angiogenesis. As shown in Table 1, angiogenesis is a feature of a limited number of physiological processes. In contrast, the etiology and pathogenesis of a much larger and increasingly expanding number of pathologic conditions (Table 2) has been shown to be a consequence of an angiogenic response that is either persistent due to the overproduction of normal or aberrant forms of angiogenic mediators, or the underproduction or decreased responsiveness of

Table 1. Physiological Angiogenesis

Chronic inflammation and wound repair
Development of the corpus luteum
Embryogenesis
Immune responses
Lactating breast
Ovulation

Table 2. Pathological Angiogenesis

Neoplasia
Solid tumors
Vascular malformations
Angiofibroma
Arteriovenous malformations
Hemangioma
Vascular adhesions
Syndromes
Dyschondroplasia with vascular hamartomas (Maffucci's syndrome)
Hereditary hemorrhagic telangiectasia (Rendu-Osler-Weber syndrome)
Von Hipple Lindau syndrome
Ocular disorders
Corneal graft neovascularization
Diabetic retinopathy
Neovascular glaucoma
Retrolental fibroplasia
Trachoma
Chronic inflammatory diseases and aberrant wound repair
Atherosclerotic plaques
Diabetes
Granulations–burns
Hemophiliac joints
Hypertrophic scars
Nonunion fractures
Osteoradionecrosis
Psoriasis
Pyogenic granuloma
Rapidly progressing adult and juvenile periodontitis
Rheumatoid arthritis
Systemic sclerosis

Note: Adapted from Moses and Langer (1991)

endothelial cells to physiological inhibitory signals. The mechanisms underlying both physiological as well as pathological angiogenesis have been the subject of considerable investigation for the past 20 years. Much of what we know today stems from studies of tumor angiogenesis. It will be shown, using tumor angiogenesis as an example, how disruption of this tightly regulated program leads to disregulated angiogenesis and disease. However, before this paradigm of angiogenesis is described, a few words must be said about the cast of characters, the mediators, that drive the angiogenic response.

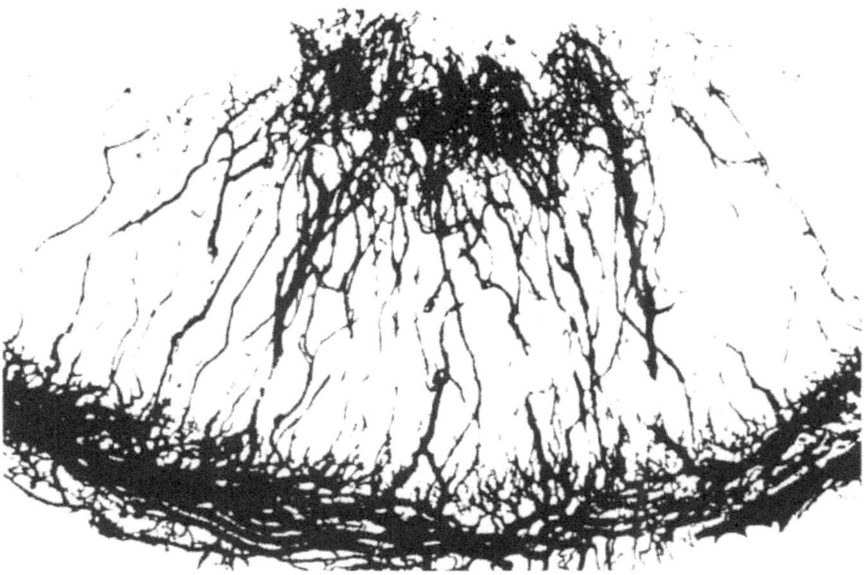

Figure 2. A neovascular response induced in the cornea of an F344 rat 7 days after implanting a Hydron pellet containing culture media (5 μg of protein) from a malignant squamous cell carcinoma. Note the brush-like ingrowth of capillary sprouts extending from the limbus toward the implant (not visible); x 30.

ANGIOGENESIS IS REGULATED BY THE BALANCED PRODUCTION OF ANGIOGENIC STIMULATORS AND INHIBITORY MOLECULES

Angiogenic mediators can be broadly divided into two groups: stimulators (Table 3) and inhibitors (Table 4) (Folkman and Klagsbrun, 1987; Klagsbrun and D'Amore, 1991; Moses and Langer, 1991). The majority of the stimulatory molecules are proteins and many of them are growth factors that induce endothelial cells to divide, migrate directionally toward the inducing stimulus, and differentiate into tubular structures. Most are secreted by a variety of cells including endothelial cells themselves, in response to exogenous or endogenous stimuli and are produced locally and function in a paracrine manner. These mediators can stimulate angiogenesis directly by interacting with receptors on the endothelial cell surface, or indirectly by attracting and activating accessory cells, i.e., inflammatory macrophages, and inducing them to produce angiogenic mediators (Polverini and Leibovich, 1986). Others such as copper may function as cofactors in key interstitial

Table 3. Proangiogenic Cytokines and Mediators

Growth factors
Acidic fibroblast growth factor (aFGF)
Basic fibroblast growth factor (bFGF)
Epidermal growth factor (EGF)
Interleukin 1(IL-1)
Interleukin 2 (IL-2)
Scatter factor/hepatocyte growth factor (SF/HGF)
Transforming growth factor alpha (TGF-α)
Transforming growth factor beta (TGF-β)
Tumor necrosis factor alpha (TNF-α)
Vascular endothelial growth factor/vascular permeability growth factor (VEGF/VPF)
Other proteins and peptides
Angiogenin
Angiotensin II
Ceruloplasm
Fibrin
Human angiogenic factor
Interleukin 8 (IL-8)
Plasminogen activator
Polyamines
Substance P
Urokinase
Carbohydrates and lipids
12(R)-hydroxyeicosatrienoic acid (Compound D)
Hyaluronan fragments
Lactic acid
Monobutyrin
Prostaglandins E1 and E2
Others
Adenosine
Angiotropin
Copper
Endothelial cell stimulating angiogenesis factor (ESAF)
Heparin
Nicotinamide

Note: Relevant reviews containing specific references: Bouck, 1990, 1993; Polverini and DiPietro, in press; Polverini, in press.

enzyme systems or, in the case of plasminogen activators can activate latent enzymes such as transforming growth factor beta to reveal its angiogenic activity. Still others play a key role in stabilizing and/or enhancing the function of stimulatory molecules normally sequestered in the extracellular matrix surrounding blood vessels, as heparin does, which when bound to basic fibroblast growth factor facilitates its interaction with high affinity receptors on the endothelial cell surface.

While the mediators responsible for the up-regulation of new capillary growth has been the subject of extensive investigation (Folkman and Klagsbrun, 1987;

Table 4. Endogenous Inhibitors of Angiogenesis

Angiostatic steroids and sulfated polysaccharides
Eosinophilic major basic protein
High molecular weight hyaluronan
γ Interferon
Interleukin-1 (IL-1)
Laminin peptides
Placental RNase (angiogenin) inhibitor
Platelet factor 4
Prostaglandin synthesis inhibitors
Protamine
Somatostatin
Thrombospondin 1
Tissue inhibitors of metalloproteinases
Vitamin A and retinoids
Vitreous fluid

Note: Relevant reviews containing specific references: Bouck, 1990, 1993; Polverini and DiPietro, in press; Polverini, in press.

Klagsbrun and Folkman, 1990; Klagsbrun and D'Amore, 1991), only recently has attention focused on the mechanisms and mediators responsible for the timely down-regulation of angiogenesis (Bouck, 1990; Klagsbrun and D'Amore, 1991; Moses and Langer, 1991) (Table 4). A common property of these compounds is that almost all of them can influence the ability of cells to produce, interact with, or degrade their extracellular matrix. Alterations in the organization and composition of the extracellular matrix has been shown to have a profound effect on the growth and function of endothelial cells and in determining whether endothelial cells will differentiate and organize into a three-dimensional capillary network.

An important feature of these opposing yet complementary mediator systems is that with rare exception none of them are endothelial cell-specific and thus unique to the process of angiogenesis. Most of these mediators have a wide range of functions and target cells. Perhaps one of the most remarkable features of the angiogenic response is its ability to respond in an identical fashion to a phylogenetically diverse range of mediators. It would appear that the angiogenic phenotype has evolved as a highly conserved response without the need for its own mediator system. Rather endothelial cells are able to utilize whatever growth stimulators and inhibitors are made available to it to produce new capillaries. For example, during embryonic development basic fibroblast growth factor (Risau, 1991) has been shown to be an important if not the major stimulator of angiogenesis. In contrast, in adult organisms, this same mediator appears to have a much more restricted role

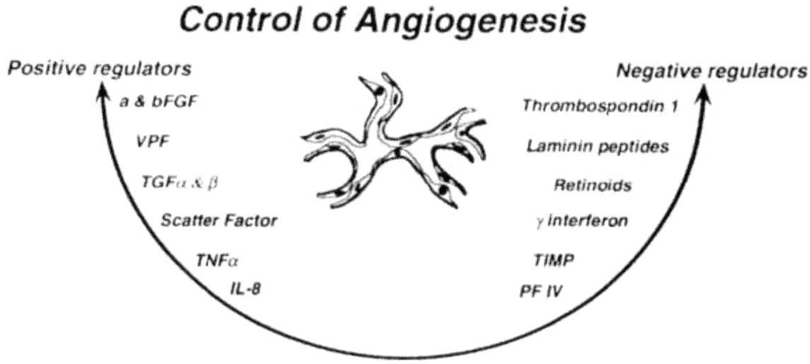

Figure 3. Diagram depicting angiogenesis as a consequence of the balanced production of positive and negative regulators.

in angiogenesis and an entirely different complement of angiogenic mediators come into play as, for example, in wound repair (Polverini et al., 1977; Polverini, 1989; Sunderkotter et al., 1991). Whether angiogenic stimulators and inhibitors are tissue- or process-specific is the subject of much speculation. Clearly the great redundancy in positive and negative regulators capable of orchestrating an angiogenic response attests to its fundamental importance in pathophysiological processes (Figure 3).

NEOPLASIA IS AN ANGIOGENESIS-DEPENDENT DISEASE

A substantial body of work published over the last 25 years (principally from the Laboratory of Judah Folkman at The Children's Hospital and Harvard Medical School in Boston) has firmly established that solid tumors are "angiogenesis dependent" (Folkman, 1985a, b; Folkman, 1989). This premise, first proposed in 1971, was based on several key observations. It was shown that tumors, implanted in isolated perfused organs, where blood vessels did not grow or when introduced into either subcutaneous transparent chambers in mice or the avascular cornea of the rabbit eye, grew ever so slowly, as small 1–2 mm^3 spheres or as thin wafers by absorbing nutrients diffusing from the surrounding tissues (Folkman et al., 1966, Gimbrone et al., 1972). The tumors were able to survive in this dormant state for an extended period of time but were unable to grow progressively. However as the advancing edge of the tumor approached adjacent microvessels, diffusible "angiogenic factors" released from the tumor stimulated endothelial cells to grow and migrate directionally toward the tumor and organize into a capillary network. This switch from the prevascular to vascular phase was accompanied by exponential growth of the tumor (Gimbrone et al., 1974). There are numerous examples where these observations have been validated in human tumors. For example, human retinoblastomas that metastasize to the vitreous or the anterior chamber of the eye remain avascular until they settle upon the richly vascular iris or retina and become vascularized. Carcinoma of the ovary metastasizes to the peritoneum as avascular spheres that fail to grow until they become vascularized. The appearance of neovascularization at the base of melanomas that enter the vertical growth phase, the red "blush" that is associated with cervical and oral carcinoma heralds the onset of rapid growth and increased metastatic potential. An increase in the number of new capillaries in certain types of breast carcinoma has been shown to correlate with malignant and metastatic potential and thus is of prognostic significance (Weidner et al., 1991).

Tumor cells can recruit new blood vessels by several different mechanisms. One of the earliest changes associated with the acquisition of tumor angiogenic activity is that tumor cells loose their ability to produce inhibitors of angiogenesis (Rastinejad et al., 1989; Bouck, 1990, 1993, 1996; Hanahan and Folkman, 1996). They also may produce diffusible angiogenic factors that directly activate endothelial cells stimulating them to sprout and grow towards the developing tumor. Tumors can recruit host cells such as macrophage (Polverini et al., 1977; Polverini and Leibovich, 1984), mast cells (Starkey et al., 1988), and neutrophils (Welch et al., 1989) which contribute a rich array of proangiogenic cytokines in an environment already enriched with angiogenic factors. They produce enzymes that release angiogenic factors sequestered in the extracellular matrix (Briozzo et al., 1991) and they stimulate adjacent normal tissues to make enzymes such as stromolysin (Basset et al., 1990) and collagenase (van den Hooff, 1988; van den Hooff, 1991) that can be activated to promote angiogenesis. Lastly, tumors can subvert host defenses that normally guard against unwarranted angiogenesis. The onset of angiogenic activity in tumor cells may occur at any time during neoplastic transformation. How and where does tumor angiogenesis fit into the multistep carcinogenic process and what is its relationship to other phenotypic traits that define the transformed phenotype is discussed below.

ANGIOGENIC ACTIVITY IS AN EARLY FEATURE OF THE MULTISTEP CARCINOGENIC PROCESS

It is now well established that tumors evolve through a series of steps that have been operationally defined as initiation, promotion, and progression (Figure 4) (Nowell, 1976; Farber and Cameron, 1980). This model, which originally evolved from studies of the mechanism of chemical carcinogen-induced liver and skin tumors in animals, is now accepted as the model for human tumor development. Initiated cells are those that have become committed to neoplastic development, presumably as a result of one or more genetic mutations that have been "locked in." During the promotion phase initiated cell populations undergo selective clonal expansion and acquire new or altered phenotypic traits indicative of transformation. While the initiation step is thought to occur within hours or days following exposure to a carcinogen, either chemical or physical in nature, promotion occurs over months or years. Lastly, during progression cells undergo frank conversion to malignancy when traits, such as invasion and metastasis, that we associate with

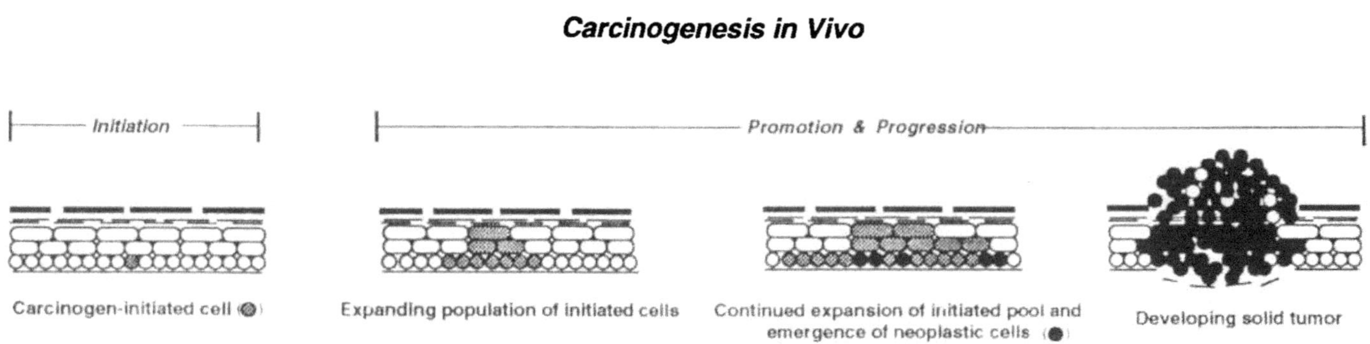

Figure 4. Diagram showing the various steps in carcinogenesis as it might occur in skin or mucosal epithelium.

malignancy emerge. It is during the multistep carcinogenic process that cells acquire altered or new phenotypic traits that enable them to evade host defense and become established as "new autonomous growths." While these steps in the carcinogenic process can be analyzed in some detail *in vitro*, and the cells in which they occur identified in some cases in experimental animal models, it is not the case for human tumors. We have yet to identify an "initiated cell population" in any human tissue that we can say unequivocally are the committed progenitors of human tumors. From a clinical standpoint the earliest lesions we are able to identify are premalignant lesions. These represent altered cell populations which show morphological and functional alterations that indicate they have acquired one or more steps in the carcinogenic process.

Although angiogenesis can be acquired at any time during the carcinogenic process, expression of angiogenic activity has been shown to be an early and predictable property of many preneoplastic cells and may represent one of the earliest indications that a cell population has become committed to malignancy (Folkman, 1985a, 1989). One of the earliest observations demonstrating the angiogenic potential of preneoplastic cells was a series of studies from Pietro Gullino's Laboratory at the National Cancer Institute. Using a series of murine breast cell lines derived from mouse strains that exhibited either a high or low incidence of spontaneous mammary tumors, these workers showed that mammary cells derived from mouse strain at risk for mammary tumors consistently were able to stimulate angiogenesis when implanted into the rabbit cornea, while cells derived from mice with a low incidence of spontaneous tumors exhibited little or no angiogenic activity (Gimbrone and Gullino, 1976a, b; Brem et al., 1977). Similarly Polverini and Solt (1988) showed that cloned populations of preneoplastic keratinocytes derived from carcinogen-treated, nontumor-bearing, hamster buccal pouches were potently angiogenic yet were not tumorigenic when injected into nude mice or neonatal hamsters. The same situation appears to hold true for some human tumors and their preneoplastic progenitors. Brem and colleagues (1978) and Jensen et al. (1982) examined a series of human breast lesions of varying malignant potential. They showed that there was a strong correlation between the angiogenic activity of a particular breast lesion and its malignant potential that preceded morphological and functional changes indicative of neoplastic transformation. Chodak, in examining the urine of patients with neoplastic and nonneoplastic lesions of the bladder for the presence of endothelial cell migration-stimulating activity, was able to distinguish between benign and premalignant conditions as well as detect the emergence of recurrent tumors (Chodak et al., 1980). Thus it would appear that expression of angiogeneic activity may be predictive of malignant potential that is acquired early in the carcinogenic process and which is independent of other transformation linked traits. It is however important to remember that while angiogenesis is necessary for tumorigenicity it is, in and of itself, not sufficient. Furthermore, the capacity of tumors to induce angiogenesis does not always correlate with either malignant potential or frank malignancy. As an example, adrenal adenoma is a benign tumor that is highly angiogenic and rarely undergoes malignant conversion.

TUMOR ANGIOGENESIS REFLECTS A SHIFT IN THE NET BALANCE BETWEEN STIMULATORS AND INHIBITORS OF ANGIOGENESIS

It is now generally agreed that growth of tumors is dependent in part on the relative concentration of proangiogenic and angiostatic mediators produced by tumor cells or tumor associated host cells (Folkman, 1985b, 1989; Liotta et al., 1991; O'Rielly et al., 1994). In a recent study from the Folkman laboratory, it was shown how a shift in the net balance between positive and negative regulators of angiogenesis can have a profound effect of tumor growth and metastasis (O'Reilly et al., 1994). There is considerable experimental and clinical evidence that a primary tumor mass can maintain its metastasis in a dormant state by inhibiting its growth. The mechanism underlying the suppressive effects of primary tumors on metastases has been the subject of much controversy and speculation for years. In an elegant series of experiments, the Folkman laboratory provided a compelling explanation for this phenomenon. They found that when a primary tumor was implanted and allowed to grow in mice, metastases were suppressed. However, upon removal of the primary neoplasm metastases underwent neovascularization and emerged. Serum from mice bearing primary tumors was found to contain a "factor" that inhibited endothelial cell proliferation in culture. Purification of this inhibitory factor revealed a 38 kDa plasminogen fragment they termed "angiostatin." Systemic administration of angiostatin potently blocked neovascularization and growth of metastases. The mechanism(s) underlying angiostatin's suppressive effect is not yet known. Nevertheless, these data further emphasize the importance of how an imbalance in a normally finely tuned regulatory system can have significant and often deleterious pathophysiological consequences.

TUMOR SUPPRESSOR GENES REGULATE THE PRODUCTION OF INHIBITORS OF ANGIOGENESIS

It is during tumor initiation, promotion, and progression when oncogenes are activated and tumor suppressor genes are lost or inactivated that angiogenic activity manifests in developing tumors. There is now compelling evidence that lesions in both of these classes of genes contributes to expression of angiogenic activity by tumor cells (Bouck, 1990, 1993). While a number of tumors derived of oncogenes and constructs capable of transforming cells encode proteins with angiogenic activity, only recently have tumor suppressor genes been implicated in angiogenesis (Rastinejad et al., 1989). Before discussing the evidence linking tumor suppressor genes with this process, I will briefly describe what these genes are and how they function.

As noted above, increasing evidence suggests that tumors result from the accumulation of genetic lesions involving two major classes of genes: oncogenes and tumor suppressor genes (Klein, 1987; Friend et al., 1988; Hansen and Cavenee, 1988; Sager, 1989; Weinberg, 1989; Bishop, 1991; Marshall, 1991; Schwechheimer and Cavenee, 1993). Proto-oncogenes, the cellular homologue of transforming oncogenes, encode a variety of proteins that regulate cell growth. These include growth factors, growth factor receptors or their functional homologues, signal transducers, and nuclear transcription factors. They can be activated and thus converted to cancer-causing oncogenes by several mechanisms including mutation, amplification, and/or translocation. This often results in structurally altered proteins, truncated gene products, or normal proteins that are overexpressed and evade appropriate feedback control mechanisms. To date, in excess of 80 oncogenes have been identified. The important point to remember is that the products of these "growth stimulatory" genes, when activated or overexpressed, function in a dominant fashion since an alteration in only one of the two copies of the gene is necessary for it to exert its effect. Tumor suppressor genes, on the other hand, function in an entirely different manner. As a group these genes negatively regulate cell growth. The protein products encoded by these genes of which about a dozen or so have been cloned include cell cycle regulatory proteins (Rb), DNA binding proteins and transcriptional activators (p53 and WT1), and signal transduction proteins (NF1). The protein encoded by several other cloned tumor suppressors as well as other candidate genes remains unidentified at this time. Of particular interest, however, are recent studies which suggest that an important target for some tumor suppressors is the regulation of angiogenesis.

A direct link between the loss and/or inactivation of a tumor suppressor gene and the expression of tumor angiogenic activity was uncovered in studies of a human/hamster hybrid tumor model system (Rastinejad et al., 1989; Good et al., 1990; Moroco et al., 1990). Bouck et al. showed that an immortal but non-tumorigenic line of hamster fibroblasts, BHK 21/cl 13 cells could be converted to malignancy and anchorage independence by loss of a functioning tumor suppressor gene. These cells were highly tumorigenic in nude mice and neonatal hamsters and were potently angiogenic *in vivo*. It is of interest that the suppressor gene that was lost during BHK transformation has a functional homologue on human chromosome 1 and whose loss has been implicated in the development of human neuroblastomas and some breast cancers. Normal BHK cells and suppressed hybrids generated by fusing transformed BHK cells with either nontransformed BHK or normal human fibroblasts are unable to induce neovascularization when cells or their concentrated CM are introduced into rat corneas, whereas transformed BHK cells and transformed segregants from the suppressed hybrids are angiogenic under the same conditions. Mixing experiments showed that normal cells elaborated an inhibitor of neovascularization whose production was blocked coincidental with suppressor loss. The presence of inhibitory activity was found to be tightly linked to the presence of an active suppressor gene in transformants and their revertants, in

Transformed Phenotypes Linked to Loss of Tumor Suppressor Genes

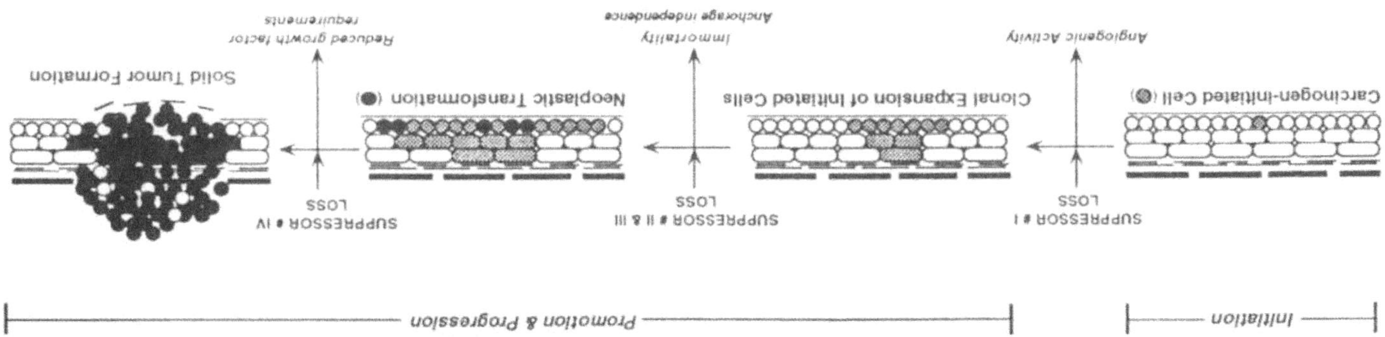

Figure 5. Diagram demonstrating the relationship between loss or inactivation of *tumor suppressor* genes and the acquisition of transformation linked phenotypes during the multistep carcinogenic process (Moroco et al., 1990).

segregating cell hybrids, and in temperature-sensitive transformants. Using endothelial cells chemotaxis as an *in vitro* corollary of angiogenesis and the rat cornea model of neovascularization, the inhibitor was purified and shown to be the well-known extracellular matrix protein thrombospondin 1 (TSP1) (Bornstein, 1992, 1995). For the first time this work established a new function for a tumor suppressor gene: the regulation of the production of a naturally occurring inhibitor of angiogenesis. More recently, Dameron et al. (1994) established a direct link between the p53 tumor suppressor gene, tumor angiogenesis, and the angiogenic inhibitor TSP1. To examine the effect of p53 on angiogenesis they used cultured fibroblasts from patients with the Li-Fraumeni syndrome. These individuals have inherited one wild-type allele and one mutant allele of the p53 gene. As a result they have an increased risk of developing sarcomas and other tumors when the second (wild-type) allele is lost. These workers found that when the wild-type allele was lost in these fibroblasts, the cells acquired potent angiogenic activity coincidental with loss of production of the angiogenesis inhibitor TSP1. Transfection assay revealed that p53 was able to stimulate the endogenous TSP1 gene and positively regulate TSP1 promoter sequences. In a related study, Van Meir et al. (1994) described a glioma-derived angiogenesis inhibitor, which did not appear to be TSP1, that was also under the transcriptional control of the p53 tumor suppressor.

Suppressor gene loss also leads to down regulation of an inhibitor of angiogenesis in keratinocytes progressing towards tumorigenicity *in vivo*. Using the hamster buccal pouch model of oral carcinogenesis, Moroco et al. (1990) were able to show that during tumor progression multiple suppressor gene controlled functions including expression of angiogenic activity, are lost and/or inactivated in mucosal keratinocytes exposed to chemical carcinogens *in vivo* (Figure 5). This same phenomenon seems to occur in human tumors, for when a tumor suppressor gene is reintroduced into cultured human neuroblastoma or osteosarcoma lines, cells begin to make an inhibitor of angiogenesis (Bouck et al., unpublished observations).

There is also indirect evidence suggesting a role for other tumor suppressors in the control of angiogenesis. Schwann cells that have lost one allele of the neurofibromatosis gene, NF-1, are angiogenic while Schwann cells from normal individuals are not (Sheela et al., 1990). Human osteosarcoma and neuroblastoma tumor cell lines which have been reverted to nontumorigenicity following the introduction of either chromosome 17 (Bader et al., 1991), the retinoblastoma, or the p53 suppressor genes become nonangiogenic and begin to produce an inhibitor of angiogenesis (Tolsma et al., unpublished observations).

FUTURE CONSIDERATIONS AND THERAPEUTIC IMPLICATIONS

It is now well established that angiogenesis is a tightly regulated process that is under the control of both growth stimulatory and inhibitory molecules. Although the complement of positive and negative regulators of angiogenesis may vary in different physiologic and pathologic settings, the recognition of this dual mechanism of control is necessary if we are to gain a more thorough understanding of this complex process and its significance in disease. The recent elucidation of the cooperative interaction among positive and negative regulatory molecules during normal physiological angiogenesis and the apparent disruption of this program in disorders such as solid tumor formation (Rastinejad et al., 1989), rheumatoid arthritis (Koch et al., 1986), and psoriasis (Folkman, 1972a) suggest that future studies of pathological angiogenesis must focus on the interaction of positive and negative regulators in this process.

The implications of these findings for the development of therapeutic strategies for the treatment of solid tumors and other angiogenesis dependent disorders are obvious. The use of inhibitors of neovascularization for the treatment of solid tumors and chronic inflammatory diseases have long been envisioned as a possible mode of therapy (Folkman, 1972b; Folkman and Klagsbrun, 1987). Several examples already exist where this approach has met with success (Orchard et al., 1989; White et al., 1991). As the tools of genetic engineering move from the laboratory to the bedside and as the molecular and biochemical basis for the functional diversity of many positive and negative regulators of angiogenesis are defined, it may be possible to up- or down-regulate angiogenic responses with exquisite precision. For example, it may be possible to customize therapy by instructing accessory cell populations such as macrophages that normally enter a focus of inflammation or infiltrate tumors to either produce or even overproduce endogenous or synthetic proangiogenic cytokines or inhibitors. Alternatively, it may be possible to block or limit production of mediators using the antisense constructs. Regardless of the diverse settings in which angiogenesis is found and the great redundancy of mediator systems that participate in this process, the sorting out of the mechanisms which control the balanced production of positive and negative regulators of angiogenesis will no doubt prove to be a fruitful area of investigation.

REFERENCES

Bader, S.A., Faschung, C., Brodeur, G.M., & Stanbridge, E.J. (1991). Dissociation of suppression of tumorigenicity and differentiation in vitro effected by transfer of single human chromosomes into human neuroblastoma cells. Cell Growth Differ. 2, 245–255.

Basset, P., Bellocq, J.P., Wolf, C., Stoll, I., Hutin, P., Limacher, J.M., Podhajter, O.L., Chenard, M.P., Rio, M.C., & Chambon, P. (1990). A novel metalloproteinase gene specifically expressed in stromal cells of breast carcinomas. Nature 348, 699–704.

Bishop, J.M. (1991). Molecular themes in oncogenesis. Cell 64, 234–248.

Bornstein, P. (1992). Thrombospondins: Structure and regulation of expression. FASEB J. 6, 3290–3293.

Bornstein, P. (1995). Diversity of function is inherent in matricellular proteins: An appraisal of thrombospondin 1. J. Cell Biol. 130, 503–506.

Bouck, N. (1990). Tumor angiogenesis:Role of oncogenes and tumor suppressor genes. Cancer Cells 2, 179–185.

Bouck, N. (1993). Angiogenesis: A mechanism by which oncogenes and tumor suppressor genes regulate tumorigenesis. In: Oncogenes and Tumor Suppressor Genes in Human Malignancy (Benz, C.C. & Liu, E.T., eds.), pp. 359–371, Kluwer Academic Publishers, Boston.

Bouck, N., Stellmach, V., & Hsu, S. (1996). How tumors become angiogenic. Adv. Cancer Res. (in press).

Brem, S.S., Gullino, P.M., & Medina, D. (1977). Angiogenesis: A marker for neoplastic transformation of mammary papillary hyperplasis. Science 195, 880–882.

Brem, S.S., Jensen, H.M., & Gullino, P.M. (1978). Angiogenesis as a marker of preneoplastic lesions of the breast. Cancer 41, 239–244.

Briozzo, P., Badet, J., Capony, F., Pieri, I., Montcourrier, P., Barritault, D., & Rochefort, H. (1991). MCF7 mammary cancer cells respond to bFGF and internalizeit following its release from the extracellular matrix: A permissive role for cathepsin D. Exp. Cell Res. 194, 252–259.

Chodak, G.W., Haudenschild, C.C., Gittes, R.F., & Folkman, J. (1980). Angiogenesis activity as a marker of neoplastic and preneoplastic lesions of the human bladder. Ann. Surg. 192, 762–771.

Dameron, K.M., Volpert, O.V., Tainsky, M.A., Bouck, N. (1994). Control of angiogenesis in fibroblasts by p53 regulation of thrombospondin-1. Science 265, 1582–1584.

DiPietro, L.A. & Polverini, P.J. (1993). Role of the macropage in the positive and negative regulation of wound neovascularization. Behring Inst. Mitt. 92, 238–247.

Engerman R.L., Pfaffenbach, D., & Davis, M.D. (1967). Cell turnover of capillaries. Lab. Invest. 17, 738–743

Farber, E. & Cameron, C. (1980). The sequential analysis of cancer development. Adv. Cancer Res. 31, 125–225.

Folkman, J., Cole, P., & Zimmerman, S. (1966). Tumor behavior in isolated perfused organs: *In vitro* growth and metastasis of biopsy material in rabbit thyroid and canine intestinal segments. Ann. Surg. 164, 491–502.

Folkman, J. (1972a). Angiogenesis in psoriasis: Therapeutic implications. J. Invest. Dermatol. 59, 40–48.

Folkman, J. (1972b). Anti-angiogenesis: A new concept for therapy of solid tumors. Ann. Surg. 175, 409–416.

Folkman, J. (1985a). Tumor angiogenesis. Adv. Cancer Res. 43, 175–203.

Folkman, J. (1985b). Perspectives Biol. Med. 9, 10–36.

Folkman, J. & Cotran, R.S. (1976). Relation of vascular proliferation to tumor growth. Int. Rev. Exp. Pathol. 16, 207–248.

Folkman, J. & Klagsbrun, M. (1987). Angiogenic factors. Science 235, 442–447.

Folkman, J. (1989). What is the evidence that tumors are angiogenesis dependent. J. Natl. Cancer Inst. 82, 4–6.

Friend, S.H., Dryja, T.P., & Weinberg (1988). Oncogenes and tumor suppressor genes. N. Engl. J. Med. 318, 618–622.

Gimbrone, M.A. Jr., Cotran, R.S., Leapman, S.B., & Folkman, J. (1974). Tumor growth and neovascularization: An experimental model using the rabbit cornea. J. Natl. Cancer Inst. 52, 413–427.

Gimbrone, M.A. Jr., Leapman, S.B., Cotran, R.S., & Folkman, J. (1972). Tumor dormancy *in vivo* by prevention of neovascularization. J. Exp. Med. 136, 261–276.

Gimbrone, M.A., Jr. & Gullino, P.M. (1976a). Neovascularization induced by intraoccular xenographs of normal, preneoplastic, and neoplastic mouse mammary tissue. J. Natl. Cancer Inst. 55, 305–318.

Gimbrone, M.A., Jr. & Gullino, P.M. (1976b). Angiogenic capacity of preneoplastic lesions of murine mammary gland as a marker of neoplastic transformation. Cancer Res. 36, 2611–2620.

Good, D.J., Polverini, P.J., Rastinejad, F., Le Beau, M.M., Lemons, R.S., Frazier, W.A., & Bouck, N.P. (1990). A tumor suppressor-dependent inhibitor of angiogenesis is immunologically and functionally indistinguishable from a fragment of thrombospondin. Proc. Natl. Acad. Sci. USA 87, 6624–6628.

Hanahan, D. & Folkman, J. (1996). Patterns and emerging mechanisms of the angiogenic switch during tumorigenesis. Cell 86, 353–364.

Hansen, M.F. & Cavenee, W.K. (1988). Tumor suppressors: Recessive mutations that lead to cancer. Cell 53, 172–173.

Jensen, H.M., Chen, I., DeVault, M.R., & Lewis, A.E. (1982). Angiogenesis induced by "normal" breast tissue: A probable marker for precancer. Science 218, 293–295.

Klagsbrun, M. & D'Amore P.A. (1991). Regulators of angiogenesis. Ann. Rev. Physiol. 53, 217–239.

Klagsbrun, M. & Folkman, J. (1990). Peptide Growth Factors and Their Receptors II. In: Angiogenesis. Handbook of Experimental Pharmacology, Vol., 95/II (Sporn, M.B. & A.B. Roberts, A.B., eds.), pp. 549–586, Springer-Verlag, Berlin, Heildelberg, Germany.

Klein, G. (1987). The approaching era of tumor suppressor genes. Science 238, 1539–1545.

Koch, A.E., Polverini, P.J., & Leibovich, S.J. (1986). Stimulation of neovascularization by human rheumatoid synovial tissue macrophages. Arth. Rheum. 29, 471.

Liotta, L.A., Steeg, P.S., & Stetler-Stevenson, W.G. (1991). Cancer metastasis and angiogenesis: an imbalance of positive and negative regulation. Cell 64, 327–336.

Marshall, C.J. (1991). Tumor suppressor genes. Cell 64, 313–326.

Moroco, J.R., Solt, D.B., & Polverini, P.J. (1990). Sequential loss of suppressor genes for three specific functions during in vivo carcinogenesis. Lab. Invest. 631, 298–306.

Moses, M.A. & Langer, R. (1991). Inhibitors of angiogenesis. Biotechnology 9, 630–634.

Nowell, P.C. (1976). The clonal evolution of tumor cell populations. Science 194, 23–28.

Orchard, P.J., Smith, C.M. II, Woods, W.G., Day, D.L., & Dehner, L.P. (1989). Treatment of hemangioendotheliomas with alpha interferon. The Lancet 2, 565–567.

O'Reilly, M.S., Holmgren, L., Shing, Y., Chen, C., Rosenthal, R.A., Moses, M., Lane, W.S., Cao, Y., Sage, E.H., Folkman, J. (1994). Angiostatin: A novel angiogenesis inhibitor that mediates the suppression of metastases by lewis lung carcinoma. Cell 79, 315–328.

Polverini, P.J., Cotran, R.S., Gimbrone, M.A., Jr., & Unanue, E.R. (1977). Activated macrophages induce vascular proliferation. Nature 269, 804–806.

Polverini, P.J. & Leibovich, S.J. (1984). Induction of neovascularization in vivo and endothelial cell proliferation in vitro by tumor–associated macrophages. Lab. Invest. 51, 635–642.

Polverini, P.J. (1989). Macrophage-induced angiogenesis: A review. In: Macrophage-Derived Regulatory Factors, (Sorg, C., ed.), pp. 54–73, S. Karger, Basel.

Polverini, P.J. (1995). The pathophysiology of angiogenesis. Crit. Rev. Oral Biol. Med. 6, 230–247.

Rastinejad, F., Polverini, P.J., & Bouck, N.P. (1989). Regulation of the activity of a new inhibitor of angiogenesis by a cancer suppressor gene. Cell 56, 345–355.

Risau, W. (1991). Vasculogenesis, Angiogenesis and Endothelial Cell Differentiation during Embryonic Development. In: The Development of the Vascular System. (Feinberg, R.N., Sherer, G.K., & Auerbach, R., eds.), pp. 58–68, S. Karger, Basel.

Sager, R. (1989). Tumor suppressor genes: the puzzle and the promise. Science 246, 1406–1410.

Schor, A.M. & Schor, S.L. (1983). Tumor angiogenesis. J. Pathol. 141, 385–413.

Schwechheimer, K. & Cavenee, W.K. (1993). Genetics of cancer predisposition and progression. Clin. Invest. 71, 488–502.

Sheela, S., Riccardi, V.M., & Ratner, N. (1990). Angiogenic and invasive properties of neurofibroma schwann cells. J. Cell Biol. 111, 645–653.

Starkey, J.R., Crowle, P.K., & Taubenberger, S. (1988). Mast-cell-deficient W/Wv mice exhibit a decreased rate of tumor angiogenesis. Intl. J. Cancer. 42, 48–52.

Sunderkotter, C., Goebeler, M., Schultze-Osthoff, K., Bhardwaj, R., & Sorg, C. (1991). Macrophage-derived angiogenesis factors. Pharmac. Ther. 51, 195–216.

Tannock, I.F. & Hayashi, S. (1972). The proliferation of capillary endothelial cells. Cancer Res. 32, 77–82.

van den Hooff, A. (1988). Stromal involvement in malignant growth. Adv. Cancer Res. 50, 159–196.

van den Hooff, A. (1991). The role of stromal cells in tumor metastasis: A new link. Cancer Cells 3, 186–187.

Van Meir, E.G., Polverini, P.J., Chazin, V.R., Huang, H-J., de Tribolet, N., Cavenee, W.K. (1994). Release of an inhibitor of angiogenesis upon induction of wild type p53 in glioblastoma cells. Nature Genet 8, 171–176.

Weidner, N., Semple, J.P., Welch, W.R., & Folkman, J. (1991). Tumor angiogenesis and metastasis-correlation in invasive breast carcinoma. N. Engl. J. Med. 324, 1–8.

Werinberg, R. (1989). Oncogenes, anticonogenes, and the molecular basis of multistep carcinogenesis. Cancer Res. 49, 3713–3721.

Welch, D.R., Schissel, D.J., Howrey, R.P., & Aeed, P.A. (1989). Tumor-elicited polymorphonuclear cells, in contrast to 'normal' circulating polymorphonuclear cells, stimulate invasive and metastatic potentials of rat mammary adenocarcinoma cells. Proc. Natl. Acad. Sci. USA 86, 5859–5863.

White, C.W., Wolf, S.J., Korones, D.N., Sondheimer, H.M., Tosi, M.F., & Yu, A. (1991). Treatment of childhood angiomatous diseases with recombinant interferon alpha. J. Pediatr. 118, 59–66.

Chapter 6

The Role of Gap Junctional Intercellular Communication in Neoplasia

RANDALL J. RUCH

Advances in Oncobiology
Volume 1, pages 119–141.

ISBN: 0-7623-0146-5

INTRODUCTION

Over the last two decades, tremendous progress has been made in defining the intracellular pathways and extracellular factors that regulate cellular division and phenotype and the changes that lead to neoplasia. Numerous oncogenes, tumor suppressor genes, and cell cycle-related genes have been discovered and their functions greatly clarified. Many growth factors, hormones, their receptors, and the signal cascades that occur after binding of ligands to receptors have also been extensively defined.

Relatively less knowledge, however, has been gleaned concerning the role of intercellular processes involved in growth control and neoplastic transformation. By this, I mean the direct interactions of cells through cell-cell contact and how this is involved in growth. Studies with cultured cells have clearly shown that cellular growth and phenotype are influenced by contact-dependent interactions with neighboring cells and the extracellular matrix. These interactions include responses of cells to (a) components on neighboring cell plasma membranes such as cadherins, (b) molecules such as laminins found within the extracellular matrices of neighboring cells or the basement membrane, and (c) signals received directly from neighboring cells via gap junction channels.

This latter mode of intercellular interaction is especially intriguing and relevant to growth regulation and neoplastic transformation because of the ubiquitous nature of gap junctions. Except for skeletal muscle cells, certain circulating blood cells, and certain neurons, all adult mammalian cells examined have gap junctions. These structures are also ancient and have been detected in such primitive invertebrates as sponges, jellyfish, and sea anemones. Similar structures known as plasmodesmata are found in plants. Thus, gap junctions are a very common feature of the majority cells in multicellular organisms and, therefore, are likely to have important functions. As will be discussed, several different lines of evidence indicate that one role for these ancient structures is the regulation of cellular growth and that the impairment of this exchange can lead to unregulated growth and neoplasia.

In this chapter, the structure of gap junctions and the mechanisms controlling their formation, degradation, and channel permeability will be reviewed. More importantly, however, the role of gap junctions in growth control and neoplasia will be described. Hopefully the reader will develop an appreciation of how this type of intercellular interaction is involved in the neoplastic phenotype.

THE STRUCTURE OF GAP JUNCTIONS

Gap junctions are one type of plasma membrane junctional complex. They are ubiquitous in the animal kingdom and have been identified in species representative of nearly all animal phyla (Loewenstein, 1981). Similar structures, the plasmodesmata, have been described in plants (Robards and Lucas, 1990). Gap junctions consist of clusters of hundreds to thousands of particles approximately 10 nm in diameter that span the plasma membranes of adjacent cells (Figure 1). Each particle contains a small pore or channel that links the interiors of the two junction-forming cells. The channels are formed by the end-to-end abutment of two hemichannels or "connexons" which are in turn comprised of six protein subunits known as "connexins." Thus, each gap junction channel contains twelve connexins.

Connexins are folded in the plasma membrane in the approximate shape of an "M" (Figure 2). The amino and carboxyl termini project into the cytoplasm while

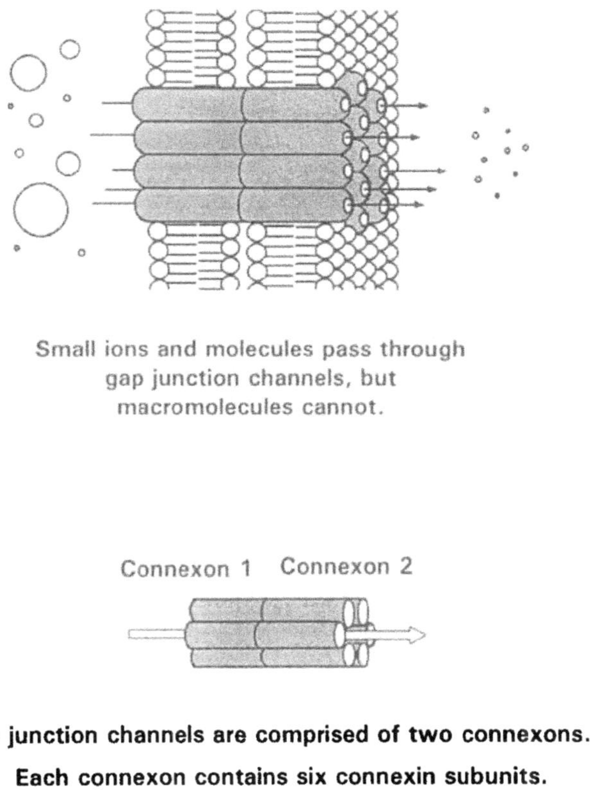

Figure 1. Model of gap junction particles embedded in the plasma membranes of two adjacent cells.

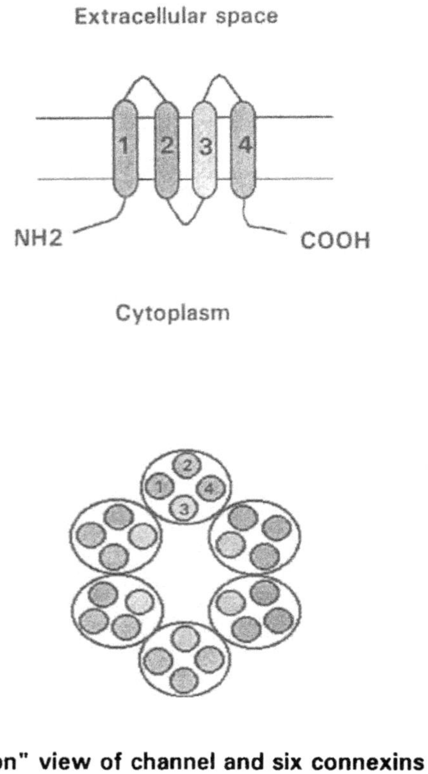

Figure 2. Diagrammatic representation of a connexin. The protein spans the plasma membrane four times and the amino (NH_2) and carboxyl tails (COOH) reside in the cytoplasm. The third membrane-spanning domain is thought to line the interior of the gap junction channel.

the rest of the molecule traverses the plasma membrane four times. These four membrane-spanning regions lie in parallel. The third region contains a high proportion of hydrophilic amino acids and is thought to line the interior of the channel. The four membrane-spanning domains and the extracellular loops are highly conserved among different connexins. More variable are the cytoplasmic regions. As will be discussed, these differences may be involved in the cellular regulation of gap junction formation and channel permeability. Connexin folding as well as connexin-connexin and connexon-connexon interactions are mediated through disulfide bonds, hydrophobic protein interactions, and other more poorly understood forces.

The diameter of vertebrate gap junction channels has been approximated using electron microscopy and X-ray crystallography and is 1.5–2 nm; invertebrate gap junction channels are slightly larger (Loewenstein, 1981). These pore sizes limit permeability through the channel to molecules up to approximately 1,000 Da (vertebrate gap junctions) and 2,000 Da (invertebrate gap junctions). Thus, ions and small molecules can traverse the channels, but macromolecules such as proteins and RNA cannot (Figure 1). Sugars, nucleotides, amino acids, fatty acids, small peptides, and even drugs and carcinogens have been shown to pass through gap junction channels into neighboring cells. However, proteins, complex lipids, polysaccharides, RNA, and other large molecules cannot. Channel passage does not require energy and appears to result from passive diffusion. This flux of ions and molecules between cells through gap junction channels is called gap junctional intercellular communication (GJIC).

One of the most significant physiological implications for GJIC is that gap junctionally-coupled cells within a tissue are not individual, discrete entities, but are highly integrated. As will be discussed, this property facilitates cellular homeostasis and permits the rapid, direct transfer of signals and coordination of cellular responses. On the other hand, the advantage of the channel size limit is that it allows cells to maintain their functional identities through cell-specific syntheses of enzymes, receptors, and other metabolic machinery involved in tissue- and cell-specific functions.

THE CONNEXIN MULTIGENE FAMILY

The channel-forming connexins comprise a multi-gene family with at least 13 rodent mammalian connexins discovered thus far (White et al., 1995). Several homologous DNAs have been identified in other vertebrate species. Several connexins and tissues in which they are highly expressed are listed in Table 1. The number associated with each connexin indicates its molecular mass. Connexins are expressed in a cell-, tissue-, and developmentally-specific manner. For instance, connexin43 is the predominant connexin expressed in cardiac muscle and was first cloned from this tissue (Beyer et al., 1987), although other connexins (connexin40, connexin45, and connexin46) have also been detected in cardiac tissue (Table 1). In adult liver, the predominant connexins are connexin32 and connexin26 (Paul, 1986; Nicholson et al., 1987) and these are expressed by adult parenchymal liver cells (hepatocytes). However, nonparenchymal liver epithelial cells, hepatic fat-storing (Ito) cells, and hepatic connective tissue cells express connexin43 (Stutenkemper et al., 1992; Ruch et al., 1994; Rojkind et al., 1995). Connexin37 and connexin40 are expressed predominantly in the lung in endothelial cells but their mRNAs have also been detected in other tissues (Willecke et al., 1991b; Hennemann et al., 1992b). The predominant connexin expressed by lung bronchial and alveolar epithelial cells is connexin43 (Guan et al., 1995; Cesen et al., 1996). In those cells where multiple connexins are expressed, gap junction channels may be comprised of more than one connexin or may be homogeneous (Nicholson et al., 1987; Sosinsky, 1995).

Table 1. Some Cloned Mammalian Connexins and Tissues That Have High Levels of Expression

Connexin (Cx)	*Tissue*
Cx26	liver, kidney
Cx30.3	skin
Cx31	skin
Cx32	liver, kidney, exocrine pancreas, brain
Cx37	lung
Cx40	lung
Cx43	heart, smooth muscle, many epithelia, brain, connective tissue, endocrine pancreas
Cx45	lung
Cx46	lens

Despite the presence of conserved sequences within connexins, the diversity of these proteins is not due to alternative splicing of one or a few precursor messenger RNAs. Instead, there appears to be one connexin gene per protein. Many connexin genes have been mapped and are located on several chromosomes (Willecke et al., 1991a) suggesting their distribution is random throughout the genome. Why there are so many connexins is not clear, but may reflect differences in their function(s) and/or the regulation of their formation and permeability.

REGULATION OF CONNEXIN GENE EXPRESSION

The structures of all connexin genes identified thus far are similar and consist of two exons separated by a long intron. In the rat connexin32 gene, the intron is approximately 6,000 base pairs (Miller et al., 1988). The first exons of connexins are quite short (about 100 base pairs) and contain no protein coding information. The ATG translational start codon is located within the first 100 base pairs of exon 2 and, thus, all of the coding information for the protein resides within this latter exon.

Little is known regarding the mechanisms regulating the expression of connexin genes, although much progress has been made within the last five years. The promoters of mouse, rat, and/or human connexins 43, 32, and 26 have been sequenced and several regulatory sites that might control expression of the respective gene have been identified (Hennemann et al., 1992a; Bai et al., 1993; Sullivan et al., 1993; Chen et al., 1995b). However, few of these sites have been examined for function. In the connexin32 and connexin26 genes, there does not appear to be a functional TATA-box, but instead transcription is most likely initiated through a

Table 2. Physiological and Pharmacological Agents That Alter Connexin Expression

Agent	*Connexin (Cx)*	*Effect*	*Tissue or cells*
	Steroid Hormones		
Estrogens	Cx43	Increase	Uterine myometrium
Progesterone	Cx43	Decrease	Uterine myometrium
Testosterone	Cx32	Increase	Spinal cord motoneurons
Glucocorticoids	Cx32, Cx26	Increase	Hepatocytes
	Cyclic AMP Agonists and Analogues		
Dibutyryl Cyclic AMP	Cx32	Increase	Hepatocytes
Forskolin	Cx43	Increase	Nonparenchymal liver cells, fibroblasts
	Retinoids and Carotenoids		
Retinoic acid	Cx43	Increase	Fibroblasts
β-carotene	Cx43	Increase	Fibroblasts

CCAAT element ("CAT-box") (Miller et al., 1988; Hennemann et al., 1992a). These elements are often characteristic of "house-keeping" genes, i.e., genes that are expressed constitutively in tissues. The expression of both Cx32 and Cx26 is inducible, however, by glucocorticoids (Ren et al., 1994) and cyclic AMP (Traub et al., 1987) and expression is dramatically reduced during liver regeneration (Dermietzel et al., 1987). There are several other pharmacological and physiological factors that enhance connexin gene expression (Table 2), but the mechanisms are not understood. Additionally, little is known how connexin genes are "turned off" or silenced in certain tissues. Clearly, a better understanding of connexin gene regulation will be beneficial to understanding their cell-specific expression and developing means to modulate their expression for the treatment of diseases such as cancer (see below).

CONTROL OF GAP JUNCTION FORMATION AND CHANNEL PERMEABILITY

Multiple mechanisms regulate gap junction formation and channel permeability. Channel formation is dependent upon levels of connexin expression, the connexin composition of the channel subunits, and the abilities of the two cells to form appropriate cell-cell contact. Connexons comprised of certain types of connexins may be incapable of forming channels with connexons containing other types (Werner et al., 1989; Rubin et al., 1992). Gap junction formation also requires that the two cells can adhere, which is dependent upon the surface properties of the cells and the extracellular matrix. In poorly communicating neoplastic cells, gap junction formation was increased when the cells were induced to express the cell adhesion molecule, E-cadherin (Jongen et al., 1991). Treatment of hepatocytes with extracellular matrix components such as proteoglycans has also resulted in increased gap junction formation (Spray et al., 1987). Therefore, the inability of cells to communicate with other cells may be related not only to decreased expression of connexins, but also to differences in connexin type, cell surfaces, and the extracellular matrix.

Rates of synthesis and degradation of connexins and removal and disassembly of gap junctions from the cell surface is also involved in the regulation of GJIC. Biochemical studies have demonstrated that some connexins are synthesized and degraded very rapidly. The half-life of connexin32 has been estimated to be only a few hours (Fallon and Goodenough, 1981; Yancey et al., 1981; Traub et al., 1989), whereas most membrane proteins have half-lives of several days. Thus, gap junction channel number and GJIC may be tightly regulated by rates of connexin turnover. Gap junctional particles aggregated in large plaques may also disaggregate in response to certain physiological or environmental cues. Following the induction of a regenerative stimulus in the rat liver, hepatocyte gap junctions decrease in number apparently due to particle dispersal (Yancey et al., 1979). We have reported that a compound from licorice root, 18β-glycyrrhetinic acid, causes the disaggregation of connexin43-containing gap junction particles and their dispersal in the plasma membrane (Guan et al., 1996). Additionally, gap junctions can be removed from the cell surface by internalization in response to physiological or environmental cues. Gap junctional internalization appears to be important in the loss of gap junctions from rabbit granulosa cells during maturation of the ovarian follicle (Larsen and Hai-Nan, 1978). Several pesticides also cause the loss of gap junctions by internalization in liver epithelial cells (Guan and Ruch, 1996). Inside the cell, the junctional proteins may be degraded in lysosomes or reutilized. The calcium-activated proteases, μ-calpain and m-calpain, can also degrade hepatocyte connexin32 (Elvira et al., 1993) which indicates there are at least two systems of connexin degradation in these cells. Thus, connexin protein turnover and gap junction assembly, disassembly, and internalization are important in the regulation of GJIC.

The permeability of gap junction channels is tightly controlled. Once formed, the channels can open and close. This gating may be modulated by connexin phosphorylation (Saez et al., 1993). Several connexins are phosphorylated on serine and threonine residues localized to the connexin cytoplasmic carboxyl tail (Figure 2). Phosphorylation by different kinases may function to either open or close the channels by altering connexin structure. Activation of cAMP-dependent protein kinase (protein kinase A) may increase the phosphorylation of connexin43 and connexin32, although it is not clear if the kinase is directly responsible. Phosphorylation of these connexins by protein kinase A usually leads to enhanced GJIC in most cells. But in other cells such as fish and turtle retina, cAMP reduces GJIC (Miyachi and Murakami, 1989). Connexin43 is also phosphorylated following the activation of calcium, phospholipid-dependent protein kinase (protein kinase C) and this usually results in reduced GJIC (Saez et al., 1993). Activation of the *src* oncogene product, pp60^{src}, which has tyrosine kinase activity, can also result in the phosphorylation of connexin43 on tyrosine residues (Crow et al., 1990). This effect is one of the first changes to occur in cells when the *src* kinase is expressed and leads to rapid channel closure. Some connexins such as connexin26 are not phosphorylated so that gating must be regulated in other ways.

Other mechanisms regulating channel gating include hydrogen and calcium ion levels, transjunctional voltage, and free radicals (Saez et al., 1993). Decreased pH or pCa lead to channel closure in cell- and connexin-specific fashion. In the case of calcium, its effect on gap junction channels may be mediated through calcium-binding proteins such as calmodulin. Transjunctional voltage effects on channels is also cell- and connexin-specific. Free radicals, which are highly reactive molecules and ions generated during normal cellular metabolism or following exposure to certain toxic agents, can also decrease channel permeability (Ruch and Klaunig, 1988). The radicals may directly attack connexins, but it is unclear whether radicals are involved in physiological regulation of channel permeability and/or gap junction turnover. The mechanisms for these various gating processes are poorly understood and controversial.

PHYSIOLOGICAL ROLES OF GJIC

Several physiological roles besides growth control have been proposed for GJIC and are briefly reviewed:

1. Homeostasis. GJIC permits the rapid equilibration of nutrients, ions, and fluids between cells. This might be the most ancient, widespread, and important function for these channels (Loewenstein, 1981).
2. Electrical coupling. Gap junctions serve as electrical synapses in electrically excitable cells such as cardiac myocytes, smooth muscle cells, and neurons (Lowenstein, 1981). In these tissues, electrical coupling permits more rapid cell-to-cell transmission of action potentials than chemical synapses. In myocytes, this enables their synchronous contraction.
3. Tissue response to hormones. GJIC may enhance the responsiveness of tissues to external stimuli (Murray and Fletcher, 1984). Second messengers such as cyclic nucleotides, calcium, and inositol phosphates are small enough to pass from hormonally activated cells to quiescent cells through junctional channels and activate the latter. Such an effect may increase the tissue response to an agonist.
4. Regulation of embryonic development. Gap junctions may serve as intercellular pathways for chemical and/or electrical developmental signals in embryos and for defining the boundaries of developmental compartments (Kalimi and Lo, 1988). GJIC occurs in specific patterns in embryonic cells

and the impairment of GJIC has been related to developmental anomalies and the teratogenic effects of many chemicals (Loch-Caruso and Trosko, 1986).

ALTERED GJIC IN GROWTH AND NEOPLASIA

More important to this discussion is that GJIC is also involved in the regulation of cellular growth and expression of the neoplastic phenotype. For some time, gap junctions have been proposed to serve as passageways for the cell-to-cell exchange of low molecular weight growth regulatory molecules (Loewenstein and Kanno, 1966). GJIC is frequently reduced in neoplastic and carcinogen-treated cells. It was hypothesized that this contributed to dysregulated cellular growth by isolating cells from their neighbors (reviewed in Loewenstein, 1979). As will be described below, there is now compelling evidence that this is true. GJIC is involved in growth control and the downregulation of GJIC facilitates abnormal growth and neoplastic transformation.

In many important biological control processes, there is redundancy by overlapping regulatory pathways. This insures that if one pathway is defective, the cell will not become dysfunctional. Similarly, GJIC should be viewed as one of several mechanisms that participate in controlling growth and phenotype. Although the details have not been worked out, GJIC undoubtedly functions in concert with more well-characterized processes such as growth factor signal cascades, cell cycle regulatory proteins such as cyclins, and pathways that activate cellular differentiation or death (apoptosis). The next decade should provide much insight into the interplay between these various processes.

Two possible schemes by which growth may be regulated by GJIC are shown in Figures 3 and 4 and are adapted from Loewenstein (1979). Both inhibitory (Figure 3) and/or stimulatory (Figure 4) low-molecular weight signals might be produced in cells and diffuse to adjacent cells through junctional channels. The negative signals would serve to inhibit cell division and maintain differentiation whereas positive signals might stimulate growth and prevent differentiation. The loss of gap junctions or reductions in channel permeability would isolate cells from inhibitory signals and/or permit the accumulation of positive signals. These effects would lead to unregulated cell division and incomplete differentiation, both of which are hallmarks of neoplasia. If the reduction of GJIC were sustained, a noncommunicating cell could expand by clonal growth into a tumor.

While this model is clearly an oversimplification of growth and tumor formation, it does provide the tissue with a potential mechanism for fine-tuning cell number and function. By regulating the quantity, frequency, and type of positive and negative signals generated, and the number and permeability of junctional channels, a steady-state of signal level(s) and cell number could be maintained in a tissue. Stresses to the system—such as the loss of cells after toxic cell death or wounding or reductions in channel number or permeability—would result in decreased

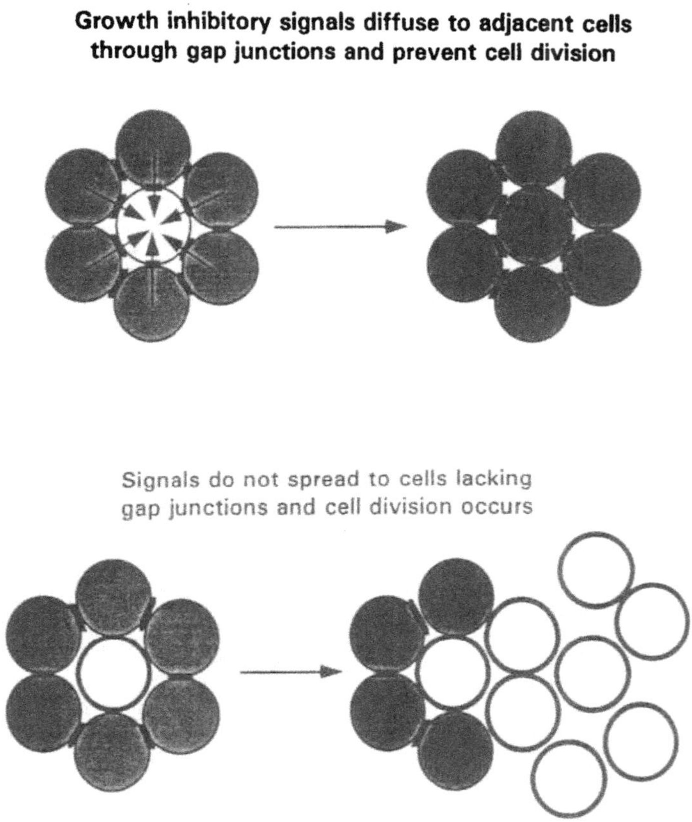

Figure 3. Gap junction model of growth control involving growth inhibitory signals. Adapted from Loewenstein (1979).

communication, altered levels of signals, and induction of tissue growth. Cell number and tissue mass would then be dependent upon the size of the communicating tissue network. Because it is a closed system, such a mechanism might be more amenable to subtle control than one involving the diffusion of regulators such as growth factors throughout extracellular spaces.

While such a model of growth regulation and neoplasia might seem plausible, is there experimental proof? The answer is affirmative and the evidence, culled from a diverse array of studies, will be briefly reviewed below.

Neoplastic Cells Have Fewer Gap Junctions

Many neoplastic cells have been examined for the presence, size, and function of their gap junctions. The vast majority of these cells have fewer gap junctions

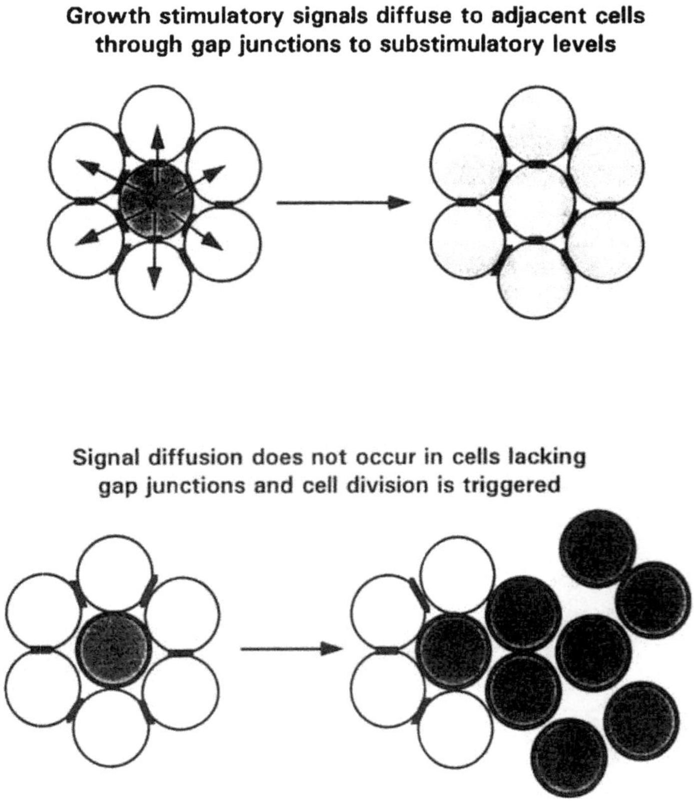

Figure 4. Gap junction model of growth control involving growth stimulatory signals. Adapted from Loewenstein (1979).

compared to homologous, nonneoplastic cells based upon ultrastructural and immunohistochemical evidence (Weinstein et al., 1976; Loewenstein, 1981, 1987). Neoplastic cells have also been evaluated for their communication levels by introduction of fluorescent or radioactive tracers into these cells and determination of tracer passage into adjacent cells. But neoplastic cells often have many abnormal features of their plasma membranes including the loss of other types of junctional complexes and defective cell adhesion molecules (Weinstein et al., 1976; Takeichi, 1990). A reduction in gap junctions does not necessarily indicate a mechanistic association with neoplasia.

Additionally, some neoplastic cells make numerous, functional gap junctions. How do these cells fit into the reduced GJIC/neoplastic transformation paradigm? Some of these cells form communicating junctions only with other neoplastic cells and not with normal cells (Yamasaki, 1990). This selective communication would effectively isolate the neoplastic cells from the regulatory influences of their surrounding neighbors and is probably related to differences in the surface properties of the normal and neoplastic cells. Alternatively, some gap junction-forming neoplastic cells do form communicating junctions with normal cells. In these instances, the tumor cells may be unresponsive to the signals received from the normal cells and would be essentially noncommunicating.

Growth Stimuli Inhibit GJIC

Growth Factors

Many growth factors inhibit GJIC when applied to cultured cells (Table 3). This effect occurs rapidly (minutes to hours) in many cells, but may also be delayed (days). The mechanism(s) of inhibition of GJIC by epidermal growth factor may be related to the stimulation of connexin phosphorylation and closure of gap junctional channels (Lau et al., 1992). How the other growth factors modulate GJIC has not been determined.

Carcinogens

A variety of carcinogens have been identified that enhance neoplastic transformation through mechanisms that do not appear to involve direct damage of DNA. Many of these so-called "nongenotoxic" carcinogens instead appear to function by selectively inducing the proliferation of preneoplastic cells (Schulte-Hermann et al., 1983). This leads to clonal expansion of the preneoplastic cell population and increased risk of subsequent genetic changes leading to full neoplastic transformation. While it is debatable how cell proliferation leads to neoplasia, it is likely that GJIC plays a role in the proliferative response. Many nongenotoxic carcinogens (over 100) have been shown to inhibit GJIC in cultured cells and tissues (Klaunig and Ruch, 1990; Budunova and Williams, 1994). A list of some of the agents that affect GJIC *in vivo* is presented in Table 4. These agents are chemically diverse and

Table 3. Growth Factors That Affect GJIC

Growth Factor	*Effect on GJIC*	*Cell Type*	*Mechanism*
Epidermal growth factor	Decrease	Fibroblasts, keratinocytes	Connexin43 phosphorylation
Platelet-derived growth factor	Decrease	Fibroblasts	Unknown
Transforming growth factor-β	Decrease	Keratinocytes, normal bronchial epithelial cells	Unknown
Transforming growth factor-β	Increase	Neoplastic bronchial epithelial cells	Unknown

Table 4. Nongenotoxic Carcinogens That Reduce GJIC and Gap Junction Number in Rodent Tissues *In Vivo*

Carcinogen	*Mechanism(s)*	*Tissue*
Pesticides		
Dichlorodiphenyltri-chloroethane (DDT)	Decreased gap junction number	Liver
Lindane	Decreased gap junction number	Liver
Phorbol Esters		
12-*O*-tetradecanoyl-phorbol-13-acetate (TPA)	Decreased gap junction number; altered connexin phosphorylation	Skin
Sedatives		
Phenobarbital	Reduced connexin32 expression and gap junction number	Liver
Hyperlipidemic Agents		
Clofibrate	Unknown	Liver
Nafenopin	Unknown	Liver

affect connexin phosphorylation and channel permeability as well as gap junction number and connexin expression (Brissette et al., 1991; Berthoud et al., 1992; Ruch et al., 1994; Matesic et al., 1994). The ability of nongenotoxic carcinogens to inhibit GJIC is one of their most common properties.

Some studies have indicated that preneoplastic cells are more sensitive than normal cells to the effects of these agents on GJIC (Klaunig et al., 1990). Such a differential response could theoretically result in proliferation and clonal expansion of preneoplastic cells at the expense of surrounding normal cells as discussed above.

In contrast to the effects of nongenotoxic carcinogens, most DNA-damaging ("genotoxic") carcinogens do not inhibit GJIC or induce cell proliferation (Ruch and Klaunig, 1986; Budunova and Williams, 1994). They instead appear to function by mutationally activating proto-oncogenes or inactivating tumor suppressor genes. Thus, carcinogens have different effects on DNA, GJIC, and growth which correlate with their mechanisms of action.

Oncogenes

Oncogenes are genes derived from normal cellular genes (proto-oncogenes) that have been mutationally activated and/or are overexpressed and that function in the transformation of a normal cell into a neoplastic one. The protein products of these genes function in signal transduction, gene regulation, growth control, and many other facets of cellular activities. Not surprisingly, the expression of several of these genes has been correlated with the reduction of GJIC in several *in vitro* systems (Table 5). Their ability to inhibit GJIC may be involved in their growth enhancing

Table 5. Oncogenes That Affect GJIC

Oncogene	*Effect on GJIC*	*Cell Type*	*Mechanism*
v-*src*, c-*src*	Decrease	Fibroblasts	Connexin phosphorylation
EJ-Ha-*ras*	Decrease	Fibroblasts, liver epithelial cells	Connexin phosphorylation and reduced expression
v-*raf*	None	Liver epithelial cells	
v-*myc*	None	Liver epithelial cells	
v-*raf* + v-*myc*	Decrease	Liver epithelial cells	Unknown
v-*fos*	None	Fibroblasts	
v-*mos*	None	Fibroblasts	
neu-1A	Decrease	Liver epithelial cells	Altered gap junction formation

and neoplastic properties. In fact, one of the first detectable events that occurs when the *src* oncogene is expressed in cultured cells is the phosphorylation of connexin43 and the reduction of gap junctional permeability (Azarnia and Loewenstein, 1984; Atkinson et al., 1981). This phosphorylation occurs on tyrosine residues of connexin43 and is mediated by activation of the *src* protein ($pp60^{src}$) which has tyrosine kinase activity (Crow et al., 1990). Other studies have shown that the inactivation of oncogene proteins using pharmacological treatments has resulted in the restoration of GJIC, a more normal cellular morphology, and reduced tumorigenicity (Ruch et al., 1993). Oncogenes can also cooperate in their ability to reduce GJIC and transform cells. Cells that expressed only *raf* or *myc* oncogenes did not have reduced GJIC and were not transformed (Table 5). However, when both oncogenes were expressed, GJIC was inhibited and the cells were highly malignant (Kalimi et al., 1992).

As mentioned above, $pp60^{src}$ appears to reduce GJIC by phosphorylating connexins. Other oncogene proteins (e.g., the p21 proteins of the *ras* oncogenes) also appear to alter connexin phosphorylation as well as to inhibit connexin expression and gap junction assembly (Brissette et al., 1991; Esinduy et al., 1995).

Growth Inhibitors Stimulate GJIC

In contrast to the effects of growth stimulatory and neoplastic agents on GJIC, many growth and cancer inhibitory agents increase GJIC and connexin expression in target cells. Some of these are listed in Table 2. Retinoids, carotenoids, dexamethasone, and cyclic AMP agonists inhibit neoplastic transformation and/or tumor cell growth and increase connexin expression and gap junction formation in target tissues (Rogers et al., 1990; Mehta et al., 1992; Zhang et al., 1992; Ren et al., 1994). Cyclic AMP agonists also increase channel permeability probably by

Table 6. Cell Cycle-Related Changes in GJIC

Cell Cycle Phase	*Tissue or Cells*
G_1/S border	Fibroblasts
S	Regenerating liver
M	Tracheal epithelial cells
M	Granulosa cells

activating protein kinases and stimulating connexin phosphorylation (Saez et al., 1993).

Cell Cycle-Related Changes in GJIC

GJIC also appears to have a role in the progression of dividing cells through the cell cycle. In several model systems, cell cycle-related changes in GJIC have been noted (Table 6). For example, a reduction in GJIC during S-phase has been observed in regenerating liver following two-thirds partial hepatectomy. The adult liver normally has a very low proportion (< 1%) of parenchymal cells (hepatocytes) that are replicating. Hepatocytes are highly coupled, however, and have over a dozen gap junctions per cell (Meyer et al., 1981; Dermietzel et al., 1987). The majority of hepatocytes are in stationary (G_0) phase. However, they can be induced to rapidly enter the cell cycle and begin dividing when two-thirds of the liver is surgically removed (partial hepatectomy). The remaining hepatocytes enter the cell cycle and undergo cell division. Analyses of hepatocyte gap junctions during this induction of cell division have shown that gap junctions were nearly completely lost in a transient manner and that this loss occurred during DNA synthesis (S-phase) (Meyer et al., 1981; Dermietzel et al., 1987). Following DNA synthesis, the junctions reappeared and remained at high levels throughout the rest of the cycle.

Several cell culture studies have also demonstrated changes in GJIC at various points in the cell cycle (Gordon et al., 1982; Shiba et al., 1990; Stein et al., 1992). Cultured cells normally replicate in an asynchronous manner, but they can be synchronized by blocking the cell cycle at a particular point such as G_1. This can be achieved by depriving the cells of nutrients or growth factors or by adding pharmacological agents that inhibit cell cycling. When released from these types of blocks, the cells will divide in a near synchronous manner through several cycles. When such cells were analyzed for GJIC, reductions were noted at the G_1/S border and in mitosis (Table 6).

Thus, both *in vivo* and *in vitro* studies have documented cell cycle-related changes in GJIC. It remains unknown how these changes are achieved and whether they contribute to cell cycle regulation.

INVOLVEMENT OF GJIC IN THE GROWTH INHIBITION OF NEOPLASTIC CELLS BY NONTRANSFORMED CELLS

Several cell culture model systems have illustrated that the growth of neoplastic cells can be inhibited by coculture with nonneoplastic cells. This effect has been attributed to noncontact-dependent and contact-dependent phenomena. For instance, normal cells may secrete growth inhibitors (e.g., TGF-β) into the culture medium that inhibit neoplastic cell growth (Massague, 1987). Contact with normal cell extracellular matrix or plasma membrane components may also trigger growth-inhibiting processes in neoplastic cells (Edelman, 1988).

The inhibition of neoplastic cell growth by normal cells also appears to involve GJIC. Our group has studied this using cultured rat liver epithelial cells (RLEC) and connexin43 as a model system (Esinduy et al., 1995). In normal RLEC, high levels of connexin43 were expressed, numerous gap junctions were formed, and the percentage of communicating cells was high (95–100%). Neoplastic transformants of these cells, however, expressed connexin43 at lower levels, formed few gap junctions, and had low incidences of communication (20–25%). When the two types of cells were cocultured, the growth of the neoplastic cells was inhibited. However, this inhibition occurred only when the two types of cells were permitted to make contact with each other and not when they were physically separated in the culture dish. This suggests that cell-cell contact, not secretion of extracellular factors was required for growth inhibition. To demonstrate this more clearly, connexin43 mutants of the normal cells were obtained that were defective in gap junction formation and did not communicate. These cells did not inhibit the growth of neoplastic cells when cell-cell contact was permitted. Interestingly, however, when GJIC was restored in the mutant cells by introduction of a functional connexin43 gene, growth inhibiting activity returned. Thus, inhibition of neoplastic cell growth by normal cells in this model system required GJIC.

CONNEXIN ANTISENSE EXPRESSION

Recently, techniques have been developed to specifically inhibit the expression of a target gene in cultured cells and animals. So-called "gene knockouts" involve the disruption of a targeted gene by the insertion of a noncoding sequence through homologous recombination. Antisense approaches entail treating cells or animals with short (usually 15–25 nucleotides), single-stranded DNA or RNA molecules that are complementary to targeted mRNA, or transfecting cells with vectors that continuously generate complementary RNA. The complementary, antisense molecules are thought to inhibit gene expression by binding to the targeted mRNA and inducing its degradation or inhibiting its translation into protein. Using such approaches, several groups have been able to inhibit connexin expression in nontransformed cells and ask how this affected cellular growth. In one study, connexin antisense-transfected, nontransformed cells lost their ability to inhibit the growth of cocultured neoplastic cells (Goldberg et al., 1994). In another study, cells treated with connexin antisense oligonucleotides grew to a much higher saturation density (i.e., maximal number of cells per dish) (Ruch et al., 1995). A connexin43 knockout mouse has been developed (Reaume et al., 1995). The mutation, which was lethal at birth, resulted in offspring that had enlarged, abnormally developed hearts. No neoplasms were evident in the embryos, possibly because of the young age at death or because of compensatory communication by the expression of other connexins. However, cell lines are being developed from the embryos and will be examined for evidence of enhanced sensitivity to neoplasia. Thus, three different approaches to specifically inhibit connexin expression have provided evidence that GJIC is involved in growth regulation.

CONNEXIN GENE THERAPY

The above discussion has indicated a role for reduced GJIC in the development and maintenance of the neoplastic cell. But can neoplasia be reversed if GJIC is restored? Several studies suggest it can. Using so-called gene therapy approaches, connexin gene expression has been enhanced in several poorly expressing malignant cell lines (Eghbali et al., 1991; Mehta et al., 1991; Zhu et al., 1991; Mesnil et al., 1995; Chen et al., 1995a; Cesen et al., 1996). This has been achieved by the introduction of active connexin genes into tumor cells by transfection with plasmids or infection with viruses that contain the connexin gene. In these "recommunicating" tumor cells, the rates of tumor cell growth *in vitro* (Mehta et al., 1991; Zhu et al., 1991; Chen et al., 1995a) and abilities to form tumors when injected into an animal (Eghbali et al., 1991; Naus et al., 1992; Rose et al., 1993; Mesnil et al., 1995; Cesen et al., 1996) were highly reduced. Unlike drugs, nutrients, vitamins, or hormones that enhance tumor cell GJIC (Table 2), but which have other effects on the cells, connexin gene introduction by plasmid transfection or viral infection is a more direct approach. The studies utilizing these newer methods have clearly linked GJIC with growth regulation and neoplastic transformation.

IMPLICATIONS AND CLINICAL RELEVANCE

While further defining the role of GJIC in growth and neoplasia has enhanced our basic understanding of normal and neoplasic cells, there is also relevance here for clinical practice. The fact that more normal cell growth and behavior of neoplastic cells has been achieved by pharmacologically or genetically enhancing GJIC, suggests such approaches may be useful for treating human cancer. Therapies specifically designed to enhance tumor cell GJIC might reduce the growth and enhance the differentiation of neoplasic cells and might be less toxic than current chemotherapies. Such applications await development, but are theoretically possible and should be considered.

SUMMARY

Gap junctions consist of clusters of channels embedded in the plasma membranes of adjacent cells that directly link the two cell interiors and that permit the cell-to-cell exchange of small (< 1,000 Da) cytoplasmic molecules and ions. This transfer has been called gap junctional intercellular communication (GJIC). The proteins forming the channels are known as connexins. Many connexin genes have been cloned, but the regulation of their expression is poorly understood.

The proposed roles of GJIC include tissue homeostasis, electrical synapsing, coordination of cellular responses to hormones, regulation of embryonic development, and regulation of cellular growth and phenotypic expression. This latter function is especially relevant to understanding dysregulated growth in the neoplastic cell. Growth regulatory factors may be exchanged between cells via gap junctions. A reduction in gap junction formation and function is commonly observed in neoplastic cells and in cells treated with certain nongenotoxic, growth-enhancing carcinogens. The inhibition of GJIC is thought to isolate cells from the growth regulatory influences of their neighbors and to facilitate dysregulated growth.

The restoration of GJIC in neoplastic cells has been achieved genetically by connexin gene transfection and pharmacologically by treatment with vitamins, hormones, and other agents. This enhancement has resulted in reduced neoplastic cell growth and tumorigenicity in experimental animals. Such results suggest that the stimulation of GJIC should be pursued as a potential antineoplastic approach.

REFERENCES

Atkinson, M.M., Menko, A.S., Johnson, R.G., Sheppard, J.R., & Sheridan, J.D. (1981). Rapid and reversible reduction of junctional permeability in cells infected with a temperature-sensitive mutant of Avian Sarcoma Virus. J. Cell Biol. 91, 573–578.

Azarnia, R. & Loewenstein, W.R. (1984). Intercellular communication and the control of growth: X. Alteration of junctional permeability by the src gene. A study with temperature-sensitive mutant Rous sarcoma virus. J. Memb. Biol. 82, 191–205.

Bai, S., Spray, D.C., & Burk, R.D. (1993). Identification of proximal and distal regulatory elements of the rat connexin32 gene. Biochim. Biophys. Acta Gene Struct. Expression 1216, 197–204.

Berthoud, V.M., Ledbetter, M.L.S., Hertzberg, E.L., & Saez, J.C. (1992). Connexin 43 in MDCK cells: Regulation by a tumor-promoting phorbol ester and Ca^{+2}. Eur. J. Cell Biol. 57, 40–50.

Beyer, E.C., Paul, D.L., & Goodenough, D.A. (1987). Connexin43: A protein from rat heart homologous to a gap junction protein from liver. J. Cell Biol. 105, 2621–2629.

Brissette, J.L., Kumar, N.M., Gilula, N.B., & Dotto, G.P. (1991). The tumor promoter 12-O-tetradecanoyl-phorbol-13-acetate and the ras oncogene modulate expression and phosphorylation of gap junction proteins. Mol. Cell. Biol. 11, 5364–5371.

Budunova, I.V. & Williams, G.M. (1994). Cell culture assays for chemicals with tumor promoting or inhibiting activity based on the modulation of intercellular communication. Cell Biol. Toxicol. 10, 71–116.

Cesen, K.T., Park, I.-Y., Malkinson, A.M., & Ruch, R.J. (1996). Transfection of the gap junction protein, connexin 43, into mouse lung carcinoma cells reduces their growth *in vitro* and tumorigenicity. Proc. Amer. Assoc. Cancer Res. 37, 39.

Chen, S.-C., Pelletier, D.B., Ao, P., & Boynton, A.L. (1995a). Connexin43 transfection reverses the phenotype of transformed cells and alters their expression of cyclin/cyclin-dependent kinases. Cell Growth Diff. 6, 681–690.

Chen, Z.-Q., Lefebvre, D., Bai, X.-H., Reaume, A., Rossant, J., & Lye, S.J. (1995b). Identification of two regulatory elements within the promoter region of the mouse connexin43 gene. J. Biol. Chem. 270, 3863–3868.

Crow, D.S., Beyer, E.C., Paul, D.L., Kobe, S.S., & Lau, A.F. (1990). Phosphorylation of connexin43 gap junction protein in uninfected and Rous sarcoma virus-transformed mammalian fibroblasts. Mol. Cell. Biol. 10, 1754–1763.

Dermietzel, R., Yancey, S.B., Traub, O., Willecke, K., & Revel, J.-P. (1987). Major loss of the 28-kD protein of gap junctions in proliferating hepatocytes. J. Cell Biol. 105, 1925–1934.

Edelman, G.M. (1988). Morphoregulatory molecules. Biochemistry 27, 3533–3543.

Eghbali, B., Kessler, J.A., Reid, L.M., Roy, C., & Spray, D.C. (1991). Involvement of gap junctions in tumorigenesis: Transfection of tumor cells with connexin 32 cDNA retards growth in vivo. Proc. Natl. Acad. Sci. USA 88, 10701–10705.

Elvira, M., Diez, J.A., Wang, K.K.W., & Villalobo, A. (1993). Phosphorylation of connexin-32 by protein kinase C prevents its proteolysis by Mu-calpain and M-calpain. J. Biol. Chem. 268, 14294–14300.

Esinduy, C., Chang, C.C., Trosko, J.E., & Ruch, R.J. (1995). *in vitro* growth inhibition of neoplastically transformed cells by nontransformed cells: requirement for gap junctional intercellular communication. Carcinogenesis 16, 915–921.

Fallon, R.F. & Goodenough, D.A. (1981). Five-hour half-life of mouse liver gap-junction protein. J. Cell Biol. 90, 521–526.

Goldberg, G.S., Martyn, K.D., & Lau, A.F. (1994). A connexin 43 antisense vector reduces the ability of normal cells to inhibit the foci formation of transformed cells. Mol. Carcinog. 11, 106–114.

Gordon, R.E., Lane, B.P., & Marin, M. (1982). Regeneration of rat tracheal epithelium: Changes in gap junctions during specific phases of the cell cycle. Exp. Lung Res. 3, 47–56.

Guan, X., Hardenbrook, J., Fernstrom, M.J., Chaudhuri, R., Malkinson, A.M., & Ruch, R.J. (1995). Downregulation by butylated hydroxytoluene of the number and permeability of gap junctions in cell lines derived from mouse lung and rat liver. Carcinogenesis 16, 2575–2582.

Guan, X., Wilson, S., Schlender, K.K., & Ruch, R.J. (1996). Gap junction disassembly and connexin43 dephosphorylation induced by 18β-glycyrrhetinic acid. Mol. Carcinog. (in press).

Guan, X. & Ruch, R.J. (1996). Gap junction endocytosis and lysosomal degradation of connexin43-P2 in WB-F344 rat liver epithelial cells treated with DDT and Lindane. Carcinogenesis (in press).

Hennemann, H., Kozjek, G., Dahl, E., Nicholson, B., & Willecke, K. (1992a). Molecular cloning of mouse connexins26 and -32: Similar genomic organization but distinct promoter sequences of two gap junction genes. Eur. J. Cell Biol. 58, 81–89.

Hennemann, H., Suchyna, T., Lichtenberg-Fraté, H., Jungbluth, S., Dahl, E., Schwarz, J., Nicholson, B.J., & Willecke, K. (1992b). Molecular cloning and functional expression of mouse connexin40, a second gap junction gene preferentially expressed in lung. J. Cell Biol. 117, 1299–1310.

Jongen, W.M.F., Fitzgerald, D.J., Asamoto, M., Piccoli, C., Slaga, T.J., Gros, D., Takeichi, M., & Yamasaki, H. (1991). Regulation of connexin 43-mediated gap junctional intercellular communication by Ca^{+2} in mouse epidermal cells is controlled by E-cadherin. J. Cell Biol. 114, 545–555.

Kalimi, G.H., Hampton, L.L., Trosko, J.E., Thorgeirsson, S.S., & Huggett, A.C. (1992). Homologous and heterologous gap-junctional intercellular communication in v-raf-, v-myc-, and v-raf/v-myc-transduced rat liver epithelial cell lines. Mol. Carcinogenesis. 5, 301–310.

Kalimi, G.H. & Lo, C.W. (1988). Communication compartments in the gastrulating mouse embryo. J. Cell Biol. 107, 241–255.

Klaunig, J.E. & Ruch, R.J. (1990). Role of intercellular communication in nongenotoxic carcinogenesis. Lab. Invest. 62, 135–146.

Klaunig, J.E., Ruch, R.J., Weghorst, C.M., & Hampton, J.A. (1990). Role of inhibition of intercellular communication in hepatic tumor promotion. In Vitro Toxicol. 3, 91–107.

Larsen, W.J. & Hai-Nan (1978). Origin and fate of cytoplasmic gap junctional vesicles in rabbit granulosa cells. Tiss. Cell 10, 585–598.

Lau, A.F., Kanemitsu, M.Y., Kurata, W.E., Danesh, S., & Boynton, A.L. (1992). Epidermal growth factor disrupts gap-junctional communication and induces phosphorylation of connexin 43 on serine. Mol. Biol. Cell 3, 865–874.

Loch-Caruso, R. & Trosko, J.E. (1986). Inhibited intercellular communication as a mechanistic link between teratogenesis and carcinogenesis. CRC Crit. Rev. Toxicol. 16, 157–183.

Loewenstein, W.R. (1979). Junctional intercellular communication and the control of growth. Biochim. Biophys. Acta 560, 1–65.

Loewenstein, W.R. (1981). Junctional intercellular communication: the cell-cell membrane channel. Physiol. Rev. 61, 829–913.

Loewenstein, W.R. & Kanno, Y. (1966). Intercellular communication and the control of growth. Lack of communication between cancer cells. Nature 209, 1248–1249.

Massague, J. (1987). The TFG-β family of growth and differentiation factors. Cell 49, 437–438.

Matesic, D.F., Rupp, H.L., Bonney, W.J., Ruch, R.J., & Trosko, J.E. (1994). Changes in gap-junction permeability, phosphorylation, and number mediated by phorbol ester and non-phorbol-ester tumor promoters in rat liver epithelial cells. Mol. Carcinog. 10, 226–236.

Mehta, P.P., Hotz-Wagenblatt, A., Rose, B., Shalloway, D., & Loewenstein, W.R. (1991). Incorporation of the gene for a cell-cell channel protein into transformed cells leads to normalization of growth. J. Memb. Biol. 124, 207–225.

Mehta, P.P., Yamamoto, M., & Rose, B. (1992). Transcription of the gene for the gap junctional protein connexin43 and expression of functional cell-to-cell channels are regulated by cAMP. Mol. Biol. Cell 3, 839–850.

Meyer, D.J., Yancey, S.B., Revel, J.-P., & Peskoff, A. (1981). Intercellular communication in normal and regenerating rat liver: a quantitative analysis. J. Cell Biol. 91, 505–523.

Mesnil, M., Krutovskikh, V., Piccoli, C., Elfgang, C., Traub, O., Willecke, K., & Yamasaki, H. (1995). Negative growth control of HeLa cells by connexin genes: Connexin species specificity. Cancer Res. 55, 629–639.

Miller, T., Dahl, G., & Werner, R. (1988). Structure of a gap junction gene: rat connexin-32. Biosci. Rep. 8, 455–464.

Miyachi, E.-I. & Murakami, M. (1989). Decoupling of horizontal cells in carp and turtle retinae by intracellular injection of cyclic AMP. J. Physiol. 419, 213–224.

Murray, S.A. & Fletcher, W.H. (1984). Hormone-induced intercellular signal transfer dissociates cyclic AMP-dependent protein kinase. J. Cell Biol. 98, 1710–1719.

Naus, C.C.G., Elisevich, K., Zhu, D., Belliveau, D.J., & Del Maestro, R.F. (1992). In vivo growth of C6 glioma cells transfected with connexin 43 cDNA. Cancer Res. 52, 4208–4213.

Nicholson, B., Dermietzel, R., Teplow, D., Traub, O., Willecke, K., & Revel, J.-P. (1987). Two homologous protein components of hepatic gap junctions. Nature 329, 732–734.

Paul, D.L. (1986). Molecular cloning of cDNA for rat liver gap junction protein. J. Cell Biol. 103, 123–134.

Reaume, A., de Sousa, P., Kulkarni, S., Langille, B., Zhu, D., Davies, T., Juneja, S., Kidder, G., & Rossant, J. (1995). Cardiac malformation in neonatal mice lacking connexin 43. Science 267, 1831–1834.

Ren, P., De Feijter, A.W., Paul, D.L., & Ruch, R.J. (1994). Enhancement of liver cell gap junction protein expression by glucocorticoids. Carcinogenesis 15, 1807–1814.

Robards, A.W. & Lucas, W.J. (1990). Plasmodesmata. Ann. Rev. Plant Physiol. Plant Mol. Biol. 41, 369–419.

Rogers, M., Berestcky, J.M., Hossain, M.Z., Guo, H., Kadle, R., Nicholson, B.J., & Bertram, J.S. (1990). Retinoid-enhanced gap junctional communication is achieved by increased levels of connexin 43 mRNA and protein. Molec. Carcinogenesis 3, 335–343.

Rojkind, M., Novikoff, P.M., Greenwel, P., Rubin, J., Rojas-Valencia, L., Campos de Carvalho, A.C., Stockert, R., Spray, D.C., Hertzberg, E.L., & Wolkoff, A. (1995) Characterization and functional studies on rat liver fat-storing cell line and freshly isolated hepatocyte co-culture system. Am. J. Pathol. 146, 1508–1520.

Rose, B., Mehta, P.P., & Loewenstein, W.R. (1993). Gap-junction protein gene suppresses tumorigenicity. Carcinogenesis 14, 1073–1075.

Rubin, J.B., Verselis, V.K., Bennett, M.V.L., & Bargiello, T.A. (1992). Molecular analysis of voltage dependence of heterotypic gap junctions formed by connexins 26 and 32. Biophys. J. 62, 183–195.

Ruch, R.J. & Klaunig, J.E. (1986). The effects of tumor promoters, genotoxic carcinogens, and hepatocytotoxins on mouse hepatocyte intercellular communication. Cell Biol. Toxicol. 2, 469–483.

Ruch, R.J. & Klaunig, J.E. (1988). Inhibition of mouse hepatocyte intercellular communication by paraquat-generated oxygen free radicals. Toxicol. Appl. Pharmacol. 94, 427–436.

Ruch, R.J., Bonney, W.J., Sigler, K., Guan, X., Matesic, D., Schafer, L.D., Dupont, E., & Trosko, J.E. (1994). Loss of gap junctions from DDT-treated rat liver epithelial cells. Carcinogenesis 15, 301–306.

Ruch, R.J., Guan, X., & Sigler, K. (1995). Inhibition of gap junctional intercellular communication and altered growth in Balb/c 3T3 cells treated with connexin 43 antisense oligonucleotides. Mol. Carcinog. 14, 269–274.

Ruch, R.J., Madhukar, B.V., Trosko, J.E., & Klaunig, J.E. (1993). Reversal of *ras*-induced inhibition of gap-junctional intercellular communication, transformation, and tumorigenesis by lovastatin. Mol. Carcinog. 7, 50–59.

Saez, J.C., Berthoud, V.M., Moreno, A.P., & Spray, D.C. (1993). Gap junctions: Multiplicity of controls in differentiated and undifferentiated cells and possible functional implications. Adv. Second Messenger Phosphoprotein Res. 27, 163–195.

Schulte-Hermann, R., Schupples, J., Timmermann-Trosiener, I., Ohde, G., Bursch, W., & Berger, H. (1983). The role of growth of normal and preneoplastic cell populations for tumor promotion in the rat. Environ. Hlth. Perspect. 50, 185–194.

Shiba, Y., Sasaki, Y., & Kanno, Y. (1990). Inhibition of gap-junctional intercellular communication and enhanced binding of fibronectin-coated latex beads by stimulation of DNA synthesis in quiescent 3T3-L1 cells. J. Cell. Physiol. 145, 268–273.

Sosinsky, G. (1995). Mixing of connexins in gap junction membrane channels. Proc. Natl. Acad. Sci. USA 92, 9210–9214.

Spray, D.C., Fujita, M., Saez, J.C., Choi, H., Watanabe, T., Hertzberg, E., Rosenberg, L.C., & Reid, L.M. (1987). Proteoglycans and glycosaminoglycans induce gap junction synthesis and function in primary liver cultures. J. Cell Biol. 105, 541–551.

Stein, L.S., Boonstra, J., & Burghardt, R.C. (1992). Reduced cell-cell communication between mitotic and nonmitotic coupled cells. Exp. Cell Res. 198, 1–7.

Stutenkemper, R., Geisse, S., Schwarz, H.J., Look, J., Traub, O., Nicholson, B.J., & Willecke, K. (1992). The hepatocyte-specific phenotype of murine liver cells correlates with high expression of connexin 32 and connexin 26 but very low expression of connexin 43. Exp. Cell Res. 201, 43–54.

Sullivan, R., Ruangvoravat, C., Joo, D., Morgan, J., Wang, B.L., Wang, X.K., & Lo, C.W. (1993). Structure, sequence and expression of the mouse Cx43 gene encoding connexin 43. Gene 130, 191–199.

Takeichi, M. (1990). Cadherins: A molecular family important in selective cell-cell adhesion. Ann. Rev. Biochem. 59, 237–252.

Traub, O., Look, J., Dermietzel, R., Brummer, F., Hulser, D., & Willecke, K. (1989). Comparative characterization of the 21-kDa and 26-kDa gap junction proteins in murine liver and cultured hepatocytes. J. Cell Biol. 108, 1039–1051.

Traub, O., Look, J., Paul, D., & Willecke, K. (1987). Cyclic adenosine monophosphate stimulates biosynthesis and phosphorylation of the 26 KDA junction protein in cultured mouse hepatocytes. Eur. J. Cell Biol. 43, 48–54.

Weinstein, R.S., Merk, F.B., & Alroy, J. (1976). The structure and function of intercellular junctions in cancer. Adv. Cancer Res. 23, 23–89.

Werner, R., Levine, E., Rabadan-Diehl, C., & Dahl, G. (1989). Formation of hybrid cell-cell channels. Proc. Natl. Acad. Sci. USA 86, 5380–5384.

White, T.W., Bruzzone, R., & Paul, D.L. (1995). The connexin family of intercellular channel forming proteins. Kidney Intl. 48, 1148–1157.

Willecke, K., Hennemann, H., Dahl, E., Jungbluth, S., & Heynkes, R. (1991a). The diversity of connexin genes encoding gap junctional proteins. Eur. J. Cell Biol. 56, 1–7.

Willecke, K., Heynkes, R., Dahl, E., Stutenkemper, R., Hennemann, H., Jungbluth, S., Suchya, T., & Nicholson, B.J. (1991b). Mouse connexin 37: Cloning and functional expression of a gap junction gene highly expressed in lung. J. Cell Biol. 114, 1049–1057.

Yamasaki, H. (1990). Gap junctional intercellular communication and carcinogenesis. Carcinogenesis 11, 1051–1058.

Yancey, S.B., Easter, D., & Revel, J.-P. (1979) Cytological changes in gap junctions during liver regeneration. J. Ultrastruct. Res. 67, 229–242.

Yancey, S.B., Nicholson, B.J., & Revel, J.-P. (1981). The dynamic state of liver gap junctions. J. Supramol. Struct. Cell. Biochem. 16, 221–232.

Zhang, L.-X., Cooney, R.V., & Bertram, J.S. (1992). Carotenoids up-regulate connexin43 gene expression independent of their provitamin A or antioxidant properties. Cancer Res. 52, 5707–5712.

Zhu, D., Caveney, S., Kidder, G.M., & Naus, C.C.G. (1991). Transfection of C6 glioma cells with connexin 43 cDNA: Analysis of expression, intercellular coupling, and cell proliferation. Proc. Natl. Acad. Sci. USA 88, 1883–1887.

Chapter 7

Cell Adhesion and Metastasis: Molecular Mechanisms

CLIVE W. EVANS

Advances in Oncobiology
Volume 1, pages 143–157.

ISBN: 0-7623-0146-5

INTRODUCTION

There are three main routes by which malignant cells spread within the body, namely via the blood, via the lymph, or by surface implantation (reviewed in Evans, 1991). Clinical evidence suggests that carcinomas tend to spread preferentially via the lymphatics whereas sarcomas spread predominantly via the blood, but there are exceptions to this generalization and the two possibilities are not mutually exclusive. Malignant melanoma is often cited as a tumor which can spread equally well by the blood or lymphatic systems, and indeed given the complex interactions between the blood and lymph it would seem that malignant cell spread for most tumors should ultimately come to involve both systems. Spreading by surface implantation typically involves the seeding of malignant cells onto internal surfaces such as those lining the pleural and peritoneal cavities. The spread of ovarian carcinomas, for example, commonly involves detachment from the primary group of cells and implantation in the lining of the peritoneal cavity.

Clinical evidence suggests that certain types of tumors have a propensity to spread to particular sites (Willis, 1973). Many such cases are attributable to anatomical considerations. Thus the liver, for example, is the dominant site (via the hepatic portal system) for blood-borne metastases from cancer of the colon. In many cases the connections are not so obvious, however, only becoming apparent on detailed anatomical examination. The seemingly anomalous spread of carcinoma of the prostate to the axial skeleton, for example, is directly attributable to the presence of the internal vertebral venous plexus which acts as a bypass for pelvic and abdominal blood, thereby allowing malignant cells from abdominal tumors to gain access to the spinal column. Despite the obvious importance of anatomical considerations in the spread of tumors to particular sites, there remain some malignancies which display patterns of spread that are not easily explainable in terms of direct anatomical connections. Tumor cells which leave the left ventricle, for example, might be expected to spread according to the pattern of arterial flow but this is often not the case. Blood-borne metastases are rare in the muscle and gut even though between them they account for more than half of the cardiac output. Specific examples of anomalous spread include clear cell carcinoma of the kidney which frequently metastasizes to the thyroid, and follicular carcinoma of the thyroid which commonly spreads to bone. Many of these unusual patterns of spread may be accounted for by the seed and soil hypothesis developed by Paget in the 19th century, which implicates specific features of the malignant cell (the seed) necessary to establish itself as a secondary tumor in the specific environment (the soil) of particular tissues or organs. As will become clear in due course, one such feature may involve selective adhesive interactions between the matastasizing cell and the endothelium or basement membrane of blood vessels coursing through particular tissues or organs.

Table 1. Major Steps in the Metastatic Cascade

a. Local invasion	e. Transport within the blood system
b. Angiogenesis	f. Lodgement at a distant site
c. Detachment from the primary	g. Extravasation
d. Invasion of a blood vessel	h. Growth

This chapter will focus predominantly on malignant cell spread via the blood system which displays the major features of what has come to be known as the metastatic cascade.

THE METASTATIC CASCADE

The spread of malignant cells in the blood has been likened to a reaction cascade in which individual steps must be accomplished before the sequence can continue. There are many advantages (especially from the experimental point of view) for using a reaction cascade paradigm, but it needs to be understood that this is an oversimplification of the metastatic process in that some steps in the cascade can be short-circuited. Invasion of a blood vessel, for example, can be achieved en masse without prior detachment from the primary. With this caveat in mind, at least eight steps, as summarized in Table 1, can be identified as key mechanistic processes involved in malignant spread after establishment of the primary tumor.

CELL ADHESION AND THE METASTATIC CASCADE

Consideration of the steps involved in the metastatic cascade indicates several stages where adhesive interactions between cells or between cells and the extracellular matrix are likely to impinge on metastatic outcome (Evans, 1992). Both local invasion and blood vessel invasion/extravasation as well as the development of new blood vessels (angiogenesis) are dependent on adhesive events since cell motility (other than by swimming) is crucially dependent on substrate interactions. Imagine trying to move on a frictionless surface. Likewise, detachment from the primary and lodgement at a distant site (usually in the wall of a blood vessel) are obviously adhesion-dependent processes, although on opposite sides of the adhesive coin. What is perhaps not so obvious is the involvement of adhesive interactions between tumor cells and host cells such as leukocytes and platelets which can occur during transport within the blood system. Such interactions may influence metastatic outcome both positively and negatively, through the formation of emboli which might occlude small vessels (thereby initiating the lodgement phase) or through interactions with defense cells (thereby promoting cytotoxicity and limiting metastatic spread).

Table 2. Major Structural Components of the Extracellular Matrix

Component	*Location*	*Major Metastatic Role*
Proteins/glycoproteins		
Collagen	diverse, including basement membrane (type IV)	cell adhesion, physical barrier
Laminin	basement membrane	adhesion
Nidogen (entactin)	basement membrane	doubtful significance
Fibronectin	vessel walls, basement membrane, plasma	adhesion
Elastin	large arteries, skin, lung	barrier?
Vitronectin	interstitial stroma, plasma	adhesion
Osteonectin (SPARC)	wide distribution including basement membrane	adhesion
Thrombospondin	wide distribution including basement membrane	adhesion, tumor cell differentiation
Proteoglycans		
Heparan sulphate	basement membrane, cell surface	adhesion
Chondroitin sulfate	cartilage, bone, muscle, skin, aorta	adhesion?
Dermatan sulfate	skin, tendon, aorta	doubtful significance
Keratan sulfate	cartilage, cornea	doubtful significance
Hyaluronic acid	vitreous, cartilage, cell surface	adhesion

Although there are a number of steps in the metastatic cascade where adhesive events are likely to play important roles, it is clear that only two categories of adhesion are involved, based on either cell-cell or cell-substrate interactions. It was at one time thought that completely different processes might be involved in the two adhesion categories, but studies of the integrin superfamily has shown that at least some principles are shared in that the integrins are involved in both cell-cell and cell-substrate interactions.

Cell-substrate adhesion in the current context largely concerns the interactions of malignant cells with components of the extracellular matrix (ECM) which comprises the interstitial stroma and basement membranes. The major components of the ECM represent an assortment of proteins, glycoproteins and proteoglycans (Table 2).

Adhesion between a cell and its substrate is believed to be mediated by molecular lock and key-type interactions in which one component acts as the ligand and the other as its receptor. There is at least one identifiable structure involved in cell-substrate adhesion and that is the hemidesmosome which plays a role in the binding of some epithelial cells to the basement membrane. The hemidesmosome bears a close resemblance to half a desmosome, but the analogy may be something of a half-truth in itself since molecular studies suggest that the two do not contain identical components. Adhesion to fibrinogen/fibrin and related products can be considered to represent a special case of cell-substrate interaction.

Cell-cell adhesive interactions of significance in metastasis include both like-like or homotypic interactions (i.e., between malignant cells) and like-unlike or heterotypic interactions (i.e., between malignant cells and platelets, leukocytes, endothelium, and others). Cell-cell adhesion can be mediated by structurally recognizable entities such as the macula adherens (spot desmosome) or the zonula adherens (belt desmosome or intermediate junction). At the molecular level, adhesion mediated by either type of adherens junction is subserved by molecular lock and key-type interactions. The major adhesive molecules of the zonula adherens have been identified as belonging to the cadherin family. Lock and key-type interactions also underlie cell-cell adhesion in the absence of recognizable structures such as cell junctions.

MOLECULAR MECHANISMS OF ADHESION

A stunning diversity of molecules have been implicated in the adhesive interactions of cells and considerably more probably remain to be discovered. Many of these molecules (Tables 3–8) are potentially involved in the metastatic process since homotypic interactions between malignant cells and heterotypic interactions between malignant cells and ECM components or between malignant cells and other cells of the body can all impinge upon metastatic outcome.

The Selectins

Members of this family have a common structure based on a N-terminal lectinlike domain, a region showing homology with the epidermal growth factor receptor, and a number of complementlike repeats. They are predominantly involved in the adhesion of leukocytes to the endothelium with a key role being played by sialylated and/or fucosylated determinants which act as their cognate ligands (Springer,

Table 3. Molecular Mechanisms of Adhesion: The Selectin Family

Determinant	*Ligand*	*Adhesive Category*
L-selectin (gp90^{MEL}, LECAM-1)	uncharacterized sialylated, fucosylated glycoprotein (GlyCAM-1?)	cell-cell (neutrophil-endothelium, lymphocyte homing to lymph nodes, rolling)
P-selectin (GMP140, PADGEM, CD62P)	sialylated glycoprotein (p150 sialyl-Lewis^{X}, CD15)	cell-cell (leukocyte rolling)
E-selectin (ELAM-1 GDG2E)	sialylated glycoprotein (sialyl-Lewis^{X})	cell-cell (leukocyte rolling? lymphocyte homing to skin)

Table 4. Molecular Mechanisms of Adhesion: The Integrin Superfamily

Determinant	*Ligand*	*Adhesive Category*
α_L/β_2 (LFA-1, CD11a-CD18)	ICAM-1, ICAM-2, ICAM-3	cell-cell (leukocyte adhesion to the endothelium, transmigration)
α_M/β_2 (Mac-1, CD11b-CD18)	ICAM-1, C3bi	cell-cell (leukocyte-endothelium, transmigration), cell-substrate?
α_D/β_2	ICAM-3	cell-cell (leukocyte-endothelium)
α_X/β_2 (p150,95; CD11c-CD18)	C3bi fibrinogen	cell-cell (leukocyte-endothelium), cell-substrate?
α_1/β_1 (VLA-1)	laminin, collagen	cell-substrate
α_2/β_1 (VLA-2, gpIa/IIa)	collagen, laminin	cell-substrate
α_3/β_1 (VLA-3)	laminin, collagen, fibronectin, epiligrin	cell-substrate
α_4/β_1 (VLA-4, LPAM-2)	VCAM-1 (INCAM-110), fibronectin (type IIICS)	cell-substrate, cell-cell (leukocyte rolling and arrest)
α_4/β_7 (α_4/β_p, LPAM-1)	MadCAM-1, VCAM-1, fibronectin (type IIICS)	cell-cell (leukocyte rolling and arrest), cell-substrate
α_5/β_1 (VLA-5, gpIc/IIa, the "fibronectin receptor")	fibronectin (RGD)	cell-substrate
α_6/β_1 (VLA-6)	laminin, cell surface located?	cell-substrate, cell-cell (lymphocyte homing to the thymus)
α_6/β_4 (TSP-180?)	laminin, epiligrin?	cell-substrate, cell-cell?
α_v/β_1	fibronectin	cell-substrate
α_v/β_{3a} (the "vitronectin receptor")	vitronectin, fibrinogen, thrombospondin, von Willebrand factor	cell-substrate
α_v/β_{3b}	vitronectin	cell-substrate
α_v/β_5	vitronectin, fibronectin	cell-substrate
α_v/β_6	?	?
αMLA/β_7 (HML-1)	?	cell-cell (lymphocyte gut retention)
α_{IIb}/β_{3a} (gpIIb,IIIa)	fibronectin, fibrinogen, von Willebrand factor, vitronectin, thrombospondin, collagen	cell-substrate, platelet aggregation (via fibrinogen binding), tumor cell-platelet interaction?
α_E/β_7	E-cadherin	cell-cell

1990). The adhesive interactions of normal leukocytes which regulate their traffic around the body obviously may be of significance in the circulation of malignant hematopoietic cells, but there is increasing evidence that the molecular processes involved may be utilized by a variety of other malignant cell types. Thus there is some support for using normal leukocyte adhesion as a paradigm to gain an understanding of the significance and molecular nature of malignant cell adhesion during the metastatic process. One feature which has become clear from the detailed

Table 5. Molecular Mechanisms of Adhesion: The Immunoglobulin Superfamily

Determinant	*Ligand*	*Adhesive Category*
N-CAM	N-CAM	cell-cell (especially neural cells)
VCAM-1 (INCAM110)	α_4/β_1, α_4/β_7	cell-cell (lymphocyte homing in inflammation)
ICAM-1, ICAM-2 (CD102)	α_L/β_2	cell-cell (lymphocyte-endothelial adhesion)
ICAM-1 (CD54)	α_M/β_2	cell-cell (lymphocyte-endothelial adhesion)
CD31 (PECAM-1)	CD31	cell-cell (platelet and lymphocyte adhesion to the endothelium), integrin activator?
MUC18 (A32)	MUC18?	cell-cell (melanoma-endothelium)

Table 6. Molecular Mechanisms of Adhesion: The Cadherin Family

Determinant	*Ligand (Homophilic)*	*Adhesive Category*
E-cadherin (uvomorulin, L-CAM, CAM 120/80)	E-cadherin	cell-cell
P-cadherin	P-cadherin	cell-cell
N-cadherin (A-CAM, N-Cal-CAM)	N-cadherin	cell-cell

Table 7. Molecular Mechanisms of Adhesion: The Proteoglycans

Determinant	*Ligand*	*Adhesive Category*
Hyaluronic acid	CD44 (gp90Hermes, Pgp-1, ECMRIII)	cell-cell, cell-substrate
Chondroitin sulphate	?	cell-substrate
Heparan sulphate	fibronectin, laminin and others	cell-cell, cell substrate

Table 8. Molecular Mechanisms of Adhesion: Miscellaneous Determinants

Determinant	*Ligand*	*Adhesive Category*
Laminin receptor (67 kD)	laminin	cell-substrate
gplb-IX	von Willebrand factor	platelet-cell (arterial), platelet-substrate
gpIV	thrombospondin, collagen	platelet-substrate
lamp-1 lamp-2	galaptin	cell-substrate

study of leukocyte adhesion is that a single cell type may utilize a variety of different adhesive mechanisms (Mackay and Imhof, 1993).

Leukocyte adhesion to the endothelium is believed to involve two steps: the first (rolling) is a fairly loose interaction involving molecules such as the selectins, whereas the second (arrest) is much stronger involving cell adhesion molecules (CAMs) such as ICAM-1, ICAM-2, and V-CAM1 on the endothelium, and the α_L/β_2 and α_4/β_1 integrins on the leukocyte. Interestingly, the switch between loose and firm adhesion may be triggered by molecules such as CD31, a member of the immunoglobulin superfamily (see below). Certain tumor cells (e.g., colorectal carcinoma) have been shown to display sialylated and/or fucosylated glycoproteins on their surfaces and thus the potential exists for the involvement of selectins in the metastatic process (Matsushita et al., 1990). Following arrest, leukocytes migrate across the endothelium in a process dependent upon the α_L/β_2 and α_M/β_2 integrins and their cognate ligands.

The Integrins

The integrins are members of a superfamily of related heterodimeric molecules in which the subunits (various α and β types) are noncovalently linked. They mediate cell adhesion to both ECM components and other cells and can act as transmembrane signalling molecules (Hynes, 1992). Different integrins may be expressed on different cell types, with the α_1/β_1, α_2/β_1, α_3/β_1, and α_6/β_1 forms being most widely expressed. The expression of integrins on malignant cells and their possible involvement in metastasis varies for different types of tumors, and there is often no simple relationship. Thus the expression of both the β_3 subunit (i.e., the vitronectin receptor) and α_4/β_1 on melanoma cells has been shown to be elevated relative to normal melanocytes, whereas expression of the β_6 subunit declines with increasing progression of this tumor type. In general, the integrin composition of progressively malignant cells remains heterogeneous, although there is a tendency towards simplification in terms of both quantity and variety. As yet, the clinical significance of these changes remains uncertain. Experimental studies, however, have provided substantial, albeit somewhat confusing, evidence in support of the notion that interactions involving the integrins contribute to metastatic outcome. At the heart of these studies lies the observation that arginine-glycine-aspartic acid (RGD)-containing peptides inhibit cell adhesion to the ECM molecules fibronectin and vitronectin. When such peptides are injected into mice along with malignant melanoma cells there is a marked reduction in the number of lung colonies seen under control conditions with a nonrelated peptide (Humphries et al., 1986). Although it is tempting to suggest that RGD-containing peptides may have interfered with melanoma cell binding to ECM components, thereby effectively reducing metastatic outcome, this is not proven and alternative explanations (e.g., reduction in interaction with platelets) remain possible. Additionally, integrin activation through ligand interaction (as might occur during adhesion or artificial peptide stimulation) may affect a number of intracellular signaling events which could influence the metastatic process in a variety of ways, including through the stimulation of cell growth (Schwartz, 1993). Clinical application of the competitive peptide approach in the treatment of metastasis is unlikely, given the quantities of peptide involved, its effective half-time, and the restricted window of potential use.

Other studies have shown the α_6 subunit to be involved in the spread of malignant melanoma cells, since antibodies against this molecule block murine melanoma cell adhesion to lung endothelium *in vitro* and inhibit lung colonization *in vivo* (Ruiz et al., 1993). Although the two identified α_6-containing integrins both bind laminin, the antibodies used in the study by Ruiz and her colleagues (1993) failed to inhibit melanoma adhesion to laminin fragments, suggesting that some other ligand may be involved. Furthermore, since the α_6-containing integrins are expressed on both the endothelium and melanoma cells, and because both sets appear to be involved in melanoma metastasis to the lungs, it would seem that any novel candidate ligand would have to be expressed reciprocally. The recent identification of a lung-specific endothelial molecule known as Lu-ECAM-1 which promotes the lung-metastasizing behavior of melanoma cells (Zhu et al., 1992) serves to reinforce the apparent redundancy in the system, thereby illustrating the point that no single adhesive mechanism is likely to underlie the metastatic process, even for a single malignant tumor type.

A number of lines of evidence point towards the involvement of platelets in the metastatic spread of malignant cells, possibly through their involvement in the promotion of intravascular trapping. Integrins such as α_{IIb}/β_{3a} may play a key role in this process via interaction with ECM components (see later). Since this integrin has now been identified on the surfaces of at least some tumor cell types it may also play a more direct role in malignant cell binding.

The Immunoglobulin Superfamily

Members of this group display a characteristic immunoglobulin domain. They include a wide variety of molecules such as carcinoembryonic antigen (CEA) and the major histocompatibility antigens as well as molecules more directly involved in adhesion (Williams and Barclay, 1988). Although many members of this superfamily are involved in leukocyte traffic, the expression of one (VCAM-1) can be induced on endothelial cells where it might act as a receptor for melanoma cells bearing increased amounts of the α_4/β_1 integrin (see above). Since many tumor cell types have been shown to release cytokines with the potential to activate endothelial cells, it is conceivable that they may be able to modulate the expression of particular endothelial determinants to promote adhesiveness prior to extravasation and establishment of secondary growth. MUC18 has been found on melanoma cells (but not on melanocytes) and on endothelial cells and may promote adhesion between them.

The Cadherins

The cadherins are Ca^{2+} -dependent adhesion molecules which bind cells via homophilic interactions (Takeichi, 1991). At least three major subclasses have been identified, namely the E-cadherins of epithelial cells; the P-cadherins of the placenta, epithelium, and other cell types; and the N-cadherins of nerve and muscle cells. Other adhesive molecules such as desmoglein (a desmosomal component) appear to belong to the cadherin family, but their precise classification within this group has yet to be established. Cellular expression of the cadherins is developmentally regulated and the molecules appear to play key roles in the maintenance of tissue structure, particularly of epithelial cells. It may thus be surmised that a breakdown of normal tissue relationships (as occurs in malignant spread) would be associated with changes in cadherin expression, and indeed the loss of E-cadherin expression has been shown to correlate with increased invasiveness in tumor cell lines. Unfortunately, available clinical data fail to show any consistent relationship between reduced cadherin expression and metastatic capacity (Eidelman et al., 1989; Shimoyama et al., 1989).

The Glycosaminoglycans and Proteoglycans

The glycosaminoglycans are saccharide chains composed of repeating dimers of uronic acid (except in keratan sulphate) and an amino sugar. Each glycoaminoglycan (except hyaluronic acid) is covalently linked to a protein core via a serine residue to form a structure known as a proteoglycan. Both hyaluronic acid and the proteoglycans are typically found in the ECM but some (heparan sulphate, chondroitin sulphate, and hyaluronic acid) are also found on the cell surface. Plasma membrane-bound heparan sulphate appears to be significant in the formation of close contacts (25–30 nm separation) between a cell and its substrate, possibly as a consequence of the interaction of this proteoglycan with the heparin-binding domain of substrate-attached fibronectin. Focal contacts (with a separating distance between the cell and the substrate of about 10–15 nm) appear to develop when the proteoglycan interaction is supplemented with the binding of the cell surface integrin α_5/β_1 with the RGD domain of fibronectin. Whereas focal contact sites are associated mainly with firmly adherent, nonmotile cells, the opposite tends to apply to close contacts. Some malignant cells have relatively few focal contact sites which may correlate with reduced cell-surface adhesiveness and enhanced motility and invasiveness.

Heparan sulphate (along with lesser amounts of the chondroitin sulphates) appears to be the major proteoglycan produced by at least some types of endothelial cells in tissue culture, and its degradation (along with that of fibronectin) is a feature of a variety of invasive malignant cell types. However, there is no simple correlation between the degradation of these molecules and metastatic capacity.

Hyaluronic acid is also found on the cell surface and in the ECM. CD44, a cell surface-located protein found on a variety of cell types including leukocytes, neuroectodermal cells, mesenchymal cells, and epithelial cells has been identified as a receptor for hyaluronic acid. It is present as a number of isoforms as a consequence of the alternative splicing involving at least 10 different exons. The smallest alternatively spliced product is the 85–90 kD form ($gp90^{Hermes}$) which plays a major role in lymphocyte homing. Interestingly, whereas CD44 antibodies inhibit the binding of lymphocytes to the endothelium (Jalkanen et al., 1987), the application of soluble hyaluronic acid or hyaluronidase treatment of target endothelial cells fails to inhibit lymphocyte adhesion (Culty et al., 1990). A possible explanation for these observations is that in some situations CD44 may recognize and bind to a determinant other than hyaluronic acid. Expression of the 85–90 kDa CD44 isoform has been shown to correlate with metastatic capacity for a number of tumors including murine malignant melanoma, and transfection of a Burkitt's lymphoma cell line with the appropriate cDNA enhanced lung colonizing ability when the transfected cells were injected into nude mice (Sy et al., 1991). Recent evidence from both laboratory (rat pancreatic carcinoma) and clinical studies (breast and colon tumors) suggests that particular CD44 variants (incorporating exon v6) may be associated with enhanced metastatic capacity, possibly via the lymphatic system (Gunthert et al., 1991; Matsamura and Tarin, 1992). Antigen-stimulated lymphocytes (in contrast to nonstimulated ones) appear to express similar types of variants, and the argument has been made that exon v6-containing CD44 variants may be of importance for both lymphocyte and malignant cell retention in the lymph node (Herrlich et al., 1993).

Miscellaneous Adhesive Molecules

Available data suggest the presence of multiple cell surface determinants which can act as receptors for the basement membrane molecule laminin. While most of these belong to the integrin superfamily, a nonrelated 67 kD plasma membrane-located protein with strong affinity for laminin has been identified on many different cell types (Mecham, 1991). Malignant cells selected for their ability to bind to laminin display increased metastatic capacity relative to unselected cells of the same type, and a specific laminin-derived peptide (YIGSR) can markedly inhibit metastatic efficiency following treatment of malignant melanoma cells prior to injection into mice (Iwamoto et al., 1987). Similar effects on a murine fibrosarcoma cell line have been shown with another laminin-derived peptide, RYVV (McCarthy et al., 1988). It is generally surmised that the inhibitory effects displayed by laminin-derived peptides are mediated via blocking of adhesive interactions between the 67 kD receptor on circulating malignant cells and laminin located within the basement membrane of blood vessel walls, but conclusive evidence that this is the limiting step is wanting. In fact, coating malignant melanoma cells with whole laminin (as against peptides) has rather perversely been shown to increase metastatic outcome (Terranova et al., 1984). The binding of laminin to receptors on the cell surface stimulates plasminogen activation through the release of tissue plasminogen activator (t-PA). The subsequent production of plasmin may aid malignant cells during invasion of the ECM (Stack et al., 1993). Recent work has shown that the 67 kD laminin receptor also binds to elastin, but the significance of this in metastasis is uncertain.

Attention has already been drawn to the possible role of platelets in tumor cell adhesion and lodgement. Glycoprotein Ib is a platelet surface determinant involved (along with another glycoprotein gpIX) in their binding to exposed subendothelial-located von Willebrand factor under the high shear conditions of the arterial system. Under relatively low shear conditions typical of the venous system, platelet adhesion is mediated largely by the integrin α_5/β_1 (gpIc,IIa) and by gpIV, a platelet-surface receptor for thrombospondin and collagen. These adhesive events are followed by a series of steps involving aggregation via the integrin α_{IIb}/β_{3a} (gpIIb,IIIa) and the formation of a platelet plug which may include fibrin. The α_{IIb}/β_{3a} integrin is also involved in the binding of platelets to a variety of ECM components, thereby strengthening the initial adhesive interactions which may develop prior to platelet activation. Since many tumor cells can bind to platelets and fibrin, the potential for the promotion of metastasis via lodgement is clear. Perhaps more significantly, there is abundant evidence that at least some tumor cells can release material which can either directly or indirectly induce platelet aggregation and fibrin deposition (Lerner et al., 1983).

Galaptin is a β-galactoside-binding lectin present in the ECM. It has been postulated to be involved in the spread by surface implantation of human ovarian carcinoma cells which have appropriate polylactosamine-containing receptors related to the lysosomal-associated membrane proteins lamp-1 and lamp-2. Lamp expression is not restricted to the lysosomal membrane, and indeed an increase in their surface expression has been correlated with increased metastatic potential (Skrincosky et al., 1993).

CELL ADHESION AND ORGAN-SELECTIVE METASTASIS

As outlined earlier, certain malignant tumors display a propensity to metastasize to particular organs or tissues. There are several reasons why this might be so. One reason has as its basis the existence of anatomical connections (e.g., the pattern of venous drainage), while another relates to the possibility that certain malignant cells may only be able to grow at particular sites (perhaps because of the need of a specific growth factor from stromal cells). A third reason may be due to selective adhesive interactions between the circulating malignant cells and the endothelium of particular organs. To some extent this parallels the trafficking of lymphocytes which can home to particular lymph nodes such as those in the gut or those of the peripheral system.

It is generally recognized that endothelial cells from the blood vessels of different organs can perform different functions. Furthermore, several lines of evidence point to the existence of organ-specific endothelial determinants and it is conceivable that such determinants might contribute to the propensity for certain malignant tumors to spread to particular organs (Auerbach et al., 1985; Pauli and Lee, 1988). The precise nature of most of these determinants and their corresponding ligands on the circulating malignant cells is as yet uncertain, although Nicolson (1988) identified a number of candidate molecules possibly involved in the selective adhesion of cells from a murine lymphoma line to the endothelium. Lu-ECAM-1, mentioned earlier, may be one such molecule involved in the adhesion of melanoma cells to lung endothelium. It perhaps needs to be reiterated here, however, that selective adhesive interactions of this type represent only one aspect of the many ways by which cell adhesiveness might influence metastatic outcome.

SUMMARY

In order to metastasize via the blood a malignant cell must leave the primary tumor, gain access to the circulatory system, enter the tissues and establish the secondary tumor, all under the watchful eye of the host defense systems. Most of these stages involve adhesive interactions of some type. Migration from the primary tumor might involve decreased homotypic adhesion to other malignant cells within the mass, while movement itself requires regulated changes in cell-substrate adhesiveness. One feasible explanation for locomotion is that the moving cell makes adhesions at its front end and breaks them down at its rear as it projects itself forward. Penetration of blood vessels may first require binding to the vessel wall (basement membrane) after which detachment (from the lining endothelium) would be necessary to utilize the medium of the circulatory system for spread to distant sites. Within the blood, collisions will take place with host cells which could include cytotoxic leukocytes and platelets. Avoidance of adhesive interactions with defense cells might enhance metastatic outcome, while promotion of binding to platelets might have a similar effect through embolus development and subsequent trapping in small vessels. Extravasation from the blood will require some form of lodgement, which may be purely mechanical (e.g., impounding of a tumor cell-platelet embolus which is of a diameter larger than that of the vessel through which it is coursing) or mediated by molecular interactions between the circulating malignant cell and the lining epithelium. Movement out of the vessel in the latter case will require detachment from the endothelium and subsequent regulation of adhesion/de-adhesion processes as the cell moves through the stroma to ultimately establish the secondary tumor.

It should be immediately clear from the scenario outlined above that the involvement of adhesion in the metastatic process is far from simple, and that at times both increased and decreased adhesiveness may be advantageous. Whether the metastasizing cell is adhering to other malignant cells or normal tissues cells (e.g., endothelium; host defense cells) or to ECM components is clearly also of importance since different interactions are likely to be involved and the metastatic outcomes can be markedly different. A comprehensive understanding of all of these aspects of the metastatic process would seem essential before valid generalizations can be made concerning the role of cell adhesion in malignant tumor spread.

ACKNOWLEDGMENTS

My thanks to Dr. Geoff Krissansen for his helpful comments on the manuscript.

REFERENCES

Auerbach, R., Alby, L., Morrissey, L.W., Tu, M., & Joseph, J. (1985). Expression of organ-specific antigens on capillary endothelial cells. Microvasc. Res. 29, 401–411.

Culty, M., & 5 others (1990). The hyaluronate receptor is a member of the CD44 (H-CAM) family of cell surface glycoproteins. J. Cell Biol. 111, 2765–2774.

Eidelman, S., Damsky, C.H., Wheelock, M.J., & Damjanov, I. (1989). Expression of the cell-cell adhesion glycoprotein cell-CAM120/80 in normal human tissues and tumors. Am. J. Pathol. 135, 101–110.

Evans, C.W. (1991). The Metastatic Cell: Behavior and Biochemistry. Chapman and Hall, London.

Evans, C.W. (1992). Cell adhesion and metastasis. Cell Biol. Intl. Reports 16, 1–10.

Gunthert, U. and 9 others (1991). A new variant of glycoprotein CD44 confers metastatic potential to rat carcinoma cells. Cell 65, 13–24.

Herrlich, P., Zoller, M., Pals, S.T., & Ponta, H. (1993). CD44 splice variants: Metastases meet lymphocytes. Immunol. Today 14, 395–399.

Humphries, M.J., Olden, K., & Yamada, K.M. (1986). A synthetic peptide from fibronectin inhibits experimental metastasis of murine melanoma cells. Science 233, 468–470.

Hynes, R.O. (1992). Integrins: Versatility, modulation and signalling in cell adhesion. Cell 69, 11–25.

Iwamoto, Y. and 6 others (1987). YIGSR, a synthetic laminin pentapeptide, inhibits experimental metastasis formation. Science 238, 1132–1134.

Jalkanen, S., Bargatze, R.F., de los Toyos, J., & Butcher, E.C. (1987). Lymphocyte recognition of high endothelium: Antibodies to distinct epitopes of an 85–95 kD glycoprotein antigen differentially inhibit lymphocyte binding to lymph node, mucosal or synovial endothelial cells. J. Cell Biol. 105, 983–990.

Lerner, W.A., Pearlstein, E., Ambrogio, C., & Karpatkin, S. (1983). A new mechanism for tumor-induced platelet aggregation. Comparison with mechanisms shared by other tumors with possible pharmacologic strategy toward prevention of metastases. Intl. J. Cancer 31, 463–469.

Mackay, C.R. & Imhof, B.E. (1993). Cell adhesion in the immune system. Immunol. Today 14, 99–102.

Matsumura, Y. & Tarin, D. (1992). Significance of CD44 gene products for cancer diagnosis and disease evaluation. Lancet 340, 1053–1058.

Matsushita, Y. & 4 others (1990). Sialyl-dimeric Lewis -X antigen expressed on mucin-like glycoproteins in colorectal cancer metastases. Lab. Invest. 63, 780–791.

McCarthty, J.B., Skubitz, A.P.N., Palm, S.L., & Furcht, L.T. (1988). Metastasis inhibition of different tumor types by purified laminin fragments and a heparin-binding fragment of fibronectin. J. Natl. Cancer Inst. 80, 108–116.

Mecham, R.P. (1991). Receptors for laminin on mammalian cells. FASEB J. 5, 2538–2546.

Nicolson, G.L. (1988). Cancer metastasis: Tumor cell and host organ properties important in metastasis to specific secondary sites. Biochim. Biophys. Acta 948, 175–224.

Pauli, B.U. & Lee, C.L. (1988). Organ preference of metastasis. The role of organ-specifically modulated endothelial cells. Lab. Invest. 58, 379–387.

Ruiz, P., Dunon, D., Sonnenberg, A., & Imhof, B.A. (1993). Suppression of mouse melanoma metastasis by EA-1, a monoclonal antibody specific for α_6 integrins. Cell Adh. Commun. 1, 67–81.

Schwartz, M.A. (1993). Signalling by integrins: Implications for tumorigenesis. Cancer Res. 53, 1503–1506.

Skrincosky, D.M., Allen, H.J., & Bernacki, R.J. (1993). Galaptin-mediated adhesion of human ovarian carcinoma A121 cells and detection of cellular galaptin-binding glycoproteins. Cancer Res. 53, 2667–2675.

Shimoyama, Y. & 6 others (1989). Cadherin cell-adhesion molecules in human epithelial tissues and carcinomas. Cancer Res. 49, 2128–2133.

Springer, T.A. (1990). Adhesion receptors of the immune system. Nature 346, 425–434.

Stack, M.S. Gray, R.D., & Pizzo, S.V. (1993). Modulation of murine B16F10 melanoma plasminogen activator production by a synthetic peptide derived from the laminin A chain. Cancer Res. 53, 1998–2004.

Sy, M.S. Guo, Y.-J., & Stamenkovic, I. (1991). Distinct effects of two CD44 isoforms on tumor growth *in vitro*. J. Exp. Med. 174, 859–866.

Takeichi, M. (1991). Cadherin cell adhesion receptors as a morphogenetic regulator. Science 251, 1451–1455.

Terranova, V.P., Williams, J.E., Liotta, L.A., & Martin, G.R. (1984). Modulation of the metastatic activity of melanoma cells by laminin and fibronectin. Science 226, 982–984.

Williams, A.F. & Barclay, A.N. (1988). The immunoglobulin superfamily-domains for cell surface recognition. Ann. Rev. Immunol. 6, 381–405.

Willis, R.A. (1973). The Spread of Tumors in the Human Body. Butterworths, London.

Zhu, D., Cheng, C.-F., & Pauli, B.U. (1992). Blocking of lung endothelial cell adhesion molecule-1 (Lu-ECAM-1) inhibits murine melanoma lung metatsasis. J. Clin. Invest. 89, 1718–1724.

Chapter 8

Ras: Processor of Vital Signals

CRYSTAL M. WEYMAN and DENNIS W. STACEY

INTRODUCTION

Each cell in an organism must correctly receive, interpret, and transmit the appropriate signal at the appropriate time to regulate and coordinate cell division, cell differentiation, and cell death (apoptosis). Cancer results when aberrations occur

Advances in Oncobiology
Volume 1, pages 159–177.

ISBN: 0-7623-0146-5

in the cellular genome which compromise normal growth constraints. A variety of precancerous states is not surprising upon realization that the pathways which control the intricate processes of proliferation, differentiation, and apoptosis are often complex and interconnected. A cell may divide inappropriately as a result of any or all of the following reasons: (a) the presence of a continuous or untimely growth-promoting signal; (b) the lack of a growth-arresting signal; (c) the lack of a signal for, or an inability to, exit the cell cycle and differentiate into a more specialized cell; and (d) the lack of a signal for, or an inability to, undergo programmed cell death. The ability of a cell which can divide inappropriately to detach from its extracellular matrix and reattach elsewhere would result in the most malignant manifestation of cancer—metastasis. Thus, a cell must circumvent a multitude of checks and balances en route to the development of malignancy.

The first goal of cancer research at the molecular level is to identify the critical alterations in the genome of various tumor types responsible for the circumvention of normal growth constraint. The assay system originally used for this purpose relied on the ability of transfected high molecular weight DNA from tumors to morphologically transform and induce unscheduled DNA synthesis in the mouse fibroblast cell line NIH 3T3 (Bowden, 1990). The dominant transforming genes (oncogenes) most frequently detected by these methods were mutated versions of the *ras* gene family. The mammalian *ras* gene family consists of three closely related members, H-*ras*, K-*ras*, and N-*ras*. Two of these had been previously observed as the transforming components of the Harvey (H-*ras*) and Kirsten (K-*ras*) tumor viruses (Lowy and Willumsen, 1993).

Both qualitative and quantitative changes in the normal *ras* gene have been found to promote its conversion to an oncogene. Qualitatively, *ras* genes frequently suffer single point mutations within their coding sequence which result in altered amino acid sequences. Naturally occurring point mutations are found most often in codon 12, 13, or 61. An additional mutation is found in codon 59 of *ras* genes that have been incorporated in retroviral genomes (Lowy and Willumsen, 1993). Qualitatively, increased expression levels of normal *ras* caused by point mutations in noncoding sequences or by generation of a high number of gene copies can induce transformation. Retroviral insertional mutagenesis can also activate *ras* when the virus integrates near the normal gene and disrupts normal transcriptional regulation (Balmain and Brown, 1988).

THE ROLE OF *ras* GENE MUTATION IN TUMOR FORMATION

ras Mutation in Human Tumors

Detection of a mutated *ras* gene varies among tumor types in humans. Almost 90% of adenocarcinomas of the exocrine pancreas, 50% of follicular and undifferentiated carcinomas of the thyroid, 30% of adenocarcinomas of the lung, 20% of melanomas, and roughly 10–20% of carcinomas of the liver, kidney, and bladder possess a mutated *ras* gene. K-*ras* gene activation predominates in adenocarcinomas of the lung, pancreas, and colon, while activation of all three *ras* genes has been observed in thyroid tumors. However, the possible correlation between tumor type and *ras* gene mutation cannot be readily explained by differences in protein function or differences in protein expression among different members of the *ras* family (Bos, 1989).

ras Mutation in Animal Models and Cultured Cells

Animal models have been extremely useful in understanding the overall process of carcinogenesis because tumor development can be empirically controlled. The fact that tumorigenesis is an ongoing process composed of many events rather than a consequence of a single event became clear from studies of chemical carcinogenesis in mouse skin. Tumorigenesis could best be achieved by a single application of carcinogen to initiate tumorigenesis, followed by repeated application of an irritant to promote its development. Carcinogens act mutagenically to create initiated cells by forming specific adducts with DNA which can lead to deletions, rearrangements, or single base mutations. Tumor initiation is thus irreversible and heritable. Promotion, on the other hand, is believed to involve the clonal expansion of an initiated cell. Such proliferation could increase the likelihood of accruing additional alterations. These tumor promoting events result in the manifestation of increasingly malignant behavior. However, while promoting agents characteristically induce proliferation, not all proliferative agents are able to promote. Promoters might therefore need also to adversely affect surrounding normal cells (Balmain and Brown, 1988).

Animal models also provide clues concerning individual genetic mutations during tumor formation. Greater than 90% of skin papillomas and carcinomas induced by treatment with the carcinogen dimethylbenzanthracene (DMBA) followed by repeated treatments with a phorbol ester irritant possess an activated H-*ras*. Similar results were obtained when mammary carcinomas were induced with nitrosomethylurea (NMU). Thymic lymphomas induced by NMU display a predominance of N-*ras* gene activation while methylcholanthrene (MCA) induced thymic lymphomas and fibrosarcomas display a prevalence of K-*ras* gene activation suggesting that these specific mutations are controlled by chromosomal structure and resulting susceptibility to mutagens. For example, fibroblast cells which have highly methylated H-*ras* treated with MCA *in vitro* give rise to tumors with activated K-*ras*. Likewise, fibrosarcomas induced by treatment of rodents with MCA also possess activated K-*ras* (Balmain and Brown, 1988).

The above results firmly established the significance of *ras* mutation in tumorigenesis. Understanding how *ras* is involved in the complex process of tumor formation has also been the focus of intense research. The immortality of NIH 3T3 cells causes them to be viewed as a precancerous state requiring only one additional event to attain morphological transformation. Although mutated *ras* genes can easily transform these cells, the transformation of primary fibroblasts by activated *ras* requires high levels of expression and that the inhibitory effect of surrounding normal cells be removed (Spandidos and Wilkie, 1984; Dotto et al., 1985). This suppressive influence of normal cells is not well understood, but can be overcome by the expression of an additional oncogene such as c-*myc*, polyoma large T (Land et al., 1983; Ruley, 1983), *jun* (Schutte et al., 1989; Binetruy et al., 1991), or by treatment with a promoter (Dotto et al., 1985). As a consequence of these observations, *ras* activation was assumed to be a late event in the process of tumor formation.

However, support for the role of *ras* activation in the initiation of tumor formation has also been firmly established. Mouse skin cells displayed no discernible neoplastic properties after application of Harvey murine sarcoma virus (harboring a transforming version of cellular H-*ras*) unless such application was followed by treatment with a promoting agent. A role for *ras* activation during initiation implies that *ras* mutation occurs as a consequence of direct interaction between a carcinogen and the *ras* gene, rather than as a consequence of a subsequent random mutation during promotion. Direct mutagenesis of the *ras* gene by the initiating agent has been inferred by correlating the DNA-binding characteristics of each initiating carcinogen with the type of *ras* mutation detected in the resulting tumor (Balmain and Brown, 1988; Bowden, 1990).

Further evidence supporting a role for *ras* activation early in malignancy comes from analyzing the frequency of *ras* mutation during the development of colon and thyroid cancer in humans and prior to the development of mammary tumors in rats treated with NMU. The fact that *ras* mutations are present in high proportions of early stage tumors in humans (Bos, 1989) and prior to tumor development in rats (Kumar et al., 1990) indicates that this mutation is an early event in carcinogenesis. These findings have been substantiated and advanced by studies performed in cultured cells. Keratinocytes with the ability to form only benign tumors were found to have a mutated cellular *ras* gene (c-Ha-*ras*). Interestingly, introduction of a mutated viral *ras* gene (v-Ha-*ras*) into keratinocytes resulted in tumors with a squamous cell carcinoma phenotype while introduction of both a v-Ha-*ras* gene and a mutated c-Ha-*ras* gene resulted in tumors with a more malignant undifferentiated carcinoma cell phenotype. These studies indicate a role for *ras* in tumor progression as well as in initiation and suggest a possible correlation between gene dosage and degree of malignancy (Greenhalgh et al., 1989).

Recently, homologous recombination was used to create Rat1 fibroblast cells with one normal *ras* allele and one activated *ras* allele, each expressed equally from its own natural promoter. These cells, which are heterozygous for activated *ras*, were not transformed, indicating that the mutant allele is not dominant over the normal allele. However, they did display a higher rate of spontaneous transformation when compared to their homozygous normal counterparts. Thus, activation of *ras* confers a predisposition to transformation and is consistent with previous studies where transformation by activated *ras* required additional events. Fourteen of the 15 spontaneously transformed cell lines derived from the nontransformed heterozygous cells had amplified the mutant allele, again suggesting the significance of gene dosage. However, one of the spontaneously transformed cell lines derived from the Rat1 fibroblasts heterozygous for *ras* and many human tumors retain only a single copy of the mutant *ras* gene without loss of the normal allele indicating that the importance of other events cannot be discounted (Finney and Bishop, 1993).

THE ROLE OF Ras IN SIGNAL TRANSDUCTION

Biochemical Properties of Ras

Biochemical studies of the Ras protein suggest how oncogenic mutation might lock Ras in a perpetually activated conformation so as to promote uncontrolled proliferation. *ras* genes are evolutionarily well conserved among eukaryotes and have been studied not only in mammals but also in fruit flies (*Drosophila melanogaster*), yeasts (*Saccharomyces cerevisiae* and *Saccharomyces pombe*), amphibians (*Xenopus laevis*), roundworms (*C. elegans*), and slime molds (*Dictoystelium discoideum*) (Bollag and McCormick, 1991). While the focus this chapter is the mammalian system, the information contributed by virtue of the ease of genetic manipulation of the other systems has been invaluable and will be noted. Mammalian *ras* genes (H-*ras*, K-*ras*, and N-*ras*) code for highly related proteins with an apparent molecular weight of 21 kD. These proteins are synthesized in the cytosol and become associated with the inner side of the plasma membrane after the posttranslational addition of a farnesyl lipid moiety to a conserved cysteine residue at position 186, near the carboxy terminus. H-Ras and N-Ras, but not K-Ras, are additionally modified by acylation with palmitic acid at an upstream cysteine (181 or 184, respectively). Ras proteins belong to a superfamily of proteins (G-proteins) which bind, exchange, and hydrolyze guanine nucleotides. As with other members of this large family, Ras is biologically active only when bound to GTP. Hydrolysis of bound GTP to GDP renders the protein inactive. In resting cells, *ras* is generally bound to GDP and therefore inactive. The cytosolic concentration of GTP, however, is much greater than GDP so that release of bound GDP from Ras and replacement by free nucleotide (nucleotide exchange) normally results in biological activation of Ras (Lowy and Willumsen, 1993).

Both the activation and the deactivation are controlled by proteins which interact directly with Ras. The rate limiting step in the exchange reaction is the release of GDP. The existence of proteins which function upstream of Ras to stimulate the release of GDP were originally identified by genetic studies in yeast (CDC25 from *S. cerevisiae* and Ste6 from *S. pombe*) and fruit flies (SOS). One mammalian homologue of CDC25 has been identified as p140Ras-GRF and three homologues of SOS have been identified one from human (hSOS) and two from mouse

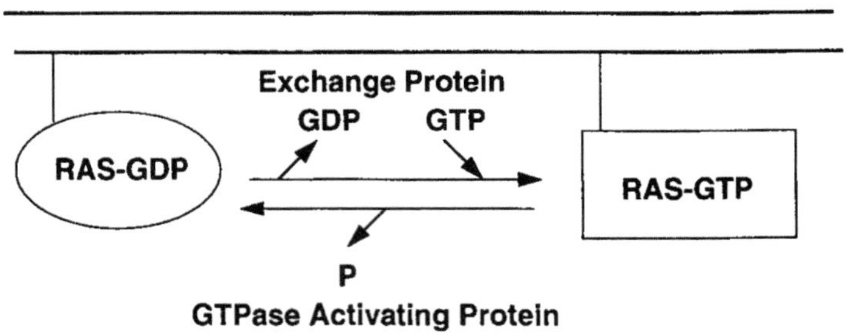

Figure 1. Mechanism of activation and deactivation of mammalian Ras proteins. Naturally occurring oncogenic mutants of Ras are insensitive to GAP and NF1 and therefore exist predominantly in the active, GTP-bound conformation.

(mSOS-1 and mSOS-2) (Feig, 1993)). The enzymatic reaction which is responsible for returning Ras to the inactive GDP-bound conformation is regulated by proteins which stimulate the intrinsic hydrolytic activity (GTPase activity) of Ras. The first of these identified was given the name GAP for GTPase-activating protein. The second protein (neurofibromin) determined to increase the intrinsic hydrolytic activity of Ras was identified as the product of the gene whose absence results in the development of neurofibromatosis type 1 (NF1). This genetic disease is characterized by both benign and malignant tumors of neural crest origin. Since it is the lack of function of neurofibromin which is responsible for tumor growth, NF1 is considered a tumor suppressor gene. Once again, the IRA genes in yeast which antagonize the exchange activity encoded by the CDC25 gene share homology with GAP and NF1. Mutations in *ras* genes which make them biologically transforming often result in proteins which are insensitive to stimulation by GAP and NF1 and therefore exist predominantly in the active GTP-bound conformation without the need for stimulated exchange (see Figure 1). Oncogenic mutations in *ras* therefore result in a constitutive gain of function (Bollag and McCormick, 1991; Lowy and Willumsen, 1993).

Proliferative Requirement for Ras

The ability of Ras to cycle between active and inactive conformations by virtue of differential nucleotide binding is a molecular mechanism shared by heterotrimeric G-proteins (so named because they consist of three subunits). These are known to be involved in transducing extracellular signals across the plasma membrane (Gilman, 1987). A distinct role for Ras as a transducer of extracellular signals is now also well established. Microinjection studies revealed the signal transduction potential of Ras proteins. Microinjection of oncogenic Ras proteins induced morphological transformation and DNA synthesis in NIH 3T3 cells within 30 hours in the absence of serum growth factors. Similar results were observed with wild-type Ras protein but required the injection of roughly 10 times as much protein.

This suggested that both normal and oncogenic Ras proteins functioned in a similar manner (Stacey and Kung, 1984). Additional information which helped to establish Ras as a key signaling molecule in the proliferative pathway came from the use of a monoclonal antibody (Y13-259) prepared against Ras, which has the ability to neutralize its transforming activity. This antibody provided the means to block Ras function within cells and observe the growth related consequences. Microinjection of cells with Y13-259 demonstrated that Ras activity was required at multiple points within the G1 phase of the cell cycle but was not involved with the actual process of DNA synthesis once initiated. It appears, therefore, that cellular Ras proteins play an essential role in the commitment to enter the replication cycle (Mulcahy et al., 1985; Dobrowolski et al., 1994).

Microinjection of anti-Ras antibody also provided one of the first clear indications that the action of Ras proteins and tyrosine kinases is functionally connected. When anti-Ras antibody was microinjected into cultured cells containing mutant, constitutively active tyrosine kinase oncogenes, the effects of these oncogenes was eliminated. The cells which had previously exhibited a transformed morphology and uncontrolled growth reverted to the appearance and growth properties of normal cells deprived of serum. This experiment, which has been duplicated by microinjection of dominant inhibitory Ras mutant protein (Stacey et al., 1991a and b), clearly indicated that tyrosine kinases of various types depend upon cellular Ras activity (Smith et al., 1986) to transmit their signal. Furthermore, microinjection studies provided some indication of the fate of the signal transduced by Ras. When anti-Ras antibodies or dominant inhibitory Ras mutant protein were microinjected into cells transformed by a third class of mutant oncogenes, the soluble serine kinases, no change was observed in their morphologic or proliferative characteristics (Smith et al., 1986). These soluble serine kinases, and other proteins to be discussed below, apparently function to receive the proliferative signal from Ras. This model of proliferative signaling where growth factors signal through Ras to soluble serine kinases has been confirmed and extended in numerous studies.

Consistent with the discoveries that Ras activity alone is sufficient to induce DNA synthesis in the absence of serum growth factors and that Ras activity is absolutely required for serum growth factor-induced DNA synthesis, individual serum growth factors have been shown to increase the proportion of Ras which is bound to GTP. These growth factors, including epidermal growth factor (EGF), platelet-derived growth factor (PDGF), and insulin, transmit signals by binding to membrane-spanning receptors. Also consistent with the discovery that tyrosine kinases signal through Ras is the fact that the cytoplasmic portion of these receptors either possess intrinsic tyrosine kinase activity or are associated with nonreceptor-type tyrosine kinase activity. This intracellular tyrosine kinase activity is stimulated by the binding of these growth factors to the extracellular portion of the receptor. As mentioned above, oncogenic mutation of these tyrosine kinases results in constitutive tyrosine kinase activity. Expression of these mutants also results in increased levels of Ras bound to GTP (Satoh et al., 1992). Furthermore, as shown in Figure 2,

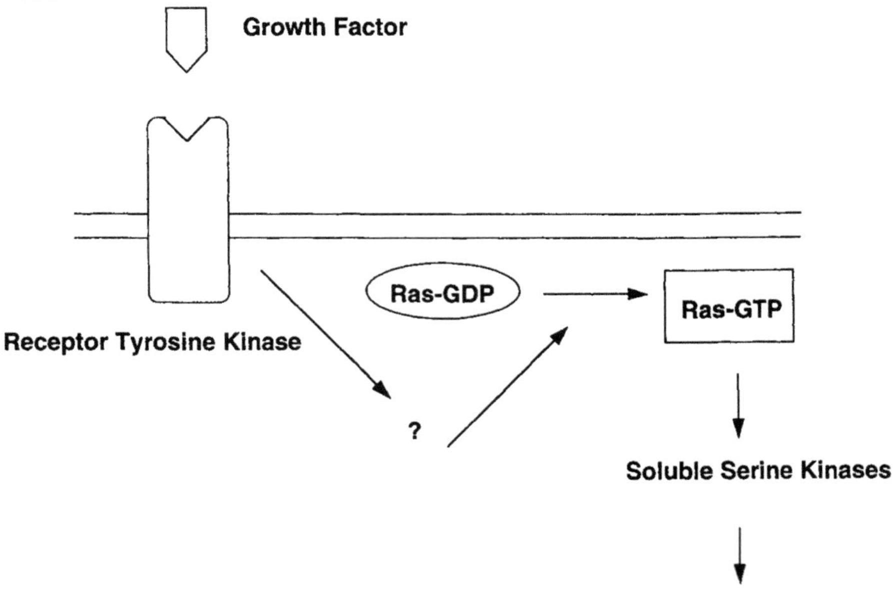

Figure 2. Generalized schematic of Ras signaling pathway. Binding of growth factor to membrane receptor and activation of tyrosine kinase activity is followed by activation of Ras, which is followed by activation of soluble serine kinases.

signaling through these pathways results in the activation of cytosolic serine kinases (Pelech and Sanghera, 1992). Both a mutant of Ras (Asn-17), which acts as a dominant negative inhibitor of Ras, and tyrosine kinase inhibitors ablate signaling through these pathways in all cases tested (Satoh et al., 1992).

The realization that such a variety of signals were all funneled through Ras revolutionized previously held beliefs that each activated transmembrane receptor initiated unique biochemical events which resulted in a specific biological response. It is staggering at first encounter to imagine mechanistically how such diverse signaling molecules could possibly converge on one molecule. This enigma has recently been resolved by the discovery of motifs that are responsible for protein-protein interactions. These motifs were originally identified in the Src protein and have been henceforth referred to as Src homology (SH2 or SH3) domains. SH2 domains couple the proteins in which they reside to other proteins which are specifically phosphorylated on tyrosine. SH3 domains couple the proteins in which they reside to other proteins which have specific proline-rich motifs. SH2 and SH3 domains have been characterized in a variety of signaling molecules such as GAP, phospholipase Cγ (PLCγ), and phosphatidylinositol-3-kinase. The SH2 domains of these molecules allow them to physically associate with specifically phosphorylated tyrosine kinase growth factor receptors which have been activated by their respective extracellular growth factors. The precise molecular consequence of this

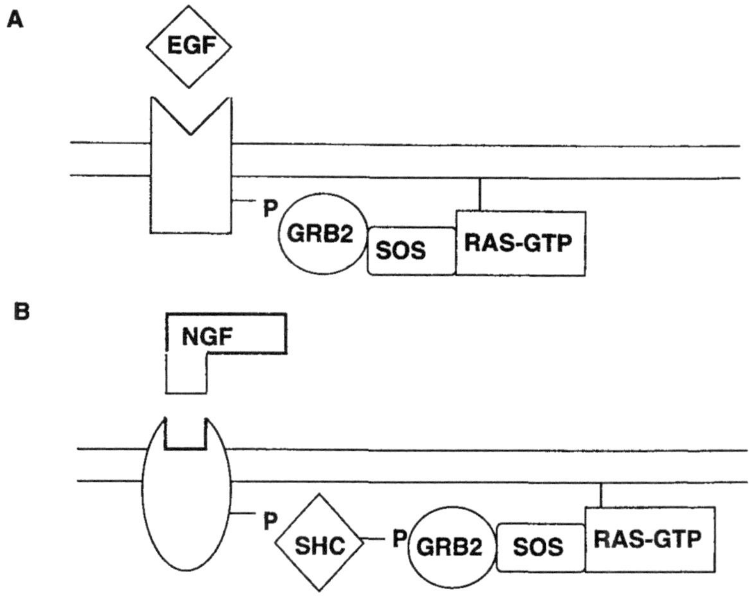

Figure 3. Possible molecular mechanisms of Ras activation by activated receptor tyrosine kinases. (**A**) Adaptor protein GRB2 associates via its SH3 domain with mSOS-1, the ubiquitous exchange factor for Ras. Association of GRB2 via its SH2 domain with tyrosine phosphorylated EGF or PDGF receptors allows localization of mSOS-1 near Ras. (**B**) Adaptor protein GRB2 associates with mSOS-1 as in (**A**). However, localization of mSOS-1 near Ras is accomplished by association of GRB2 via its SH2 domain with the tyrosine phosphorylated adaptor protein SHC. SHC associates via its SH2 domain with tyrosine phosphorylated NGF receptors. Activation of Ras by heterotrimeric-G-protein coupled receptors appears to occur via tyrosine phosphorylation of SHC and its subsequent association with GRB2-mSOS-1. Calcium induced activation of Ras occurs through p140-GRF.

association is currently the focus of intense research. SH2 and SH3 domains have also been characterized in proteins which apparently have no other functional domain. These include Shc, Crk, and Nck which are considered to be adapter proteins, able to couple tyrosine phosphorylated proteins (probably receptors) to proteins with proline-rich regions (Carpenter, 1992; Pawson and Gish, 1992; Mayer and Baltimore, 1993; McCormick, 1993).

A direct molecular link from the activation of tyrosine kinase receptors to the activation of Ras is completed by the discovery of Grb2, an adapter protein which binds via its SH3 domains to the proline-rich carboxy-terminal tail of mSOS-1, the ubiquitous exchange factor for Ras (see Figure 3A) and which binds via its SH2 domains directly to activated EGF receptors and PDGF receptors. Although mSOS-1 in the cytosol of unstimulated cells is active and found complexed with Grb2, translocation of mSOS-1 to the membrane via the interaction of the Grb2-mSOS-1 complexes with activated tyrosine kinase receptors could increase exchange activity by positioning mSOS-1 near Ras at the plasma membrane (Egan et al., 1993; Feig, 1993).

Several variations of this model also exist. In contrast to the association of Grb2 with activated EGF and PDGF receptors in fibroblasts, Grb2 is not found associated with the insulin receptor in myoblast cells treated with insulin (Skolnik et al., 1993a and b) or with the nerve growth factor (NGF) receptor in PC12 cells induced to differentiate with NGF (Suen et al., 1993). However, Grb2 is found associated with Shc in these cells and in cells transformed by v-*src*. Shc is a tyrosine phosphorylated, SH2 containing, protein capable of transforming fibroblasts. Shc can also induce the differentiation of PC12 cells. Since this induction can be blocked by the dominant inhibitory Ras mutant, Shc function is localized upstream of Ras and may also serve to position Grb2-mSOS-1 near Ras and activated tyrosine kinases (see Figure 3B) (Egan et al., 1993).

Providing additional flexibility in the pathways which lead to Ras activation are the presence of tissue specific exchange factors. p140-Ras-GRF is brain specific and lacks the proline-rich region found in mSOS-1 responsible for Grb2 binding. p140-Ras-GRF appears to mediate the activation of Ras by calcium influx (Farnsworth et al., 1995). Heterotrimeric G-protein-coupled receptors can activate Ras either through p140-Ras-GRF (Crespo et al., 1994) or through phosphorylation of Shc and consequent positioning of Grb2-mSOS-1 near Ras (Biesen et al., 1995).

It is also clear that Ras activity is controlled to some extent by alteration of GTPase activity in some situations. Inhibition of GAP activity and increased levels of Ras bound to GTP has been demonstrated in T cells treated with phorbol ester (which activates the serine/threonine kinase, PKC) or T cell receptor agonist (Downward et al., 1990; Izquierdo et al., 1992) and in erythroleukemia cells treated with erythropoietin (Torti et al., 1992). Stimulation of GAP activity has been observed in PC12 cells following NGF stimulation (Li et al., 1992) and in fibroblast cells grown to high density (Zhang et al., 1992).

The means by which GTPase activity can be controlled is a matter of investigation. It has been observed that one of the first and most dramatic biochemical events which occurs in mitogen stimulated cells is the metabolism of various phospholipids. This observation led to the suggestion that Ras or another oncogene might control phospholipid metabolism (Macara, 1985). However, it was demonstrated that lipid metabolism must generally occur before the action of Ras since anti-Ras antibody completely abolished the activity of lipid related mitogens. This observation suggested that lipids might be involved in the control of Ras activity (Yu et al., 1988). Accordingly, a search was initiated to identify a lipid able to control Ras activity which led to the observation that certain lipids are able to completely block the activity of GAP and of neurofibromin. The active lipids are related to arachidonic acid or one of its metabolites (Tsai et al., 1989b), physically associate with GAP (Serth et al., 1991; Tsai et al., 1991), and have been shown to block the activity of the GTPase activating proteins of several Ras related proteins (Tsai et al., 1989a).

The same types of experiments which were instrumental in revealing the molecular transmission from signals outside the cell to Ras have also begun to reveal how these signals are transmitted from Ras to the interior of the cell. A group of point mutations in oncogenic *ras* which abolish its transforming activity but do not alter the protein biochemically have identified a region of Ras considered to be the effector binding domain (amino acids 32–40). It is assumed that a downstream target of Ras should associate preferentially with this region in Ras-GTP. The first molecules speculated to be candidates by these criteria were GAP and NF1. NF1 stably and specifically associates with Ras in the GTP-bound conformation (DiBattiste et al., 1993). NF1 functions as a negative regulator of Ras signaling independent of its GTPase-accelerating ability (Johnson et al., 1994). No data has yet been obtained to indicate that NF1 is responsible for transmitting a signal from Ras. A few situations exist where GAP may function downstream of Ras. However, overexpression of GAP inhibits the activity of normal Ras. This clearly demonstrates a role for GAP as negative regulator, and indicates that GAP cannot function simply as a downstream effector (Stacey et al., 1991; Lowy and Willumsen, 1993).

In addition to the anti-Ras antibody injection experiments already discussed, several other types of experiments suggest that the Raf family of serine/threonine kinases (Raf-1, B-Raf, and A-Raf) functions downstream of Ras, possibly as the Ras effector molecule. Expression of oncogenic Ras activates Raf while expression of the dominant inhibitory Ras mutant (Asn-17) blocks extracellular signal-induced activation of Raf. These studies place Raf downstream of Ras in both the proliferation of NIH 3T3 cells (Morrison, 1990) and the differentiation of PC12 cells (Wood et al., 1993). Raf-1 has been localized downstream of Ras in experiments using antisense *raf* or dominant inhibitory Raf mutants (Kolck et al., 1991). Evidence placing Raf downstream of Ras has been strengthened by studies which demonstrate the physical association of Raf with Ras. This association is selective for Ras in the GTP-bound conformation and requires an intact effector binding domain (Moodie et al., 1993; Van Aelst et al., 1993; Vojtek et al., 1993). Phosphatidylinositol-3-OH kinase (Rodriguez-Viciana et al., 1994) and Ral GDS (guanine nucleotide dissociation stimulator) (Hofer et al., 1994) have also been shown to physically associate specifically with Ras-GTP (see Figure 4).

Another family of cytosolic serine/threonine kinases has been localized downstream of both Ras and Raf. These are referred to as MAP kinases which serve as an acronym for either mitogen-activated protein kinases or microtubule-associated protein kinases because microtubule-associated proteins (MAPs) are a preferred substrate *in vitro*. Studies performed with antisense RNA or kinase-deficient mutants have established that MAP kinases are absolutely required for proliferation (Pagès et al., 1993). MAP kinases are themselves directly regulated by phosphorylation on both tyrosine and threonine (Pelech and Sanghera, 1992; Thomas, 1992). The dual action kinase responsible for this activation is referred to as either MAP

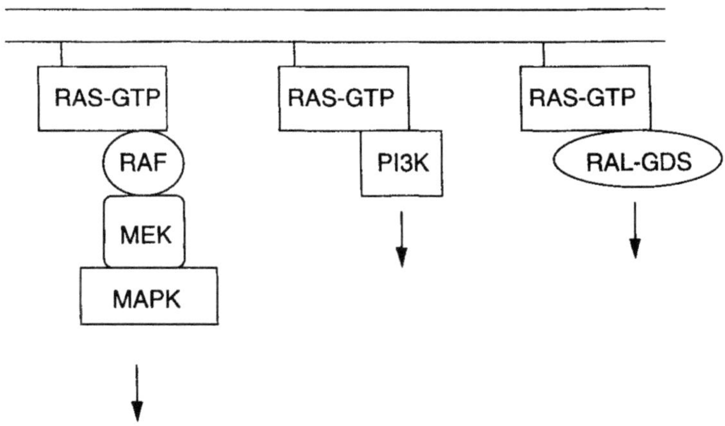

Figure 4. Identification of Ras effector molecules. Activated Ras stably associates with NF-1, Raf-1, B-Raf, PI3-Kinase, and Ra1-GDS. Association with B-Raf is necessary for activation of MEK which is responsible for activating MAP kinase.

kinase kinase or MEK (Rossomando et al., 1992). Expression of oncogenic Ras can result in constitutive activation of MAP kinases. Expression of Asn-17 Ras blocks the growth factor-induced activation of MAP kinases (De Vries-Smits et al., 1992). Finally, activated MEK and MAP kinase activity are found physically associated with Ras. This association is selective for Ras in the GTP-bound conformation and requires an intact effector binding domain (Moodie et al., 1993; Van Aelst et al., 1993) and the presence of B-Raf, but does not require the presence of Raf-1 (Moodie et al., 1993).

The discovery that Raf activity can induce MEK activity (Dent et al., 1992; Howe et al., 1992; Kyriakis et al., 1992) seemed to complete a tidy linear cascade from receptor tyrosine kinases to MAP kinases. However, anomalies do exist. Expression of an activated Ras in PC12 cells results in constitutive activation of MAP kinase while expression of activated Raf does not (Wood et al., 1992; Blenis, 1993). Neither activated Ras nor Raf activates MAP kinase in Rat1a cells (Gupta et al., 1992) and alternative pathways have been demonstrated in NIH 3T3 cells (Mitra et al., 1993). These anomalies are perhaps a reflection of the robust nature of signal transduction. They are certainly indicative of the flexible, albeit complex, signaling made possible by the existence of isozymes with potentially different functions. Ras and Raf each exist as three isozymes (Storm et al., 1990). There are two forms of MEK and as many as four forms of MAP kinase (Pelech and Sanghera, 1992). In addition to MAP kinase activation, oncogenic Ras also activates JNK (*jun* NH2-terminal kinase) (Hibi et al., 1993). JNK may play a role in tumor promotion since it is also activated by ultraviolet light (Devary et al., 1992). When it comes to understanding a role for MAP kinases in cancer, clues can be found historically in the fact that MAP kinases acquired notoriety for their possible role in the cell cycle (Pelech and Sanghera, 1992). Critical targets for MAP kinases and JNKs remain unclear, but certainly extracellular signals transmitted through Ras to these kinases ultimately regulate nuclear transcription factors (Hill and Treisman, 1995).

The importance of the Ras signaling pathway is underscored by the realization that in addition to the Ras protein itself, many of the other proteins in the pathway have been relatively conserved throughout evolution. An adaptor-type protein (Sem5) has also been characterized in *C. elegans*. It too is composed of one SH2 domain flanked by two SH3 domains. Genetic analysis has determined that this protein functions upstream of a Ras homologue (let-60) and downstream of the *C. elegans* homologue of the EGF receptor tyrosine kinase (let23) (Clark et al., 1992). Also, Ras and Raf are both necessary for signaling by let23 (Han et al., 1993). Drk, the *Drosophila* counterpart to Sem5, binds through its SH2 domains to the activated receptor tyrosine kinase, sevenless. This is consistent with earlier genetic studies in *Drosophila* which demonstrated that Sos (Son of sevenless) functioned downstream of the sevenless receptor tyrosine kinase and the *Drosophila* homologue of the EGF receptor tyrosine kinase (DER) and upstream of the Ras homologue (Dras1) (Simon et al., 1992, 1993; Olivier et al., 1993). *raf* function is required downstream of the sevenless receptor tyrosine kinase and *ras* in *Drosophila* (Dickson et al., 1992) and upstream of the *Drosophila* homologue of MEK (Perrimon, 1993).

Ras AS A CLINICAL TARGET

The information gained to date is now being put to practical use. Because mutated *ras* can be detected in bodily fluids (Brandt-Rauf, 1991; Berthelemy et al., 1995; Malats et al., 1995) and feces (Hasegawa et al., 1995), it is being considered as a potential diagnostic and prognostic biomarker. Ras is also a target for specific immunotherapy (Jung and Schluesener, 1991; Skipper and Stauss, 1993; Peace et al., 1993) and gene therapy (Aoki et al., 1995). Because Ras must be associated with the cell membrane in order to be transforming, compounds which inhibit the transfer of the farnesyl lipid moiety to Ras, and thus its association with the cell membrane, represent potential therapeutic agents (Hara et al., 1993; James et al., 1993; Kohl et al., 1993; Gibbs et al., 1994; Kohl et al., 1994; Prendergast et al., 1994).

Ras IN NONPROLIFERATIVE PATHWAYS

The focus of this discussion has been on the role of Ras in proliferative signal transduction. The highest level of Ras protein expression, however, is in the brain and therefore presumably not related to proliferation. In cells of neuronal origin Ras activity induces differentiation (Bar-Sagi and Feramisco, 1985; Satoh et al., 1987), and both Ras induction (Gartner et al., 1995) and MAP kinase activity have been implicated directly in the pathology of Alzheimer's disease (Drewes et al.,

1992). Ras activity also induces the differentiation of adipocytes (Benito et al. 1991) but inhibits the differentiation of myoblasts (Alema and Tato, 1987) and mammary epithelial cells (Jehn et al., 1992). Previously, mention has been made of Ras homologues, in *Drosophila* and *C. elegans*. In both of these cases Ras functions to control differentiation as opposed to proliferation. In addition, even in cells which normally respond mitogenically to growth factors, the pathway in which Ras participates has other actions. It has been shown by microinjection of anti-Ras antibody that Ras is needed from the start and continuously through the process of migration (Fox et al., 1994).

SUMMARY

The pivotal role of Ras as a transducer of vital signals is well established. As such, aberrant Ras signaling is associated with cancer, neurofibromatosis, and Alzheimer's disease. This realization could make detection of mutated *ras* a useful diagnostic and prognostic tool. Exploiting the biochemical properties of Ras has made interfering with Ras activity a potential clinical option and dissection of the relative contribution of each signaling pathway emanating from Ras will certainly provide additional targets for clinical intervention.

REFERENCES

Alema, S. & Tato, F. (1987). Interaction of retroviral oncogenes with the differentiation program of myogenic cells. Adv. Cancer Res. 49, 1–28.

Aoki, K., Yoshida, T., Sugimura, T., & Terada, M. (1995). Liposome-mediated in vivo gene transfer of antisense K-ras construct inhibits pancreatic tumor dissemination in the murine peritoneal cavity. Cancer Res. 55, 3810–3816.

Bar-Sagi, D. & Feramisco, J. (1985). Microinjection of the ras oncogene protein into PC12 cells induces morphological differentiation. Cell 42, 841–848.

Balmain, A. & Brown, K. (1988). Oncogene activation in chemical carcinogenesis. Adv. Cancer Res. 51, 147–182.

Benito, M., Porras, A., Nebreda, A., & Santos, E. (1991). Differentiation of 3T3-L1 fibroblasts to adipocytes induced by transformation of ras oncogenes. Science 253, 565–568.

Berthelemy, P., Bouisson, M., Escourrou, J., Vaysse, N., Rumeau, J., & Pradayrol, L. (1995). Identification of K-ras mutations in pancreatic juice in the early diagnosis of pancreatic cancer. Ann. Intern. Med. 123, 188–191.

Biesen, T., Hawea, B., Luttrell, D., Krueger, K., Touhara, K., Porfiri, E., Sakaue, M., Luttrell, L., & Lefkowitz, R. (1995). Receptor-tyrosine-kinase- and Gβγ-mediated MAP kinase activation by a common signaling pathway. Nature 376, 781–784.

Binetruy, B., Smeal, T., & Karin, M. (1991). Ha-Ras augments c-Jun activity and stimulates phosphorylation of its activation domain. Nature 351, 122–127.

Blenis, J. (1993). Signal transduction via the MAP kinases: Proceed at your own RSK. Proc. Natl. Acad. Sci. USA 90, 5889–5892.

Bollag, G. & McCormick, F. (1991). Regulators and effectors of *ras* proteins. Ann. Rev. Cell Biol. 7, 601–632.

Bos, J. (1989). *ras* oncogenes in human cancer: A review. Cancer Res. 49, 4682–4689.

Bowden, G.T. (1990). Oncogene activation during multi-stage carcinogenesis. Mutation and the Environment Part D, 1–12.

Brandt-Rauf, P.W. (1991). Advances in cancer biomarkers as applied to chemical exposures: The ras oncogene and p21 protein and pulmonary carcinogenesis. J. Occup. Med. 33, 951–955.

Carpenter, G. (1992). Receptor tyrosine kinase substrates: *src* Homology domains and signal transduction. FASEB J. 6, 3283–3289.

Clark, S.G., Stern, M.J., & Horvitz, H.R. (1992). *C. elegans* cell signalling gene *sem-5* encodes a protein with SH2 and SH3 domains. Nature 356, 340–344.

Crespo, P., Xu, N., Simonds, W., & Gutkind, J. (1994). Ras-dependent activation of MAP kinase pathway mediated by βγ subunits. Nature 369, 418–420.

DiBattiste, D., Golubic, M., Stacey, D., & Wolfman, A. (1993). Differences in the interaction of p21 c-Ha-ras-GMP-PNP with full-length neurofibromin and GTPase-activating protein. Oncogene 8, 637–643.

De Vries-Smits, A.M.M., Burgering, B.M.T., Leevers, S.J., Marshall, C.J., & Bos, J.L. (1992). Involvement of $p21^{ras}$ in activation of extracellular signal-regulated kinase 2. Nature 357, 602–604.

Dent, P., Haser, W., Haystead, T.A.J., Vincent, L.A., Roberts, T.M., & Sturgill, T.W. (1992). Activation of mitogen-activated protein kinase by v-raf in NIH 3T3 cells and in vitro. Science 257, 1404–1407.

Devary, Y., Gottlieb, R., Smeal, T., & Karin, M. (1992). The mammalian ultraviolet response is triggered by activation of Src tyrosine kinases. Cell 71, 1081–1091.

Dickson, B., Sprenger, F., Morrison, D., & Hafen, E. (1992). Raf functions downstream of Ras1 in the Sevenless signal transduction pathway. Nature 360, 600–603.

Dobrowolski, S., Harter, M., & Stacey, D.W. (1994). Cellular ras activity is required for passage through two separate points in the GO-G1 phase of Balb/c 3T3 cells. Molec. Cell Biol. 14, 5441–5449.

Dotto, G.P., Parada, L.F., & Weinberg, R.A. (1985). Specific growth response of *ras* - transformed embryo fibroblasts to tumour promoters. Nature 318, 472–475.

Downward, J., Graves, J.D., Warne, P.H., Rayter, S., & Cantrell, D.A. (1990). Stimulation of $p21^{ras}$ upon T-cell activation. Nature 346, 719–723.

Drewes, G., Lichtenberg-Kraag, B., Döring, F., Mandelkow, E.-M., Biernat, J., Goris, J., Dorée, M., & Mandelkow, E. (1992). Mitogen activated protein (MAP) kinase transforms tau protein into an Alzheimer-like state. EMBO J. 11, 2131–2138.

Egan, S.E., Giddings, B.W., Brooks, M.W., Buday, L., Sizeland, A.M., & Weinberg, R.A. (1993). Association of Sos Ras exchange protein with Grb2 is implicated in tyrosine kinase signal transduction and transformation. Nature 363, 45–51.

Farnsworth, C., Freshney, N., Rosen, L., Ghosh, A., Greenberg, M., & Feig, L. (1995). Calcium activation of Ras mediated by neuronal exchange factor Ras-GRF. Nature 376, 524–527.

Feig, L.A. (1993). The many roads that lead to ras. Science 260, 767–768.

Finney, R. & Bishop, J.M. (1993). Predisposition to neoplastic transformation caused by gene replacement of H-*ras* 1. Science 260, 1524–1527.

Fox, P.L., Dobrowolski, S., Gaurisankar, S., & Stacey, D.W. (1994). Cellular ras activity is required for endothelial cell motility. Oncogene 9, 3519–3526.

Gartner, U., Holzer, M., Heumann, R., & Arendt, T. (1995). Induction of p21 ras in Alzheimer pathology. Neuroreport 6, 1441–1444.

Gibbs, J., Oliff, A., & Kohl, N. (1994). Farnesyltransferase inhibitors: Ras research yields a potential cancer therapeutic. Cell 77, 175–178.

Gilman, A.G. (1987). G proteins: Transducers of receptor generated signals. Ann. Rev. Biochem. 56, 615–649.

Greenhalgh, D.A., Welty, D.J., Strickland, J.E., & Yuspa, S.H. (1989). Spontaneous Ha-*ras* gene activation in cultured primary murine keratinocytes: Consequences of Ha-*ras* gene activation in malignant conversion and malignant progression. Molec. Carcin. 2, 199–207.

Gupta, S.K., Gallego, C., Johnson, G.L., & Heasley, L.E. (1992). MAP kinase is constitutively activated in *gip2* and *src* transformed rat 1a fibroblasts. J. Biol. Chem. 267, 7987–7990.

Han, M., Golden, A., Han, Y., & Sternberg, P.W. (1993). *C. elegans lin-45 raf* gene participates in *let-60 ras*-stimulated vulval differentiation. Nature 363, 133–140.

Hara, M., Kazuhito, A., Akinaga, S., Okabe, M., Nakano, H., Gomez, R., Wood, D., Uh, M., & Tamanoi, F. (1993). Identification of ras farnesyltransferase inhibitors by microbial screening. Proc. Natl. Acad. Sci. USA 90, 2281–2285.

Hasegawa, Y., Takeda, S., Ichii, S., Koizumi, K., Maruyama, M., Fujii, A., Ohta, H., Nakajima, T., Okuda, M., Baba, S., & Nakamura, Y. (1995). Detection of K-ras mutations in DNAs isolated from feces of patients with colorectal tumors by mutant-allele-specific amplification (MASA). Oncogene 10, 1441–1445.

Hibi, M., Lin, A., Smeal, T., Minden, A., & Karin, M. (1993). Identification of an oncoprotein- and UV-responsive protein kinase that binds and potentiates the c-Jun activation domain. Genes Dev. 7, 2135–2148.

Hill, S. & Treisman, R. (1995). Transcriptional regulation by extracellular signals: Mechanisms and specificity. Cell 80, 199–211.

Hofer, F., Fields, S., Schneider, C., & Martin, G. (1994). Activated ras interacts with the ral guanine nucleotide dissociation stimulator. Proc. Natl. Acad. Sci. USA 91, 11089–11093.

Howe, L.R., Leevers, S.J., Gómez, N., Nakielny, S., Cohen, P., & Marshall, C.J. (1992). Activation of the MAP kinase pathway by the protein kinase raf. Cell 71, 335–342.

Izquierdo, M., Downward, J., Graves, J.D., & Cantrell, D.A. (1992). Role of protein kinase C in T-cell antigen receptor regulation of $p21^{ras}$: Evidence that two $p21^{ras}$ regulatory pathways coexist in T cells. Mol. Cell. Biol. 12, 3305–3312.

James, G.J., Goldstein, J.L., Brown, M.S., Rawson, T.E., Somers, T.C., McDowell, R.S., Crowley, C.W., Lucas, B.K., Levinson, A.D., & Marsters, J.C. (1993). Benzodiazepine peptidomimetics: Potent inhibitors of ras farnesylation in animal cells. Science 260, 1937–1941.

Jehn, B., Costello, E., Marti, A., Keon, N., Deane, R., Li, F., Friis, R., Burri, P., Martin, F., & Jaggi, R. (1992). Overexpression of mos, ras, src, and fos inhibits mouse mammary epithelial cell differentiation. Mol. Cell. Biol. 12, 3890–3902.

Johnson, M., DeClue, J., Felzmann, S., Vass, W., Xu, G., White, R., & Lowy, D. (1994). Neurofibromin can inhibit ras-dependent growth by a mechanism independent of its GTPase-accelerating function. Mol. Cell Biol. 14, 641–645.

Jung, S. & Schluesener, H.J. (1993). Human T lymphocytes recognize a peptide of single point-mutated, oncogenic ras proteins. J. Exp. Med. 173, 273–276.

Kohl, N.E., Mosser, S.D., deSolms, S.J., Giuliani, E.A., Pompliano, D.L., Graham, S.L., Smith, R.L., Scolnick, E.M., Oliff, A., & Gibbs, J.B. (1993). Selective inhibition of ras-dependent transformation by a farnesyltransferase inhibitor. Science 260, 1934–1937.

Kohl, N., Wilson, F., Mosser, S., Giuliani, E., DeSolms, S., Conner, M., Anthony, N., Holtz, W., Gomez, R., Lee, T., Smith, R., Graham, S., Hartman, G., Gibbs, J., & Oliff, A. (1994). Protein farnesyltransferase inhibitors block the growth of ras-dependent tumors in nude mice. Proc. Natl. Acad. Sci. USA 91, 9141–9145.

Kolck, W., Heidecker, G., Lloyd, P., & Rapp, U.R. (1991). Raf-1 protein kinase is required for growth of induced NIH/3T3 cells. Nature 349, 426–428.

Kumar, R., Saraswati, S., & Barbacid, M. (1990). Activaton of *ras* oncogenes preceding the onset of neoplasia. Science 248, 1101–1104.

Kyriakis, J.M., App, H., Zhang, X., Banerjee, P., Brautigan, D.L., Rapp, U.R., & Avruch, J. (1992). Raf-1 activates MAP kinase-kinase. Nature 358, 417–421.

Land, H., Parada, L.F., & Weinberg, R.A. (1983). Tumorigenic conversion of primary embryo fibroblasts requires at least two cooperating oncogenes. Nature 304, 596–602.

Li, B.Q., Kaplan, D., Kung, H., & Kamata, T. (1992). Nerve growth factor stimulation of the ras-guanine nucleotide exchange factor and GAP activities. Science 256, 1456–1459.

Lowy, D.R. & Willumsen, B.M. (1993). Function and regulation of ras. Annu. Rev. Biochem. 62, 851–891.

Macara, I. (1985). Oncogenes, ions, and phospholipids. Am. J. Physiol 17, c3–c11.

Malats, N., Porta, M., Pinol, J., Corominas, J., Rifa, J., & Real, F. (1995). K-ras mutations as a prognostic factor in extrahepatic bile system cancer. J. Clin. Oncology 13, 1679–1686.

Mayer, B.J. & Baltimore, D. (1993). Signalling through SH2 and SH3 domains. Trends Cell Biol. 3, 8–13.

McCormick, F. (1993). Signal transduction: How receptors turn ras on. Nature 363, 15–16.

Mitra, G., Weber, M., & Stacey, D.W. (1993). Multiple pathways for activation of MAP kinases. Cell. Mol. Biol. Res. 39, 517–523.

Moodie, S.A., Willumsen, B.M., Weber, M.J., & Wolfman, A. (1993). Complexes of ras•GTP with Raf-1 and mitogen-activated protein kinase kinase. Science 260, 1658–1661.

Morrison, D.K. (1990). The Raf-1 kinase as a transducer of mitogenic signals. Cancer Cells 2, 377–382.

Mulcahy, L.S., Smith, M.R., & Stacey, D.W. (1985). Requirement for ras proto-oncogene function during serum-stimulated growth of NIH 3T3 cells. Nature 313, 241–243.

Olivier, J.P., Raabe, T., Henkemeyer, M., Dickson, B., Mbamalu, G., Margolis, B., Schlessinger, J., Hafen, E., & Pawson, T. (1993). A Drosophila SH2-SH3 adaptor protein implicated in coupling the sevenless tyrosine kinase to an activator of Ras guanine nucleotide exchange. Sos. Cell 73, 179–191.

Pagès, G., Lenormand, P., L'Allemain, G., Chambard, J.-C., Meloche, S., & Pouysségur, J. (1993). Mitogen-activated protein kinases $p42^{mapk}$ and $p44^{mapk}$ are required for fibroblast proliferation. Proc. Natl. Acad. Sci. USA 90, 8319–8323.

Pawson, T. & Gish, G.D. (1992). SH2 and SH3 domains: From structure to function. Cell 71, 359–362.

Peace, D.J., Smith, J.W., Disis, M.L., Chen, W., & Cheever, M.A. (1993). Induction of T cells specific for the mutated segment of oncogenic p21ras protein by immunization in vivo with the oncogenic protein. J. Immunotherapy 14, 110–114.

Pelech, S.L. & Sanghera, J.S. (1992). Mitogen-activated protein kinases: versatile transducers for cell signaling. Trends Biochem. Sci. 17. 233–238.

Perrimon, N. (1993). The torso receptor protein-tyrosine kinase signaling pathway: An endless story. Cell 74, 219–222.

Prendergast, G., Davide, J., DeSolms, S., Guiliani, E., Graham, S., Gibbs, J., Oliff, A., & Kohl, N. (1994). Farnesyltransferase inhibition causes morphological reversion of ras-transformed cells by a complex mechanism that involves regulation of the actin cytoskeleton. Mol. Cell. Biol. 14, 4193–4202.

Rodriguez-Viciana, P., Warne, P., Dhand, R., Vanhaesebroeck, B., Gout, I., Fry, M., Waterfield, M., & Downward, J. (1994). Phosphatidylinositol-3-OH kinase as a direct target of ras. Nature 370, 527–532.

Rossomando, A., Wu, J., Weber, M.J., & Sturgill, T.W. (1992). The phorbol ester-dependent activator of the mitogen-activated protein kinase $p42^{mapk}$ is a kinase with specificity for the threonine and tyrosine regulatory sites. Proc. Natl. Acad. Sci. USA 89, 5221–5225.

Ruley, H.E. (1983). Adenovirus early region 1A enables viral and cellular transforming genes to transform primary cells in culture. Nature 304, 602–606.

Satoh, T., Nakafuku, M., & Kaziro, Y. (1992). Function of Ras as a molecular switch in signal transduction. J. Biol. Chem. 267, 24149–24152.

Satoh, T., Nakamura, S., & Kaziro, Y. (1987). Induction of neurite formation in PC12 cells by microinjection of proto-oncogenic Ha-ras protein preincubated with guanosine-5′-0-(3-thiophosphate). Mol. Cell. Biol. 7, 4553–4556.

Schutte, J., Minna, J.D., & Birrer, M.J. (1989). Deregulated expression of human c-*jun* transforms primary rat embryo cells in cooperation with an activated c-Ha-*ras* gene and transforms Rat 1a cells as a single gene. Proc. Natl. Acad. Sci. USA 86, 22257–22261.

Serth, J., Lautwein, A., Frech, M., Wittinghofer, A., & Pingoud, A. (1991). The inhibition of the GTPase activating protein—Ha-*ras* interaction by acidic lipids is due to physical association of the C-terminal domain of the GTPase activating protein with micellar structures. EMBO J. 10, 1325–1330.

Simon, M.A., Carthew, R.W., Fortini, M.E., Gaul, U., Mardon, G., & Rubin, G.M. (1992). Signal transduction pathway initiated by activation of the *sevenless* tyrosine kinase receptor. Cold Spring Harbor Symp. Quant. Biol. 57, 375–380.

Simon, M.A., Dodson, G.S., & Rubin, G.M. (1993). An SH3-SH2-SH3 protein is required for $p21^{Ras1}$ activation and binds to sevenless and Sos proteins in vitro. Cell 73, 169–177.

Skipper, J. & Stauss, H.J. (1993). Identification of two cytotoxic T lymphocyte-recognized epitopes in the ras protein. J. Exp. Med. 177, 1493–1498.

Skolnik, E.Y., Batzer, A., Li, N., Lee, C.-H., Lowenstein, E., Mohammadi, M., Margolis, B., & Schlessinger, J. (1993a). The function of GRB2 in linking the insulin receptor to Ras signaling pathways. Science 260, 1953–1955.

Skolnik, E.Y., Lee, C.-H., Batzer, A., Vicentini, L.M., Zhou, M., Daly, R., Myers, M.J., Jr., Backer, J.M., Ullrich, A., White, M.F., & Schlessinger, J. (1993b). The SH2/SH3 domain-containing protein GRB2 interacts with tyrosine-phosphorylated IRS1 and Shc: Implications for insulin control of *ras* signalling. EMBO J. 12, 1929–1936.

Smith, M.R., DeGudicibus, S.J., & Stacey, D.W. (1986). Requirement for c-ras proteins during viral oncogene transformation. Nature 320, 541–543.

Spandidos, D.A. & Wilkie, N.M. (1984). Malignant transformation of early passage rodent cells by a single oncogene. Nature 310, 469–475.

Stacey, D.W., Feig, L.A., & Gibbs, J.B. (1991a). Dominant inhibitory Ras mutants selectively inhibit the activity of either cellular or oncogenic Ras. Mol. Cell. Biol. 11, 4053–4064.

Stacey, D.W. & Kung, H.F. (1984). Transformation of NIH 3T3 cells by microinjection of Ha-ras p21 protein. Nature 310, 508–511.

Stacey, D.W., Roudebush, M., Day, R., Mosser, S.D., Gibbs, J.B., & Feig, L.A. (1991b). Dominant inhibitory Ras mutants demonstrate the requirement for Ras activity in the action of tyrosine kinase oncogenes. Oncogene 6, 2297–2304.

Storm, S.M., Cleveland, J.L., & Rapp, U.R. (1990). Expression of raf family proto-oncogenes in normal mouse tissue. Oncogene 5, 345–351.

Suen, K.-L., Bustelo, X.R., Pawson, T., & Barbacid, M. (1993). Molecular cloning of the mouse *grb2* gene: Differential interaction of the Grb2 adaptor protein with epidermal growth factor and nerve growth factor receptors. Mol. Cell Biol. 13, 5500–5512.

Thomas, G. (1992). MAP kinase by any other name smells just as sweet. Cell 68, 3–6.

Torti, M., Bencke Marti, K., Altschuler, D., Yamamoto, K., & Lapetina, E.G. (1992). Erythropoietin induces $p21^{ras}$ activation and p120GAP tyrosine phosphorylation in human erythroleukemia cells. J. Biol. Chem. 267, 8293–8298.

Tsai, M.H., Hall, A., & Stacey, D.W. (1989a). Inhibition by phospholipids of the interaction between R-ras, rho, and their GTPase-activating proteins. Mol. Cell. Biol. 9, 5260–5264.

Tsai, M.H., Roudebush, M., Yu, C.L., Gibbs, J.B., & Stacey, D.W. (1991). Ras GTPase-activating protein physically associates with mitogenically activated phospholipids. Mol. Cell Biol. 11, 2785–2793.

Tsai, M.H., Yu, C.L., Wei, F.S., & Stacey, D.W. (1989b). The effect of GTPase activating protein upon *ras* is inhibited by mitogenically responsive lipids. Science 243, 522–526.

Van Aelst, L., Barr, M., Marcus, S., Polverino, A., & Wigler, M. (1993). Complex formation between RAS and RAF and other protein kinases. Proc. Natl. Acad. Sci. USA 90, 6213–6217.

Vojtek, A.B., Hollenberg, S.M., & Cooper, J.A. (1993). Mammalian ras interacts directly with the serine/threonine kinase raf. Cell 74, 206–214.

Wood, K.W., Qi, H., D'Arcangelo, G., Armstrong, R.C., Roberts, T.M., & Halegoua, S. (1993). The cytoplasmic *raf* oncogene induces a neuronal phenotype in PC12 cells: A potential role for cellular *raf* kinases in neuronal growth factor signal transduction. Proc. Natl. Acad. Sci. USA 90, 5016–5020.

Wood, K.W., Sarnecki, C., Roberts, T.M., & Blenis, J. (1992). *ras* Mediates nerve growth factor receptor modulation of three signal-transducing protein kinases: MAP kinase, Raf-1, and RSK. Cell 68, 1041–1050.

Yu, C.L., Tsai, M.H., & Stacey, D.W. (1988). Cellular ras activity and phospholipid metabolism. Cell 52, 63–71.

Zhang, K., Papageorge, A.G., & Lowy, D.R. (1992). Mechanistic aspects of signaling through Ras in NIH 3T3 cells. Science 257, 671–674.

Chapter 9

Human Cytokines

BRYANT G. DARNAY and BHARAT B. AGGARWAL

Advances in Oncobiology
Volume 1, pages 179–205.

ISBN: 0-7623-0146-5

INTRODUCTION

It is well known that cells respond to extracellular stimuli in order to adapt to their dynamic environment. Evidence for this responsive capability dates back to the early 1900s to experiments of both Erdman and Burrows who demonstrated that sustained *in vitro* growth of bone marrow-derived cells required the presence of blood plasma. With the discovery of interferon in the 1950s by Issacs and Lindermann, the identification of specific molecules that alter biological responses of a cell had begun. In 1969, Dumonde and colleagues coined the word "lymphokines" for factors produced by lymphocytes that could regulate the growth or mobility of a variety of leukocytes. About 10 years later, Cohen and co-workers employed the less restrictive term "cytokine" since not only lymphocytes but also other leukocytes, fibroblasts, keratinocytes, and transformed cell lines could produce such factors. It was only in 1980 that interferon was for the first time purified to homogeneity and its structure determined, demonstrating its uniqueness from all the previously known polypeptide hormones. Even today, we are only beginning to understand the functions and mechanisms of action for this diverse class of molecules.

Cytokines are a family of polypeptide hormones produced by a multitude of cell types in response to diverse stimuli including attack by foreign antigens and tissue damage after myocardial infarction. Since cytokines generally are produced at low levels, their isolation, purification, and characterization has been impeded by technical problems. Additional difficulties were encountered since many effects by the same factor in a bioassay could act in an additive, synergistic, or suppressive manner. Fortunately, while the cytokine field was emerging, the fields of recombinant DNA technology, hybridoma technology, and protein microsequencing were also making rapid strides. Use of these techniques has allowed identification and characterization of a vast array of cytokines.

CYTOKINES—A BRIEF OVERVIEW

The term "cytokine" is currently applied to polypeptide hormones that are produced primarily by cells of the immune system in response to specific stimuli to influence the state of a target cell (Figure 1). Cytokines generally act as paracrine or autocrine

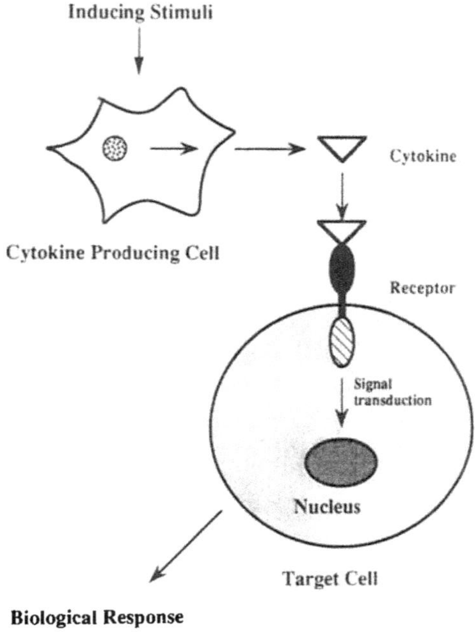

Figure 1. General elements of cytokine action.

signals (some even act as endocrine factors) within local tissues and only following excessive potent stimulation in the circulation, will they cause systemic symptoms such as fever and other acute phase responses (Figure 2). In contrast, endocrine polypeptide hormones and neuropeptides are generally produced by specialized glands, are constantly present in the circulation, act on neighboring as well as distant organs, serve to maintain homeostasis, and mobilize the "flight or fight" response of stress. Unlike hormones, cytokines are generally not constitutively expressed or stored within the cells as are some neuropeptides. The term autocrine pertains to the action of a cytokine on the same cell that produces it. In contrast, paracrine pertains to the production of a cytokine from one cell type and the action of this cytokine on the other. The target cells may be located in close vicinity to the producer cell or be very distant from the cells which produce the cytokine, suggesting a relationship to typical polypeptide hormones. Because cytokine production is induced or suppressed by specific stimuli rarely at a constant rate, their survival in the circulation is short lived and their effects are well regulated. This plays an important role in controlling their activities *in vivo*. Thus, these proteins transmit signals between tissues and their surroundings. Some functional characteristics demonstrated by these cytokines include regulation of the immune

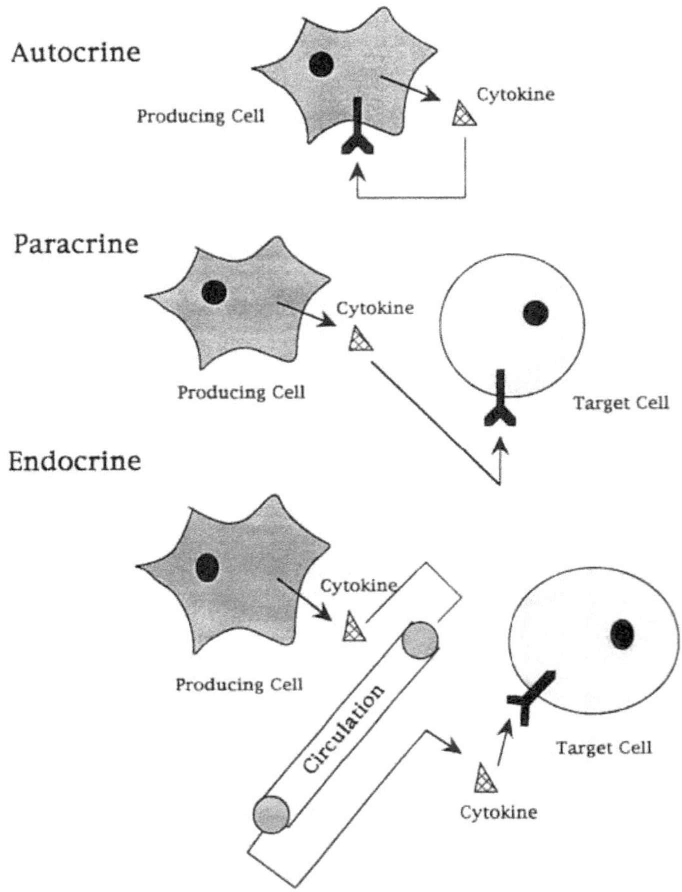

Figure 2. Cytokine and cell interactions.

system, inflammation, tissue remodeling, embryonic development, and cell growth and differentiation.

Cytokines share many common characteristics as follows:

1. They are low molecular weight proteins (< 80 kD) that are secreted and sometimes glycosylated.
2. They regulate amplitude and duration of biological response.
3. They are produced transiently and act locally in a combined autocrine and paracrine rather than endocrine manner.
4. They are extremely potent even at picomolar concentrations.
5. Their receptors exhibit low density (10 to 10,000/cell) but high-affinity.

6. Their cell surface binding leads to induction of different genes resulting in intracellular changes.
7. Several different cytokines may exhibit overlapping and sometimes redundant actions.
8. The response to them is dependent upon local concentration, cell type, and other regulatory factors to which they are exposed.
9. They interact in a network by inducing or suppressing the expression of each other.
10. They are capable of transmodulating their own or another cytokine cell-surface receptor.
11. Different cytokines may exhibit synergistic, additive, and antagonistic interaction on cell function.

Because the field of cytokines has proliferated enormously in the last decade, it is not possible to outline all the cytokines and their effects in detail that have been identified thus far. Therefore, this chapter will instead outline some of the basic principles of cytokines and their actions. This chapter is divided into three general sections: (a) cytokines; their structural and biological aspects, (b) cytokine receptors and signal transduction, and (c) cytokine function in health and disease.

CYTOKINES

Most, but not all, cytokines identified to date have been characterized based on their biological activities as found in cell culture assays. Therefore, cytokines are grouped by their *in vitro* biological activities. However, large quantities of highly purified protein is needed to elucidate their activities *in vivo*. As such, the structure of most cytokines is known not only at the protein level but also at the gene level. Thus, the production of large quantities of cytokines by recombinant DNA technology has enabled the elucidation of their activities both *in vitro* and *in vivo*. The major groups described below include interferons, interleukins, colony stimulating factors, transforming growth factors, tumor necrosis factors, and chemokines. Though most cytokines, based on their structural characteristics, are classified as members of a given family, they may possess activities that overlap with those of other family members. Cytokines are produced by a variety of cells and act upon many types of target cells (Figure 3).

Interferons

Interferons (IFNs), whose name is derived from the word interference, were originally identified as agents produced by cells in response to viral infections for self-protection. Some general characteristics of interferons are shown in Figure 4. Besides antiviral properties, interferons exhibit other activities which include antiproliferative effects on certain cells, cell differentiation, and modulation of the

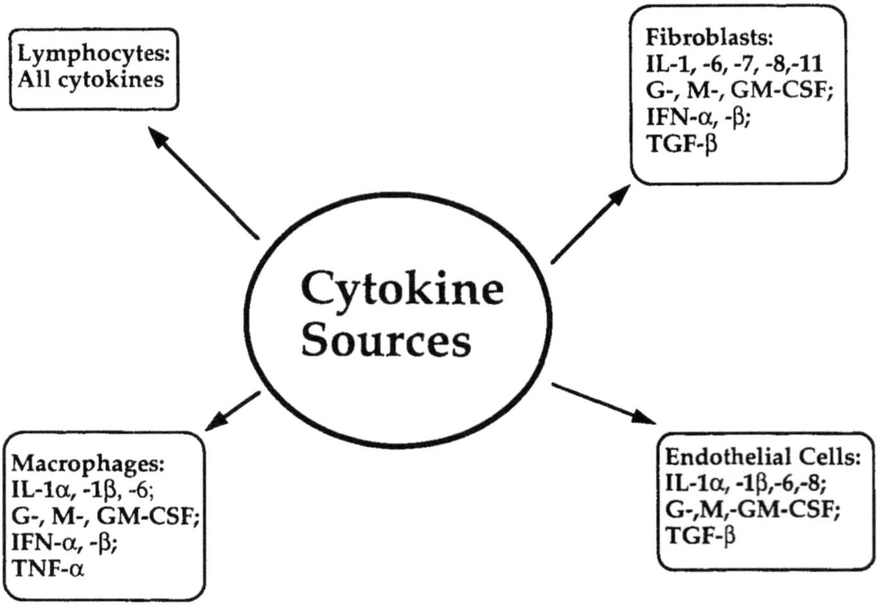

Figure 3. Different types of cells produce cytokines.

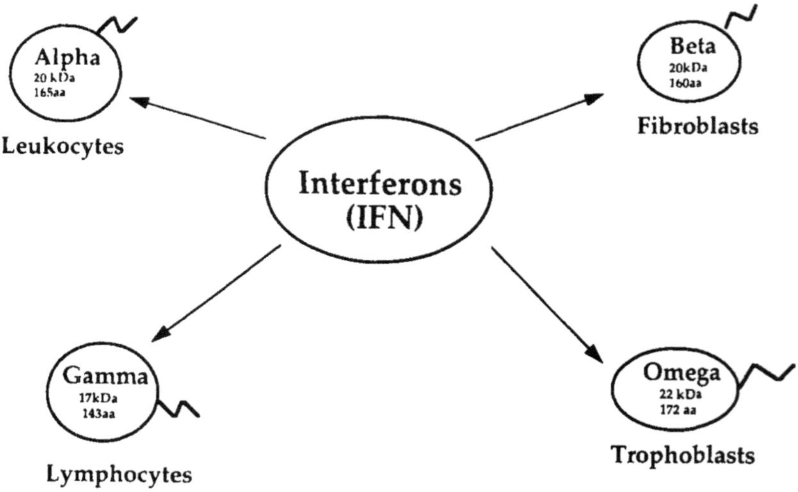

Figure 4. Production of different types of interferons by different cell types.

immune system. Interferon-α (IFN-α) is a 19–26 kD, acid-stable protein produced by T and B cells and monocytes. There are 16 different subtypes of IFN-α genes which encode for proteins between 166 and 172 amino acids long and the sequence of most of these subtypes is highly conserved. The production of IFN-α is induced by double-stranded RNA (dsRNA), which facilitates the cell's response to viral infection.

Interferon-β (IFN-β) is a protein produced by fibroblasts, is 20 kD in size and 160 amino acids in length. Like IFN-α, IFN-β aids in resistance to viral infection, is induced by dsRNA, and shares a common receptor of 130 kD with IFN-α. An additional effect of IFN-α and -β is activation of natural killer cells that may be responsible for destroying virus-infected cells or tumor cells *in vivo*. In addition, an increase in expression of major histocompatibility complexes (MHC) on many cells and stimulation of immunoglobulin-Fc receptor expression on macrophages are also observed following IFN treatment. The Fc receptor is needed for antibody-dependent cellular cytotoxicity of macrophages. IFN-α and -β treatment of B-lymphocytes induce low levels of antibody production.

Interferon-γ is produced by T-lymphocytes and its primary role is activation of macrophages which provides protection from microorganisms. Only one gene is known for IFN-γ and it encodes for a protein of 143 amino acids with a molecular mass of 15 kD. Unlike the other IFNs, IFN-γ binds to a distinct cell surface receptor.

While the biochemical and biological properties of IFN-α, IFN-β, and IFN-γ have been extensively investigated over the past decade, little is known about interferon-ω (IFN-ω). The latter shares about 60% sequence homology with IFN-α, but only 29% with IFN-β. IFN-ω is a 172 amino acid polypeptide with an apparent molecular mass varying between 22–24 kD due to the glycosylation. In sheep and cattle, this cytokine is secreted by the trophoblast during early pregnancy, resulting in extension of the luteal lifespan (i.e., maternal recognition of pregnancy). However, its identification in humans is still uncertain. Similar to IFN-α, the expression of IFN-ω mRNA is induced after viral infection of leukocytes and lymphoma cells. *In vitro* and *in vivo* characterization of IFN-ω has been restricted so far due to the lack of pure IFN-ω, or specific antibodies to it. Overall, interferons are an important part of the body's defense mechanism against foreign organisms and also perhaps against cells with deregulated growth.

Interleukins

Mediators that are produced by and act upon cells of the immune system are known as lymphokines. This class has many characteristics in common with other known types of cytokines, both with respect to their production and mechanism of action. The name interleukin, was coined in 1979 for cytokines that are produced by activated T-lymphocytes and that act upon other lymphocytes to generate different biological activities. The interleukin family represents a class of 14 distinct

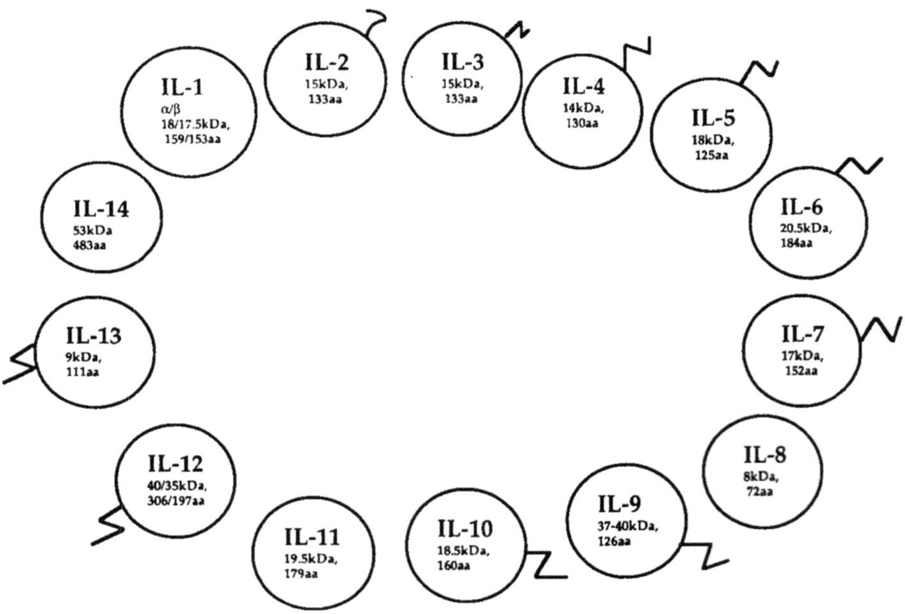

Figure 5. The interleukin family.

proteins whose molecular masses range between 8 and 75 kD and have varying lengths of amino acids (72 to 306) in the mature form (Figure 5).

Interleukin-1 (IL-1) is produced primarily by monocytes and macrophages upon induction by lipopolysaccharide (LPS) or other stimuli (hence the name monokine). There are two types of IL-1, viz., IL-1α which is produced mainly by fibroblasts and IL-1 β which is produced by activated macrophages. *In vitro*, IL-1 exhibits autostimulatory activity suggesting a positive feedback mechanism. This cytokine stimulates the immune response by induction of other lymphokines. IL-1 has also been shown to be one of the key mediators of fever and inflammation response via an induction of prostaglandin.

Interleukin-2 (IL-2) is produced almost exclusively by T-lymphocytes in response to antigenic as well cytokine stimulation. IL-2 was originally described as a T-lymphocyte growth factor, suggesting its importance in regulation of T cell proliferation. In cell culture systems, this stimulation is nonspecifically mimicked by a plant lectin, phytohemagglutinin. In addition, IL-2 causes proliferation of B cells and natural killer cells, and is responsible for differentiation of B cells. The production of IL-2 and its high-affinity receptors allow autocrine-stimulation, which is largely responsible for the state of active proliferation found in T cells.

Interleukin-3 (IL-3) is a hematopoietic growth factor mainly produced by activated T-lymphocytes. It is highly glycosylated at several possible sites. IL-3 also targets many nonlymphoid cell types within several hematopoietic lineages. IL-3 regulates growth and differentiation of pluripotent stem cells, progenitors of mast cells, neutrophils, macrophages, and others. IL-3 stimulates proliferative responses in cells of both the erythroid and myeloid lineages, suggesting overlapping activities with granulocyte-macrophage colony stimulating factor (GM-CSF). However, IL-3 may exhibit a broader specificity which includes early multipotential cells but not primitive stem cells. Though IL-3 can be detected in the circulation and not within the bone marrow, the physiological relevance of its activities on early multipotential cells remains unclear. In addition, IL-3 stimulates the production of histamine-containing mast cells during an allergic response.

Interleukin-4 (IL-4) is also a product of T-lymphocytes. Even though IL-4 is a glycoprotein, glycosylation does not appear to be required for biological activity. It exhibits activities on a wide range of target cells including participation in the proliferation of B and T cells, natural killer cells, thymocytes, and mast cells. Its principal role *in vivo* probably is to promote proliferation of B-lymphocytes that have been activated by antigen binding or other cell surface-mediated events. In addition, IL-4 can enhance the production of IgE, implicating its role in the development of an allergic response. While most of its activities are growth stimulatory, IL-4 antagonizes the stimulatory effects of IL-2 on lymphokine-activated killer (LAK) cells. Conversely, IFN-γ antagonizes IL-4 effects on B-lymphocytes.

Interleukin-5 (IL-5) was first discovered as a cytokine that can activate B lymphocytes and induce differentiation of eosinophils. Thus, the stimulation of eosinophil maturation by IL-5 is probably crucial for antibody-dependent control of parasitic, bacterial, and viral infections of these cells. In contrast, interleukin-6 (IL-6) was first defined as interferon (IFNβ2) because it is induced by viruses, LPS, IL-1, and phorbol esters. Additionally, IL-6 was described as a B cell growth and differentiation factor, as well as a factor which causes antibody production, and that which stimulates production of acute-phase proteins by liver in response to injury or inflammation. IL-6 is produced by several cell types including T-lymphocytes, monocytes, fibroblasts, and endothelial cells. IL-6 and IL-1 constitute the most ubiquitous cytokines. IL-6 serves as a signal enhancer for various T-lymphocyte activities such as IL-2 production and cell proliferation, and IL-3 dependent development of hematopoietic precursors in the bone marrow.

Unlike other lymphokines which act on the terminally differentiated target cells (such as IL-5 and IL-6), interleukin-7 (IL-7) is responsible for proliferation of early B-cell progenitors and of pro-B cells prior to immunoglobulin gene rearrangements. This interleukin is produced by the bone marrow stromal cells.

The macrophage-derived factor which is chemotactic for neutrophils is called interleukin-8 (IL-8). This is a protein that is the smallest in size of all the interleukins having a molecular mass of only 8 kD. This cytokine attracts neutrophils and also activated T cells to sites of tissue damage or infection, thus enhancing the phagocytic and cytolytic activity of the neutrophils.

Produced mainly by T-lymphocytes, interleukin 9 (IL-9) was first identified as a growth factor for proliferation of T-helper cell clones. Like most other cytokines, IL-9 exhibits synergistic activity with other cytokines. IL-9 synergizes with IL-2, IL-3, and erythropoietin (EPO) in the proliferation of mast cells, fetal thymocytes, and erythroid precursor cells, respectively. Since IL-9-transfected murine helper T cells become tumorigenic and lymph nodes from patients with Hodgkin's disease or large-cell anaplastic lymphomas express constitutively IL-9, the suggestion has been made that IL-9 plays a role in tumorigenesis. In addition, IL-9 supports the maturation of burst-forming units erythroid (BFU-E) of adult and fetal origin and the growth of megakaryoblastic cell lines.

Interleukin 10 (IL-10) was originally described as a cytokine synthesis inhibitory factor (CSIF) that can also inhibit arachidonic acid and prostaglandin synthesis. Production of cytokines, such as IFN-γ and IL-2 produced by Th2 cells, are inhibited by the action of IL-10. Synergistically, IL-10 and IL-4 are thought to mediate susceptibility to infection via immunosuppression of cell-mediated immune response. Similar to transforming growth factor β (TGF-β), IL-10 also inhibits the function of macrophages. Thus, it has been suggested that IL-10 functions to balance the actions of stimulating and inhibiting factors, thereby affecting lymphocytes secondarily. However, IL-10 activates the proliferation of mast cells and thymocytes assisting in antibody production for humoral immunity.

A cytokine known as interleukin 11 (IL-11) is a stromal cell-derived cytokine that activates proliferation of IL-6-dependent plasmacytoma cells. It stimulates the production of IgG-secreting B cells in spleen cell cultures and augments the IL-3-dependent development of megakaryocyte colonies in bone marrow cell clonal cultures. IL-11 functions by increasing T cell-dependent antibody production in B cells.

Interleukin 12 (IL-12), produced by macrophages and B lymphocytes, plays a major role in the effective elimination of microbial pathogens. When a host macrophage interacts with a pathogen, IL-12 serves as a messenger in the development of a cellular immune response of T cells. Pathogen-activated macrophages are stimulated to release IL-12 and TNF-α. IL-12 synergizes with TNF-α to stimulate natural killer cells to produce IFN-γ, which activates macrophages to a microbicidal state, thus restricting the spread of infection. In addition, IL-12 induces the differentiation of uncommitted T cells to the Th1 pathway. Thus, the Th1 effector cells generate IFN-γ, IL-2, and TNF-β to maximize macrophage microbicidal response.

Interleukin 13 (IL-13) was identified as a product of activated T-lymphocytes that inhibits inflammatory cytokine production induced by LPS in human peripheral blood monocytes, a function shared with IL-4 and IL-10. Its anti-inflammatory functions could be crucial for clinical applications, such as septic shock or rheumatoid arthritis. It appears to synergize with IL-2 in regulating IFN-γ synthesis in large granular lymphocytes. IL-13 increases the proliferation of B-lymphocytes and the expression of CD23 surface antigen.

A factor initially identified as a high-molecular-weight B cell growth factor (HMW-BCGF), has now been termed interleukin 14 (IL-14). Produced mainly by T-lymphocytes and some malignant B cells, IL-14 mainly causes the proliferation of normal and malignant B cells. IL-14 lacks the ability to stimulate resting B cells and to induce antibody synthesis or secretion by B cells. Among blood leukocytes and tumors derived from them, only B cell lineages have receptors for IL-14.

From a brief description of interleukins given above, it is clear that many of them exhibit overlapping activities. Besides most, but not all, are produced by the lymphoid cell lineages as was initially thought.

Hematopoietic Growth Factors

Cytokines which stimulate the formation of "colonies" of cells *in vitro* are referred to as colony-stimulating factors (CSFs). When such factors were initially discovered, the term CSF was used to describe factors involved in the growth and differentiation of specific blood cell lineages in semi-solid agar. To understand the necessary role of these factors, one must understand how hematopoietic cell lineages arise from a small population of multipotential stem cells in the bone marrow. Stem cells undergo continual self-renewal, providing a pool of cells that become committed to any of the hematopoietic lineages. While the committed cells are still proliferating, the number of progenitors increase before they eventually

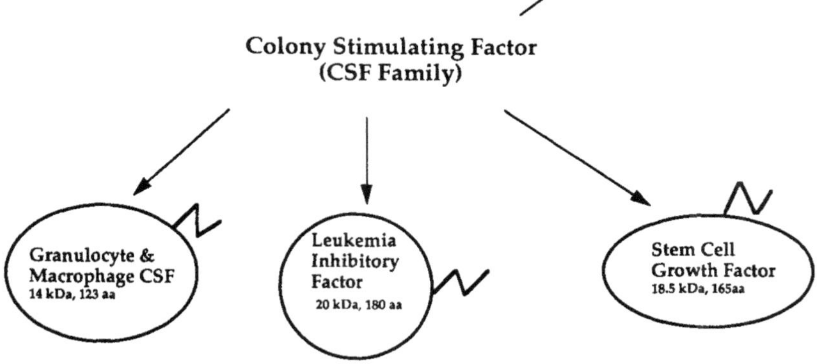

Figure 6. Production of different colony stimulating factors by immune cells.

undergo terminal differentiation to produce the various mature blood cells. Thus, it is during these stages that the CSFs exert their effects *in vivo* and, as described later, have proven to be most successful in the clinic. The CSF family includes granulocyte-macrophage CSF (GM-CSF), macrophage CSF (M-CSF), granulocyte CSF (G-CSF), erythropoietin (EPO), leukocyte-inhibitory factor (LIF), and stem cell factor (SCF). CSFs range between 127–224 amino acids and have a molecular mass of 18–45 kD (Figure 6).

GM-CSF supports the proliferation of bone marrow-derived progenitors of both granulocyte and macrophage lineages. This polypeptide is produced by a number of different cell types including activated T-lymphocytes, fibroblasts, endothelial cells, and monocytes. GM-CSF and IL-3 exert overlapping activities in promoting proliferation of other hematopoietic lineages, such as erythroid and megakaryocyte precursors; however, though such lineages may be sensitive to GM-CSF, IL-3 is probably the more efficient cytokine *in vivo*. In addition, IL-1 and IFN-γ regulate the production of GM-CSF via responses to antigenic stimulation, suggesting a role of GM-CSF in host defense by increasing the phagocytic activation of granulocytes and macrophages. It has been shown that stromal cells produce GM-CSF and this regulates the development of early hematopoietic progenitor cells.

Similar to GM-CSF, G-CSF is also produced by the same cells, but not by T-lymphocytes. This cytokine exerts its effects on both granulocyte precursors and mature cells of this hematopoietic lineage. Its inducers include bacterial LPS, IL-1, and other mitogens. As compared to GM-CSF, G-CSF is highly specific in activating neutrophils for phagocytosis and superoxide production to fight bacterial and other parasitic infections. It also enhances antibody-mediated cytotoxicity against tumor cells and is chemotactic for neutrophil migration towards sites of inflammation. Furthermore, both GM-CSF and G-CSF have the ability to promote the growth of their target cells and to activate them in response to specific stimuli.

Although there is considerable information on M-CSF, its clinical role, unlike GM-CSF and G-CSF, is less clear. It is a protein produced by normal monocytes which also exhibit its receptors and respond to it. M-CSF activates macrophages to produce products such as IL-1, prostaglandin E, and peroxide that serve in the activation of phagocytosis and cell killing.

Both the generation of normal levels of erythrocytes and the adaptation of these levels to changes in the physiological requirements are essential in humans. The cytokine that regulates this process is known as erythropoietin (EPO). It is synthesized primarily by kidney glomerular cells in response to the level of oxygenated erythrocytes. This protein stimulates the proliferation and differentiation of erythroid precursor cells in the bone marrow, and its receptor is found only on the surface of erythroid precursor cells. EPO, like many of the other CSFs, only exert effects on cells that have been committed to a particular lineage. Differentiation of BFU-E colonies to CFU-E is induced by exposure to EPO causing the production of erythroblasts that synthesize hemoglobin. These effects of EPO can be synergistic with other lymphokines such as IL-1 and IL-3. However, even though EPO and other CSFs control these processes, it must be emphasized that these effects *in vivo* possibly require a complex interaction between multiple factors, and that no single factor is capable of eliciting all of the essential effects.

Transforming Growth Factors

TGF-β comprises a family of cytokines involved in cell growth and differentiation. In contrast, TGF-α is a member of the epidermal growth factor (EGF) family. While these two factors share the same name, neither TGF-α nor TGF-β share any structural or functional relationships. Though TGF-a consists of only a single polypeptide, there are three related TGF-βs: TGF-β1, β2, and β3 (Figure 7). TGFβs are disulfide-linked homodimers of approximately 25 kD polypeptides with each subunit having 112 amino acids and are produced by megakaryocytes, macrophages, lymphocytes, and bone cells. TGF-β is a family of related proteins with more than 20 members that control growth, differentiation, and morphogenesis. This family includes activin and inhibin, which act on the endocrine system, and the Müllerian inhibiting substance (MIS) which induces regression of the female genital primordium and the Müllerian duct early in male gonadal tissue development.

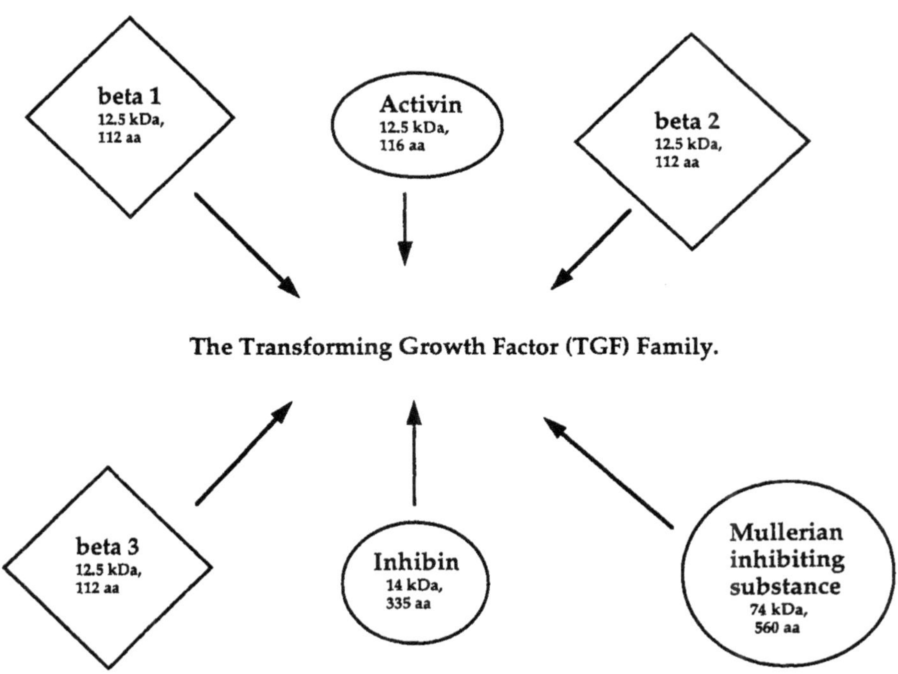

Figure 7. The transforming growth factor (TGF) family.

The actions of TGF-β are numerous and diverse. Though this cytokine was originally thought to be a growth factor, it is apparent now that it has very little proliferative activity on most cells. In fact, it exerts antiproliferative action. TGF-β regulates differentiation of myoblasts, preadipocytes, osteoblasts, chondroblasts, hematopoietic progenitors, and other cell types. TGF-β is also known to increase the expression of cell adhesion molecules (CAMs) resulting in cell-to-cell adhesion. In addition, TGF-β increases the expression of extracellular matrix molecules such as fibronectins, collagens, small secretory proteoglycans, laminin, and others. Along with the extracellular matrix proteins, TGF-β attracts other cell types (such as monocytes) by chemotaxis, which contributes to wound healing and tissue repair. TGF-β antagonizes the effects of growth-promoting factors such as EGF and fibroblast growth factors (FGFs). As such, TGF-β acts as a negative regulator of cell growth in an autocrine manner, thus limiting the uncontrolled proliferation of cells.

Tumor Necrosis Factors

Tumor necrosis factors (TNFs) represent a family of related polypeptides among which are TNF-α, TNF-β (also referred to as lymphotoxin), CD40L, CD30L,

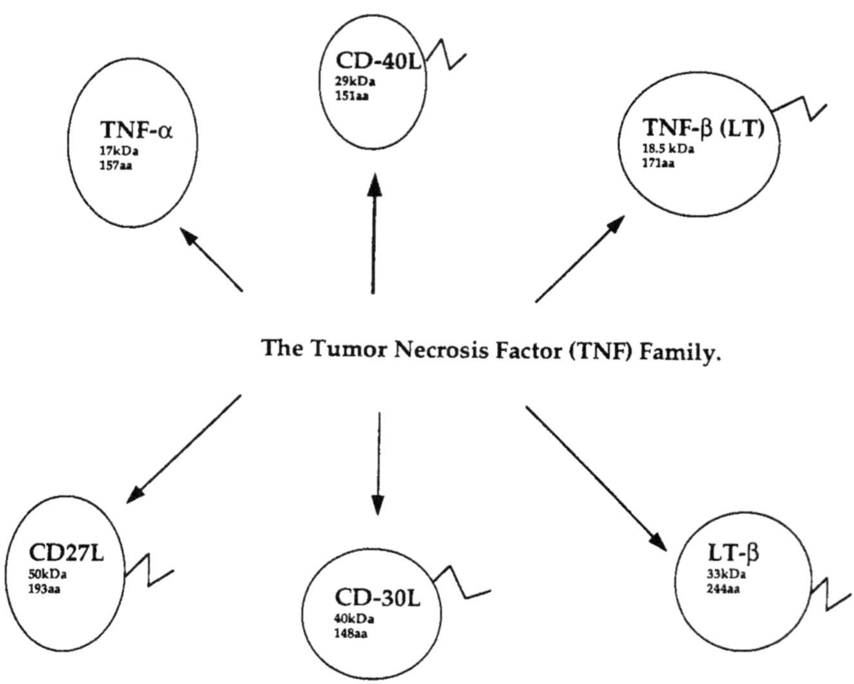

Figure 8. The tumor necrosis factor (TNF) family.

CD27L, and lymphotoxin β (LT β) (Figure 8). Two of these factors, TNF-α and TNF-β, were identified almost 10 years ago, and since then a great deal of information has been gathered on their biological effects. TNF-α, is a 17 kD protein, produced mainly by macrophages in response to a wide variety of agents, in particular LPS, IL-1, and IL-2. Though TNF-α can cause regression of tumor cells, this is just one of its many functions. TNF-α induces cachexia as indicated by tissue wasting, negative nitrogen balance, and loss of body weight. It also promotes proliferation of fibroblasts and is involved in the host defense against infection and inflammation. TNF-α and IL-1 are emerging as the primary mediators of the inflammatory response.

TNF-β, a 20–25 kD polypeptide, also known as lymphotoxin, is produced mainly by T-lymphocytes but also by certain B cells. Even though both TNF-α and -β share the same receptors, several systems have been reported where they may display differing effects. This cytokine is known for its antiproliferative effects against certain tumors and normal hematopoietic cells. Other possible activities include its ability to induce monocyte/macrophage differentiation and induction of other cytokines. TNFs also activate granulocytes and eosinophils leading to an increase in phagocytic activity and other mechanisms of cell killing.

Chemokine Family

Cytokines which possess chemotactic activity are known as chemokines. Members of this family are responsible for mobilizing inflammatory and immune cells to regions of inflammation and immunological reactions. Chemokines range from 8 to 11 kD in molecular mass and are produced by a wide variety of cells (Figure 9). They are induced by exogenous stimuli and irritants and by endogenous mediators such as IL-1, IL-2, TNF-α, IFN-γ, and platelet-derived growth factor (PDGF). The chemokines, unlike most other cytokines, have limited capacity to induce cytokines, exhibit more specialized functions in inflammation and repair, and at present, appear to be less pleiotropic than the first order proinflammatory cytokines. Some of the chemokines can be assigned to a subset based on their location on chromosome 4 and on the fact that the first two of their four cysteines are separated by one amino acid (C-X-C). This chemokine-α family group consists of IL-8, melanoma-derived growth stimulatory activity (MGSA/GRO), platelet factor 4 (PF4), β thromboglobulin (βTG), neutrophil attracting peptide (NAP-2) IP-10, and ENA-78. The chemokine-β subgroup is located on chromosome 17, has no intervening amino acid between the first two cysteines (C-C), and includes macrophage chemotactic and activating factor (MCAF/MCP-1), RANTES, LD78 which is also known as human MIP1α, ACT-2 or huMIP-1β, and I-309. The major difference between the -α and -β chemokines is that they primarily attract different types of cells during an inflammatory response. The chemokine-α subfamily attracts primarily neutrophils, basophiles, and fibroblasts; in contrast, the -β family recruits mostly monocytic and lymphocytic mononuclear cells.

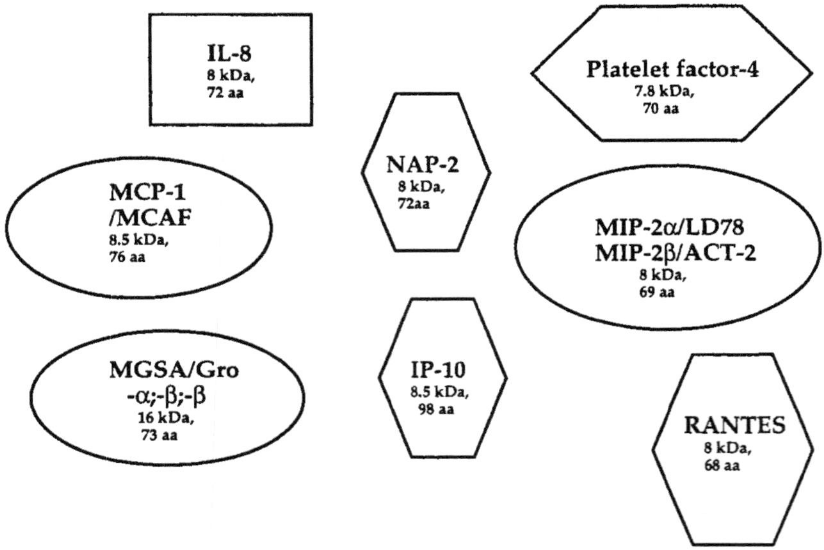

Figure 9. The chemokine family.

CYTOKINE RECEPTORS AND SIGNAL TRANSDUCTION

Cytokines, like hormones, bind specifically to receptors on the cell surface to elicit their responses. In many instances a single cytokine is able to interact with more than one type of cell. The same cytokine may also interact with more than one type of receptor. In the past few years, an enormous amount of information has been accumulated concerning the identification of specific receptors, not only at the protein level but also at the DNA level. Since it is the interaction between the cytokine and its receptor that initiates a signal transduction pathway, it is essential to understand the relationship between the structure of the receptor to its function.

A typical receptor consists of extracellular, transmembrane, and intracellular domains (Figure 10). The extracellular domain contains all the necessary information for ligand binding. The transmembrane region consists of approximately 25 residues, most of which are hydrophobic that allow interaction with the lipid bilayer. The intracellular domain may exhibit enzymatic activity (i.e., protein kinase activity) and also may serve as a contact surface for binding to intracellular molecules. Some receptors have been shown to lack an intracellular domain, but they may interact with other receptors or polypeptide chains to form a functional receptor.

The interaction between a ligand and its receptor is relatively strong with dissociation constants (Kd) approaching the nanomolar and even picomolar range. It is not uncommon that the stoichiometry between a ligand and its receptor is

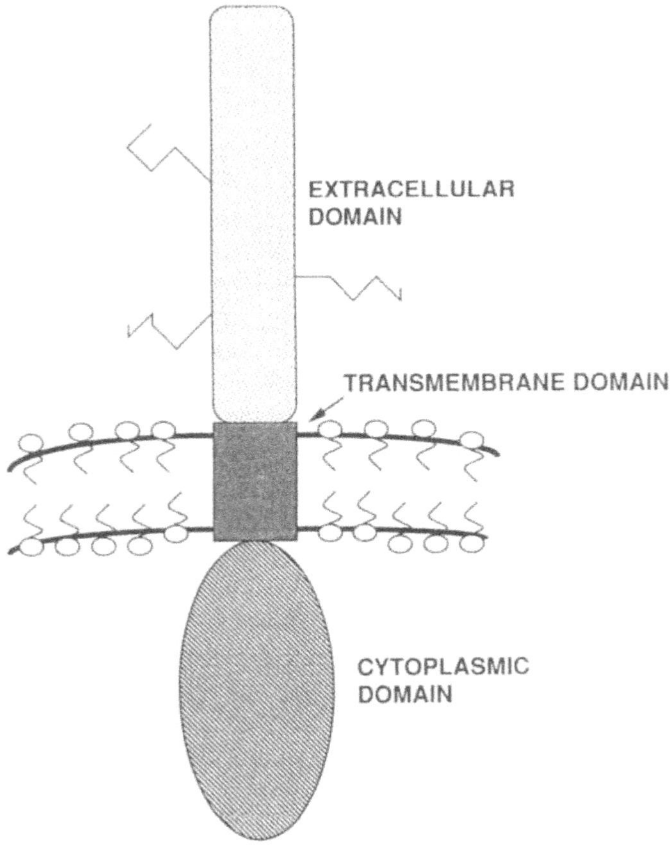

Figure 10. Structural features of a typical cytokine receptor.

greater than 1:1. The strong affinity between a cytokine and its receptor has enabled the identification and subsequent cloning of several cytokine receptors. Most of the cytokine receptors have been identified by chemical crosslinking experiments using radiolabeled cytokines or by protein purification and antibody selection. Scatchard analysis is typically used to determine the number of receptors found on the cell surface and the affinity of the receptor for its cytokine.

After ligand binding, the receptor/ligand complex is internalized, dissociated, and rapidly degraded or the receptor may be recycled back to the cell surface. The rate of synthesis and turnover of the receptor may regulate the activity of some cytokines. The synthesis of their receptor is regulated at both the transcriptional and posttranscriptional levels, but it is less certain whether receptor recycling can also be modulated. In some instances, the endocytosis of the receptor/ligand complex may constitute an essential part of the mechanism by which the cytokine produces its biological response.

Cytokine Receptor Families

Similar to their ligands, cytokine receptors are also grouped into families based upon their structural similarities (Figure 11). It has become apparent that the signal-transducing molecules found in the cytoplasm are shared by a number of different receptors. Hence, these observations have led to the classification of receptors that display structural similarities and share common transducing proteins. These insights have provided evidence to support the signal-transducing mechanisms that mediate the pleiotropic and sometimes redundant cytokine activities.

The cytokine receptors fall into five different families. The hematopoietic cytokine receptor family, which is the largest among all the cytokine receptors, is characterized by the presence of an immunoglobulin-like domain in its extracellular portion. This feature of hematopoietic receptors is also shared with growth factor

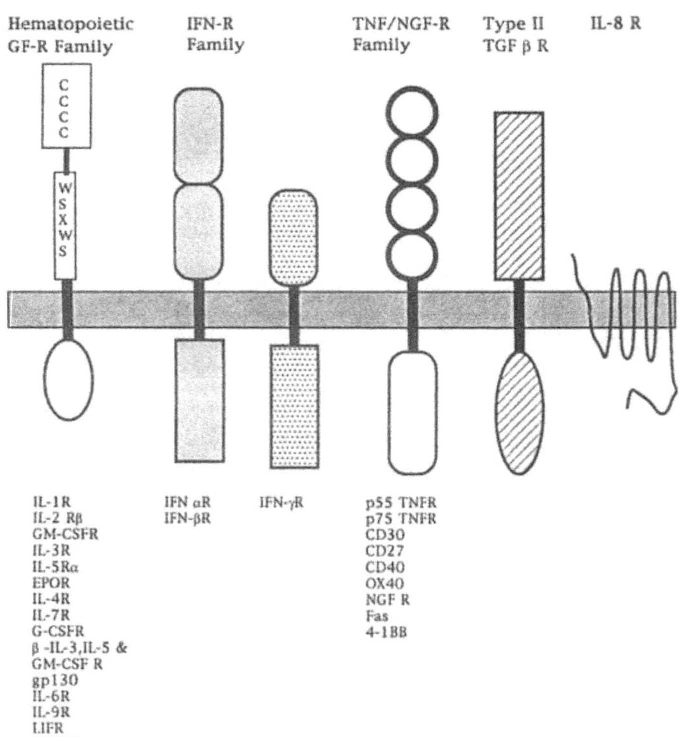

Figure 11. Cytokine receptor families.

receptors. However, unlike the growth factor receptors that display protein kinase activity, members of the hematopoietic receptor family do not have any intrinsic kinase activity. The extracellular domain of these receptors shares about a 200-residue region characterized by four conserved cysteine residues in the amino-terminal half of this region, and a Trp-Ser-Xaa-Trp-Ser (WSXWS) motif at its carboxyl-terminal end. In addition, the ligands for these receptors also share certain structural motifs, such as four α-helices interconnected by small peptide loops. Among all the members of the hematopoietic cytokine receptor family, the intracellular domains of only M-CSF receptor and SCF receptor exhibit tyrosine kinase activity.

The second of the five cytokine receptor families, the interferon receptors, include two types of receptor that are distinct from each other. The extracellular region of IFN-α/β receptor contains two repeat domains of approximately 200 residues each. This region also includes two cysteine residues separated by seven residues and another set of cysteines separated by 20 residues. Interestingly, the extracellular domain of IFN-γ receptor possesses only one of these cysteine regions. Thus, these two receptors are structurally unrelated to the hematopoietic cytokine receptor family.

The third cytokine receptor family, the TNF/nerve growth factor receptors are characterized by cysteine rich regions in their ligand-binding domain. This domain contains four repeats of cysteine-rich regions in which each repeat consists of approximately six cysteines. Members of this family share very little sequence homology in their cytoplasmic domains. Again, like the hematopoietic cytokine receptor family, most of the ligands which interact with this type of receptor share common structural motifs.

TGF-β receptors comprise the fourth family of cytokine receptors. There are three distinct high affinity cell surface receptors for TGF-β. Type I (55 kD) and II receptors (80 kD) bind the ligand and mediate suppression of cell growth and gene activation. Type III (280 kD) is a transmembrane proteoglycan receptor and has a short cytoplasmic domain with no apparent signaling motif. The type II TGF-β receptor, however, is a transmembrane serine/threonine protein kinase whose cytoplasmic domain undergoes autophosphorylation. The TGF-β receptor type I is also a transmembrane serine-threonine kinase with a shorter extracellular domain than TGF-βRII. This evidence is novel in that most growth factor receptor kinases are tyrosine kinases and not serine/threonine kinases. Unlike the TNF/NGF receptor family, members of this receptor family show striking sequence similarities in their cytoplasmic domains but not in their extracellular domains.

The IL-8 receptor belongs to a distinct class of cytokine receptors and it is characterized by the presence of a seven α-helical, hydrophobic membrane-spanning region, the cytoplasmic domain of which is coupled to guanine nucleotide-binding protein (G-protein). G-proteins serve to couple signals from the receptors to enzymes such as adenylylate cyclase and phospholipase C.

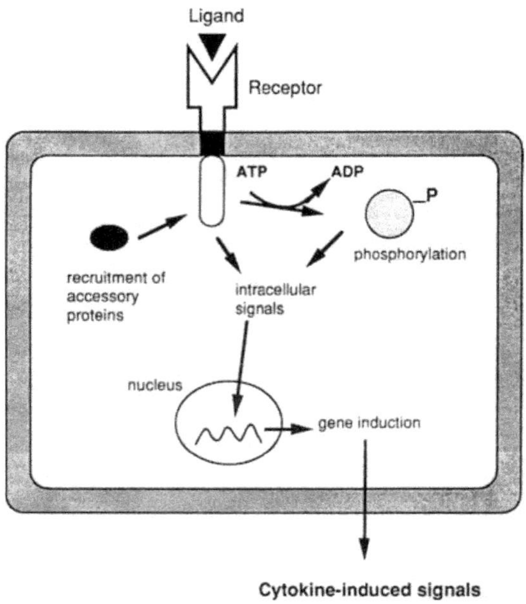

Figure 12. Ligand-induced signal transduction cascade.

Signal Transduction

Unlike the growth factor receptor families, most cytokine receptors contain no intrinsic kinase activity in their cytoplasmic regions. However, after binding their respective ligand, cytokine receptors seem to recruit intracellular molecules (usually protein kinases, G-proteins, and other enzymes) as a means of initiating intracellular signals. The kinase-mediated protein phosphorylation leads to activation of various genes resulting in biological responses (Figure 12). It is noteworthy to mention here that a single cytokine can transduce both negative and positive signals, depending on the type and developmental state of the target cell as well as the presence of other molecules. Conversely, different cytokines can work through different receptors yet evoke the same biological responses, thus suggesting that even though cytokine signaling pathways inside a cell are nonlinear, they can form a framework with multiple cross-talk among different cytokine pathways.

CYTOKINE FUNCTION IN HEALTH AND DISEASE

As the previous sections have indicated, cytokines control a wide range of biological responses on a variety of target cells *in vitro*. Because of the pleiotropic action, *in vivo*, cytokine actions are tightly regulated. Though understanding of the mechanisms underlying the *in vivo* actions of cytokines is still unclear, correlation between the *in vitro* and *in vivo* effects are being investigated. Since cytokines possess many overlapping biological activities, it is probable that one or more cytokines is involved in each of these processes. These include tissue growth and turnover, generation of different blood cell types and wound repair, and possibly many others. Diseases with abnormal growth patterns or inappropriate states of cell differentiation often include disorders of cytokine function, for example, all types of cancer, nonmalignant abnormalities of tissue growth, defects in healing or repair, and many kinds of hematopoietic deficiencies. Thus, individuals would in essence possess a consort of these pleiotropic molecules, thus giving rise to the cytokine network.

Role of Cytokines in Hematopoiesis

Perhaps the best understood function of cytokines is their ability to control hematopoiesis, i.e., the growth and differentiation of blood cells (Figure 13). For instance, the multipotential stem cells, from which all blood cells are derived, must be maintained in their number and in their commitment to differentiate along the possible cell lineages. This process is regulated by several cytokines which include

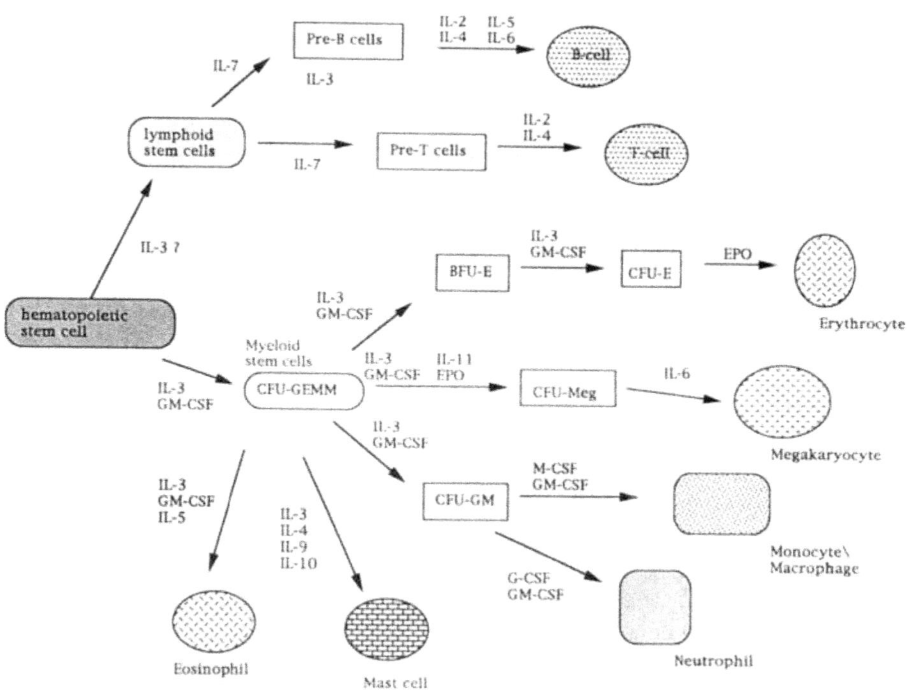

Figure 13. Cytokines involved in hematopoiesis.

CSFs, almost 12 different ILs, LIF, SCF, and EPO. Some cytokines such as G-CSF and M-CSF display very tight cell specificity, whereas others like GM-CSF have been shown to act upon both granulocytes and macrophages. Similarly, EPO has been found to be a specific growth factor for the formation of erythrocytes.

In addition, IL-1, IL-4, IL-5, and IL-6 play various roles in the differentiation pathways of hematopoietic cells. For instance, IL-3 participates early on in the differentiation of CFU-GEMM cells (colony-forming units for granulocytes, erythrocytes, monocytes, and megakaryocytes) into downstream lineages of erythroid and myeloid cells; it also promotes growth of later single-lineage progenitor cells. Conversely, IL-1 exhibits enhancing effects in synergy with IL-3 and M-CSF. The mechanism by which IL-1 synergizes with IL-3 and M-CSF to promote differentiation of multipotential stem cells may be through induction of receptors for these cytokines. IL-1 also induces other cytokines such as GM-CSF and M-CSF, thus affecting cell growth differentiation indirectly.

Several cytokines have been described which upregulate hematopoiesis, but there are very many molecules which downmodulate this process. These include TGF-β, macrophage inflammatory protein-1a (MIP-1α), TNF-α and IFNs. TGF-β is inhibitory upon marrow precursor cells in the initial stages of the stem cell lineage. MIP-1α also decreases proliferation of stem cells in the bone marrow. Additionally, TNF-α acts as a negative regulator of proliferation of myeloid precursor cells whereas interferon inhibits the proliferation of CFUs of all cell lineages.

Hematopoiesis in the bone marrow is controlled by the cytokines produced by the heterogenous stromal cells in a paracrine manner. Because the progenitor cells are closely associated with the stromal cells, these cytokines need not be transported through the circulation. In some cases when the cytokine is membrane bound (i.e., IL-1α), a direct cell-to-cell contact is necessary to illicit the response. Under situations when a rapid response is required, cytokines are recruited from outside the marrow. For instance, under conditions of acute infection, antigenic stimulation, or inflammation, the immune system assists in producing CSFs, IL-3, and IL-5; and these cytokines increase hematopoietic activities.

Role of Cytokines in Modulation of the Immune System

Immunological responses are initiated when a host encounters foreign antigens by the actions of B- and T-lymphocytes of the immune system. Before antigens are presented to B cells, they are first processed by the T-lymphocytes and reexpressed on its cell surface. In conjunction with T cell antigens, the class I and II MHC molecules are then recognized by cytotoxic T cells and helper T cells, respectively. It is the interaction between the antigen-presenting cells (macrophages) and antigen-recognizing cells (T cells) that leads to the increased production of cytokines. Almost all the cytokines are involved directly or indirectly in growth, differentiation, or activation of various immune cells.

Although much is known of cytokines and their effects upon B- and T-lymphocytes *in vitro*, the understanding of this complex process *in vivo* is much less clear. Though most of the cytokines are able to stimulate the immune system, some cytokines such as TGF-β, IL-4, IL-6, and IL-10 cause its suppression. These positive and negative actions depend upon the cytokine, cell type, and stage of differentiation of the cell. As noted above, the activities of cytokines noted *in vitro* may only suggest their role in regulating the immune system *in vivo*. In addition to pleiotropic effects of these molecules, sometimes synergistic, antagonistic, and overlapping actions of these cytokines make it difficult to predict their true role *in vivo*.

Role of Cytokines in Inflammation

Inflammation results in response to different stimuli (such as tissue damage and/or invasion of pathogens) and it involves a complex interaction between different cells and their mediators. Several cytokines have been directly implicated in this pathway, in particular TNF, IL-1, IL-6, IL-8, and monocyte chemoattractant protein-1 (MCP-1). Most of these cytokines contribute to the recruitment of various cells (i.e., neutrophils) to migrate towards the site of inflammation. IL-8 specifically recruits neutrophils, whereas MCP-1 recruits only monocytes. GM-CSF may also act in the inflammation process by activating specific granulocyte effector functions, such as antibody-dependent cytotoxicity, phagocytosis, chemotaxis, and oxidative metabolism. Furthermore, inflammation not only causes the enhancement of cytokines and specific cells, but also the release of other mediators like prostaglandin, histamine, and leukotrienes from mast cells.

Role of Cytokines in Wound Healing

Wound healing involves cellular migration and inflammation for several days, proliferation of fibroblasts with new collagen synthesis for about three weeks, and last, the remodeling of the scar for one month to one year. Several cytokines, like IL-1, TNF, and TGF-β, have been shown to play an important role in different phases of wound healing. Even several growth factors such as TGF-α, PDGF, EGF, and FGFs may also be involved. For example, TNF causes fibroblasts to grow and to release collagen for building up strength at the site of tissue damage. Though TGF-β can antagonize these effects, it is also capable of promoting biochemical events for the laying down of connective tissue. Thus, the acceleration of the process of wound healing by growth factors suggests the clinical use of cytokines.

Role of Cytokines in Bacterial, Parasitic, and Viral Infections

Several cytokines have been shown to exhibit activities against different kinds of infection. This includes IFNs, TNF-α, TNF-β, IL-1, IL-6, and IL-12. These cytokines combat the invading organisms through both immune and nonimmune systems. For instance, some cytokines enhance antibody production from plasma cells which inactivate viral particles while others activate granulocytes and macrophages, leading to increase in phagocytosis of microorganisms and also release of agents that destroy virus-infected cells. Moreover, cytotoxic T-lymphocytes kill infected cells by recognizing foreign antigens on their cell surfaces. IFNs directly induce a state of resistance to viral replication in various target cells. This process is known to require induction of gene expression. More recently, IL-12 was identified as one of the major players in the signaling pathway of microbial pathogenesis. Since the temporal viral resistance is a consequence of rapid increase of IFNs, the long-term defense is mediated by the induction of MHC-dependent, cell-mediated, and humoral immunity along with the killing and phagocytosis of virally or bacterially infected cells.

Role of Cytokines in Cancer

Many cancers are the result of deregulated cell growth and differentiation. This may be due to disorder in the signaling pathways of cytokines. In many instances, growth factor receptors have been altered in such a way that they continue to signal even in the absence of the cytokine. It is even possible that proteins farther downstream of the receptor-ligand complex are disrupted so as to cause uncontrolled proliferation. Overexpression of genes that are normally under the strict control of cytokines could cause a state of disorder. Some of these genes—now referred to as proto-oncogenes—are normal genes that have undergone certain alterations in transformed cells. Interestingly, it has become apparent that some of the proto-oncogenes are components of cytokine signaling pathways.

Therapeutic Uses for Cytokines

It is widely recognized that cytokines play a diverse role in many biological processes. Their essential role in the regulation of the disease demonstrates their potential use in cytokine therapy. Some diseases are caused by a disruption of the normal cytokine activity; thus, cytokines could be used clinically to intervene against the progress of disease or even to alleviate the symptoms or side effects of other therapies. Since the production of pure, recombinant cytokines is no longer a problem, opportunities for testing many of these cytokines in the clinic have increased and produced many valuable results. However, most cytokines have multiple actions *in vitro*, and thus it is not always possible to predict their outcomes *in vivo*.

Some obvious criteria to evaluate before allowing the use of cytokines for therapy are dosage, side effects, and efficacy of the cytokine in question. It is known that administration of cytokines in high doses is not always beneficial and may produce unpleasant or dangerous side effects. Thus, preclinical testing has served as a guide in selecting the appropriate cytokine and patient for evaluating the potential uses of cytokines as therapeutic agents. Cytokines which have been tested in patients thus far include IFN-α, IFN-β, IFN-γ, IL-1, IL-2, IL-3, TNF-α, G-CSF, GM-CSF,

M-CSF, and EPO. In addition, IL-4, IL-5, IL-6, and IL-7 are in the early phases of preclinical testing.

The first cytokine to be used in the treatment of malignancies was IFN-α. Initial studies were carried out with a crude preparation of IFN, but later a more pure protein obtained from recombinant technology was administered to patients with cancers such as leukemia; melanomas; and renal cell, breast, and colorectal carcinomas. While not all of these malignancies were cured, some results were found highly encouraging. For instance, the best results were noted in patients with chronic myeloid leukemia and hairy-cell leukemia in whom remission rates were as high as 90%. Thus, IFN-α is now considered an effective therapy for this leukemia. The precise mechanism of action of IFN-α in this disease, however, remains unclear. The antiviral properties of IFN-α have also been evaluated in patients infected with viruses such as hepatitis B. In such studies, 48% of patients responded positively. In addition, IFN-β has been used successfully in the treatment of multiple sclerosis.

The IFN-γ has also been used clinically, but with mixed results. It has been used less successfully in patients with solid carcinomas since response rate is less than 10%. However, this cytokine has become the treatment of choice for a rare disorder of neutrophil cell function, childhood chronic granulomatous disease, in which the defective production of reactive oxygen intermediates predisposes patients to recurrent and severe pyrogenic infections.

While IL-1 at low dosages has a wide variety of activities that operate by inducing other cytokines like IL-6, TNF-α, and CSFs, it seems to be toxic at high dosages causing hypotension, increased systemic vascular resistance, depressed myocardial function, and pulmonary congestion. Potential therapeutic uses of IL-1 include radioprotection, enhancement of hematopoiesis, and stimulation of T and NK cells. Due to the role of IL-2 in activating T and NK cell functions, it has found its use in certain types of cancers, immunodeficiency, and chronic infections. The effect of IL-2 on peripheral lymphocytes is to cause the production of certain cytotoxic cells, such as LAK cells, which are involved in tumor regression. Clinically, the patients with melanomas and renal carcinomas treated with a combination of both IL-2 and LAK cells have shown tumor regression. Furthermore, the clinical efficacy of IL-2 has been hindered by many side effects including fever, malaise, nausea, vomiting, and chills. Recently, IL-2 has been approved for treatment of renal cell carcinoma.

In vitro, recombinant TNF-α was shown to exhibit antiproliferative effects against a wide variety of tumor cell lines, in addition to synergism with IFNs. It was also found to be a potent modulator of the immune system and an inhibitor of angiogenesis. These observations prompted many clinical trials for developing TNF-α as an antitumor agent. Both single- and multiple-dose phase I trials have confirmed that recombinant TNF, like other cytokines, causes severe side effects including hypotension, fever, nausea, vomiting, and other symptoms inherent with endotoxin shock. Some of these side effects have been circumvented by localized administration of the cytokine to the cancer patients which has resulted in impressive antitumor effects.

EPO serves as a regulator for the generation of normal levels of red blood cells. Clinical studies have already shown it to be advantageous in patients who are EPO deficient such as those with renal disease and those with rheumatoid arthritis. Treatment of patients with recombinant EPO reduces the severity of red cell anemia and diminishes dependence on blood transfusions. In addition, hematopoietic differentiators, like G-CSF, GM-CSF, and IL-3, show promise in accelerating the recovery of damaged bone marrow during bone marrow transplantation or chemotherapy. Among other cytokines, G-CSF, GM-CSF, and IL-3 have thus far shown very little of the toxicity and thus have also become widely used for cancer treatment.

FUTURE PROSPECTS FOR CYTOKINE THERAPY

Recombinant technology has had a great influence thus far in making available cytokines as clinical therapeutics. Programs to produce recombinant cytokines, soluble cytokine receptors as "anticytokines," as well as growth factor antagonists are well underway. Some possible developments for the use of cytokines and, related molecules for clinical medicine are summarized below. Mutant and chimeric cytokines with specific properties (i.e., competition with endogenous cytokines, presentation as toxic or radioactive conjugates, linkage to cell-specific antibodies) are being developed such that specific target cells bind the engineered cytokine. With the help of structural biochemists, the three-dimensional structure of several cytokines and their receptors have been elucidated and this has resulted in the design of specific cytokine antagonists. Soluble ligand-binding domains of the receptors appear to act as potent and specific inhibitors of cytokines. Specific inhibitors are also being designed which interfere with the cytokine signaling pathways. These inhibitors are being linked to specific cytokines to insure delivery to the appropriate cell type. Furthermore, the known conventional chemotherapeutic regimens are being combined with cytokines for combating many diseases.

SUMMARY

Human cytokines represent a new class of polypeptide hormones which are produced by and targeted to not only the cells of the immune system but many others. Cytokines are key players involved in the modulation of the immune system, the maintenance of homeostasis, and hematopoiesis. Most cytokines possess many overlapping activities as evident by their *in vitro* activities as well as their *in vivo* activities. In the past few years, the identification of new factors, the characterization of their activities *in vitro*, and the identification of their receptors has increased rapidly. The detection of cytokines in different disease states may hold promise in diagnosis. The role of these molecules in the therapy of the disease is already evident. The utilization of human cytokines and their receptors in rational drug design has widened the scope of the current biotechnology industry.

ACKNOWLEDGMENTS

This research was conducted, in part, by The Clayton Foundation for Research and was supported, in part, by new program development funds from The University of Texas M.D. Anderson Cancer Center.

REFERENCES

Aggarwal, B.B. & Gutterman, J.U. (eds.) (1996). Human Cytokines: A Handbook for Basic and Clinical Researchers, Volume II. Blackwell Science, Boston.

Aggarwal, B.B. & Gutterman, J.U. (eds.) (1992). Human Cytokines: A Handbook for Basic and Clinical Researchers. Blackwell Scientific Publications, New York.

Aggarwal, B.B. & Pocsik, E. (1992). Cytokines: From clone to clinic. Arch. Biochem. Biophysics 292, 335–359.

Aggarwal, B.B. & Puri, R.K. (eds.) (1995). Human Cytokines: Their Role in Disease and Therapy. Blackwell Science, Boston.

Cohen, S. & Bigazzi, P.E. (1980). Lymphokines, cytokines, and interferons. Interferon 2, 81–95.

Clemens, M.J., Read, A.P. & Brown, T. (eds.) (1991). Cytokines. Bios Scientific Publishers, Oxford.

Dumonde, D.C., Wolstencroft, R.A., Panayi, G.S., Matthew, M., Morley, J., & Howson, W.T. (1969). Lymphokines: Non-antibody mediators of cellular immunity generated by lymphocyte activation. Nature 224, 38–42.

Elsasser-Beile, U. & von Kleist, S. (1993). Cytokines as therapeutic and diagnostic agents. Tumor Biology 14, 69–94.

Galvani, D.W. & Cawley, J.C. (eds.) (1992). Cytokine Therapy. Cambridge University Press, New York.

Miyajima, A., Kitamura, T., Harada, N., Yokota, T., & Arai, K. (1992). Cytokine receptors and signal transduction. Ann. Rev. Immunology 10, 295–331.

Chapter 10

Immunity to Cancer:
CYTOTOXIC LYMPHOCYTES, INTERLEUKIN-2, AND THE TUMOR NECROSIS FACTOR SUPERFAMILY

MICHAEL J. ROBERTSON and JEROME RITZ

Advances in Oncobiology
Volume 1, pages 207–219.

ISBN: 0-7623-0146-5

INTRODUCTION: THE IMMUNE SYSTEM

The immune system defends multicellular organisms from infection by microbial pathogens. The cardinal feature of the immune system is its ability to discriminate self from other. The immune system of an individual does not normally react against the tissues of that individual, but will mount a vigorous response to eliminate foreign substances. Substances capable of being recognized by the immune system are called antigens. Vertebrate immune responses depend on the cooperation of several different types of cells, all of which are probably derived from a common hematopoietic stem cell in the bone marrow. The predominant mediators of the immune response are leukocytes (white blood cells). Based on their morphology, leukocytes are classified as granulocytes, monocytes, or lymphocytes. Lymphocytes can be further subdivided into T cells, B cells, and natural killer (NK) cells based on their function and their expression of certain cell surface antigens (Figure 1). The surface antigens of human leukocytes have been extensively studied using monoclonal antibodies and are described using the "cluster of differentiation" (CD) nomenclature. More than 100 CD antigens have been defined in five sequential International Workshops on Human Leukocyte Differentiation Antigens (Table 1).

T cells develop in the thymus from immature bone marrow-derived progenitor cells. Mature T cells constitute ~65 to 75% of human peripheral blood lymphocytes, and almost all express CD2, CD3, CD5, CD6, and CD7. Each T cell expresses a unique T cell receptor (TCR) on its surface that determines its specificity for foreign antigen. Binding of the TCR to foreign antigen stimulates T cells to proliferate and to exhibit specialized immune functions. T cell receptors, however, do not interact with intact foreign molecules, but rather with small peptides that are bound to major histocompatibility (MHC) antigens on the surface of other cells. There are two major kinds of MHC antigens: class I MHC antigens are expressed by almost all cell types, whereas class II MHC antigens are expressed by specialized antigen-

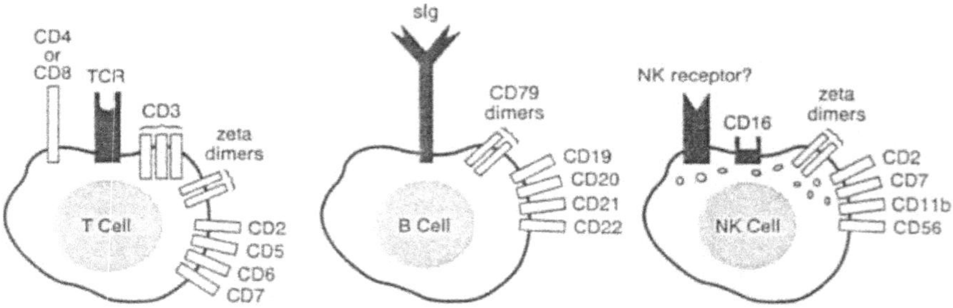

Figure 1. The three major subsets of lymphocytes. T cells express antigen-specific TCR complexes and B cells express antigen-specific B cell receptor (BCR) complexes. NK cells are recognized by the absence of TCR and BCR as well as the presence of characteristic surface antigens, such as CD11b, CD16, and CD56. Other antigens typically expressed by each of the lymphocyte subsets are also shown.

Table 1. Characteristics of Selected Human Leukocyte Antigens*

Cluster of Differen-tiation	*Other Names for Antigen*	*Molecular Weight kD*	*Main Cellular Expression*	*Additional Comments*
CD2	T11, SRBC R	50	T, NK	Adhesion molecule
CD3	T3 complex (γ, δ, ε)	26 (γ,) 20 (δ, ε)	T	Part of TCR complex
CD4	T4, class II MHC R	59	T subset, Mono	Receptor for HIV
CD5	Tp 67	67	T, B subset	
CD6	T12	100	T	
CD7		40	T, NK	
CD8	T8, class I MHC R	32	T subset, NK subset	α and β subunits
CD11a	LFA-1 α subunit	180	All leukocytes	Adhesion molecule
CD11b	Mo1, Mac1, CR3	165	NK, Mono, Gran	Complement receptor
CD16	Fc γ R III	50–70	NK, Mono, Gran	Receptor for ADCC
CD18	β_2 integrin	95	All leukocytes	Common β subunit for CD11a, CD11b, CD11c
CD19	B4	90–95	B	
CD20	Bp 35, B1	35–37	B	
CD21	CR2, B2	140	B,DC	Receptor for EBV
CD22	Bgp 135, B3	135	B	
CD27		55 (110 dimer)	T, B subset, NK subset	Member of TNF R superfamily
CD28	Tp 44	44 (90 dimer)	T	Costimulatory receptor for T cells
CD30	Ki-1	105–120	Act lymphocytes	Member of TNF R superfamily
CD40	gp 50	50	B cells, DC	Member of TNF R superfamily
CD54	ICAM-1	85–90	Mono, Act lymphocytes	Adhesion molecule
CD56	NKH-1	140–220	NK, Neural cells	Isoform of NCAM
CD58	LFA-3	55–70	Many cell types	Adhesion molecule
CD70	CD27 ligand	Uncertain	Act T, Act B	Homology to TNF
CD79a	mb-1, Ig α	33	B cells	Part of BCR complex
CD79b	B29, Ig β	40	B cells	Part of BCR complex
CD80	B7, BB1	60	Act B, Act mono	Ligand for CD28
CD95	APO-1, Fas	40–50	Many cell types	Member of TNF R superfamily

Notes: *Selected from more than 150 clusters and subclusters defined in five International Workshops on Human Leukocyte Differentiation Antigens. Abbreviations: T, T cell; B, B cell; NK, NK cell; Mono, monocyte/macrophage; Gran, granulocyte; Eo, eosinophils; DC, dendritic cell; Act, activated cell; R, receptor; SRBC, sheep red blood cell; TCR, T cell receptor; HIV, human immunodeficiency virus; CR, complement receptor; EBV, Epstein-Barr virus; NCAM, neural cell adhesion molecule; BCR, B cell receptor complex.

presenting cells. Antigen-presenting cells take up foreign proteins, digest them into small peptides, bind the small peptides to class II MHC antigens, and express foreign antigen/class II MHC complexes on their surface. Antigen-presenting cells also express costimulatory ligands that are required for full activation of T cells. One such ligand, CD80 (B7/BB1), binds to the CD28 protein on helper T cells. In the absence of such CD28 binding, engagement of the TCR causes helper T cells become inactive (anergic) rather than activated. This appears to be one of the ways that the immune system avoids reactivity against self antigens. There are two major subsets of mature T cells, CD4 T cells and CD8 T cells. CD4 T cells (also called helper T cells) express CD4 but not CD8, recognize foreign peptides that are bound to class II MHC antigens, and assist other lymphocytes in generating effective immune responses. CD8 T cells express CD8 but not CD4, recognize foreign peptides bound to class I MHC antigens, and kill infected cells or regulate the immune response.

B cells develop in the bone marrow and then migrate predominantly to lymph nodes and other secondary lymphoid tissues. B cells represent ~5 to 10% of blood lymphocytes and express CD19, CD20, CD21, and class II MHC antigens. Each B cell expresses a unique immunoglobulin (antibody) on its surface that determines its specificity for foreign antigen. Unlike T cell receptors, B cell surface immunoglobulins can bind to intact antigens, including proteins, carbohydrates, and lipids. Binding of surface immunoglobulin to foreign antigen stimulates B cells to proliferate and differentiate into plasma cells, although the participation of CD4 helper T cells is generally required. Plasma cells secrete antibodies that bind to and inactivate foreign antigens.

NK cells constitute ~10 to 15% of peripheral blood lymphocytes. NK cells are derived from bone marrow progenitor cells, but the ontogeny of NK cells is less well understood than that of T and B-lymphocytes. NK cells do not rearrange TCR or immunoglobulin genes and do not express TCR or BCR complexes on their surface. Like T cells, most NK cells express CD2 and CD7 and a subset expresses CD8. However, NK cells do not express other typical T cell antigens (including CD3, CD4, CD5, and CD6) or B cell antigens (including CD19, CD20, and CD21). Moreover, NK cells express the CD16 and CD56 antigens, which are absent from B cells and the vast majority of T cells.

The growth and differentiation of activated lymphocytes is regulated by small soluble proteins called cytokines. Some of these factors have also been called lymphokines (factors produced by lymphocytes), monokines (factors produced by monocytes), or interleukins (factors mediating communication between leukocytes). Like endocrine hormones, cytokines permit communication and cooperation between different cells. Unlike hormones, however, most cytokines have very short half-lives and are not usually detectable in the blood. Rather, cytokines tend to act locally at sites of leukocyte differentiation (e.g., bone marrow, thymus, and lymph node) and at sites of infection or inflammation.

CYTOTOXIC LYMPHOCYTES AND TUMOR IMMUNITY

Cytotoxic lymphocytes (or killer cells) are defined by their ability to destroy other cells, including those infected with microorganisms or that have undergone malignant transformation. Killer cells often possess abundant cytoplasm containing prominent granules and have therefore been called large granular lymphocytes or LGL. The two major types of cytotoxic lymphocytes are cytotoxic T lymphocytes (CTLs) and NK cells. CTLs are usually CD8 T cells, although under some circumstances CD4 T cells behave as killer cells. Fully differentiated, active CTLs are not present in an animal that has never previously encountered the foreign antigen that the CTL recognizes. Instead, CTLs are generated from CTL precursors, which are small, agranular, quiescent T lymphocytes. These resting CD8 T cells are stimulated by target cells that express foreign antigen complexed with class I MHC antigens. With the assistance of helper T cells, activated CD8 T cells differentiate into fully functional CTLs that possess cytotoxic granules and can kill target cells. The cytotoxic activity of CTLs is triggered by the same signals that initiated their differentiation from CTL precursors (i.e., interaction of their TCRs with foreign antigens bound to class I MHC on the surface of the target cell) (Figure 2A). Most CTLs are therefore both antigen-specific and MHC-restricted; target cells that express either the "wrong" foreign antigen or the "wrong" class I MHC antigen (or no MHC antigen) will not activate the CTL and will not be killed.

Antigen-specific CTLs can reject transplanted tissues, including large organs such as the kidney and liver. Although graft rejection is a serious and undesirable complication of clinical transplantation, this phenomenon demonstrates that the human immune system can destroy large masses of human cells if the cells are perceived as foreign. Most human cancer cells probably contain mutant genes that encode aberrant proteins, so malignant tumors should theoretically express antigens that could be recognized as foreign by the immune system. Such immunogenic tumors might then be eradicated by an effective immune response. Indeed, specific immune elimination of tumors has been demonstrated in animal models. For example, malignant murine tumors that have been induced by chemicals or ultraviolet light can be rejected by antigen-specific, MHC-restricted murine CTLs. Although antibodies to tumors have also been detected in some animals, it appears that cell-mediated immune responses are predominantly responsible for tumor rejection.

Despite these theoretical considerations and the animal model data, it has proved difficult to demonstrate tumor-specific antigens in most human cancers. Potential reasons that human tumors do not appear to be immunogenic include: (a) most human cancers do not express tumor-specific antigens; (b) tumor-specific antigens exist but are not presented in a fashion that evokes an effective immune response; and (c) the malignant tumor produces immunosuppressive factors that prevent an effective immune response. There is experimental evidence to support the latter two hypotheses. Class I MHC antigens are expressed on some cancers at very low or

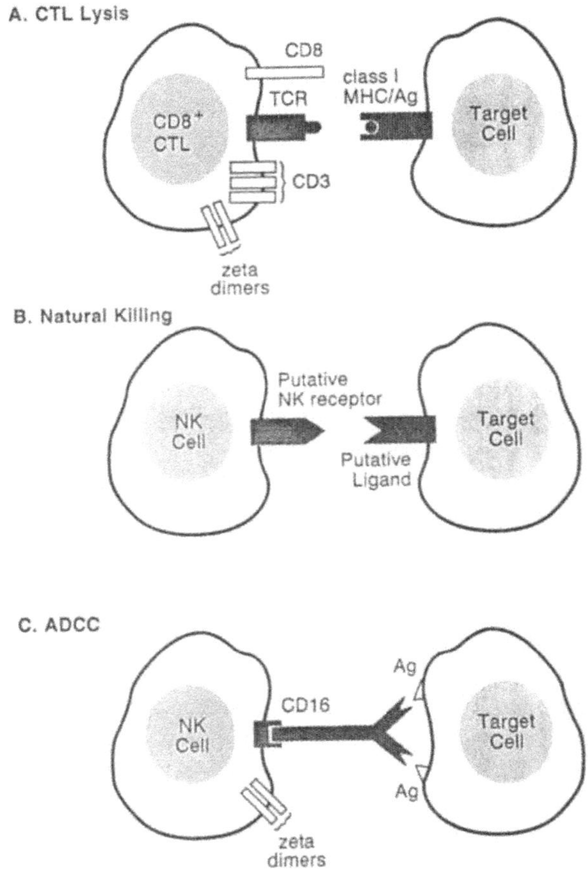

Figure 2. Receptors on killer cells that trigger cytotoxicity. (**A**) $CD8^+$ CTL express TCR complexes, consisting of variable TCR heterodimers, invariant CD3 ($\gamma\ \delta\ \varepsilon$) complexes, and dimers of ζ family proteins. Engagement of the TCR complex by antigenic peptide bound to class I MHC molecules on the target cell surface activates the CTL. (**B**) NK cells express putative receptor(s) for natural killing; these receptors and their putative ligands on NK-sensitive target cells have not been fully characterized. (**C**) NK cells express an antibody Fc receptor complex consisting of CD16 and dimers of ζ family proteins. Engagement of this complex by the invariant portion of IgG that is bound to specific antigen on target cells triggers ADCC.

undetectable levels so that mutant peptides from these tumors would not be effectively presented to $CD8^+$ CTLs. Moreover, many malignant tumors do not express costimulatory ligands, such as CD80 (B7/BB1), that are required for the full activation of helper T cells. Some tumors have been shown to produce cytokines, such as TGF-β, that can suppress immune responses.

Malignant cells tend to be genetically unstable and to undergo frequent mutations. Thus, a tumor mass may be composed of malignant cells exhibiting considerable antigenic heterogeneity. Even if some of these cells express tumor-specific antigens that are presented appropriately and evoke an immune response, a subpopulation of cancer cells lacking such antigens could survive and grow. Thus, some human tumors could appear "nonimmunogenic" when they are clinically detected precisely because their immunogenic subclones have already been eliminated by effective immune responses. In those relatively rare instances in which immune responses to human tumors have been successfully demonstrated, the candidate "tumor antigens" have been found to be expressed by some normal tissues as well as cancer cells. Thus, the immune responses to these antigens represent a type of autoimmunity rather than true tumor-specific immunity.

Because of the difficulty in developing tumor-specific CTLs for cancer therapy, some investigators have attempted to exploit the antitumor activity of NK cells. NK cells differ from CTLs in several ways. NK cells have cytotoxic granules and can spontaneously kill certain tumor cells and virus-infected cells in the absence of known prior stimulation. Moreover, NK cells are not MHC-restricted because they can lyse target cells that express no MHC antigens or MHC antigens different than those of the host. NK cells do not appear to have exquisite antigen specificity. Whereas a CTL specific for a given tumor will not generally lyse cancer cells from a different tumor, individual clones of NK cells can lyse several different types of malignant cells. The structures expressed by target cells that trigger natural killing have not been well-characterized, nor have the receptors on the NK cells that interact with putative target structures.

Natural killing or NK activity refers to the spontaneous, MHC-unrestricted lysis of target cells (Figure 2B). NK cells can also kill target cells by a different process known as antibody-dependent cell-mediated cytotoxicity (ADCC). In ADCC, antibody bound to specific antigen on the surface of a target cell triggers an antibody receptor on the NK cell, causing the NK cell to lyse the target cell (Figure 2C). The receptor for ADCC on NK cells is a complex consisting of the CD16 (Fcγ RIII) antigen associated with putative signal transducing molecules of the ζ family. The CD16 molecule binds to the invariant (Fc) portion of IgG, whereas the target antigen binds to the variable (Fab) portion of the antibody. Granulocytes and macrophages also possess receptors for immunoglobulin and can mediate ADCC.

A small subset of CTLs exhibits MHC-unrestricted, NK-like killing. These CTLs represent less than 5% of peripheral blood T cells, and like most CTLs they express CD3, CD5, and CD8. However, they also express the CD56 antigen, which is absent from other CTLs but is expressed by almost all NK cells. These unusual lymphocytes are known as non-MHC-restricted CTLs or $CD56^+$ CTLs; their function in immune responses is presently unclear.

MECHANISMS OF CELL-MEDIATED CYTOTOXICITY

Regardless of the type of cytotoxic lymphocyte involved, the cytotoxic process can be divided into several stages (Figure 3). First, the cytotoxic lymphocyte binds weakly to a potential target cell. This cell contact is mediated by surface adhesion molecules, such as LFA-1 (CD11a/CD18) and CD2 on the killer cell and ICAM-1 (CD54) and LFA-3 (CD58) on the target cell. If the target cell does not express surface structures recognized by the lymphocyte, this binding is weak and transient and the lymphocyte will move on to make contact with another potential target cell.

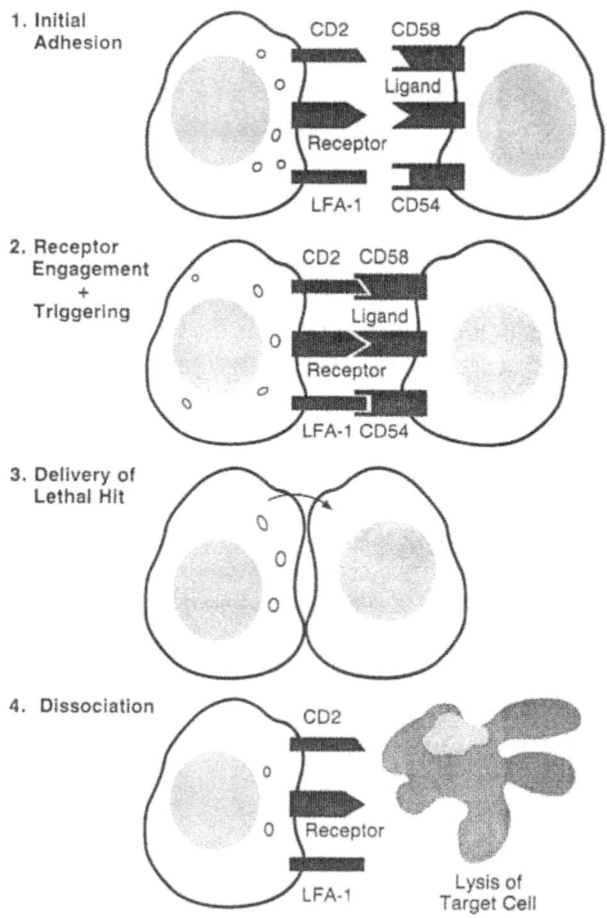

Figure 3. Cell-mediated cytotoxicity. Although the receptors that trigger the cytolytic activity of CTL and NK cells differ, both types of lymphocytes appear to use similar mechanisms to bind and kill target cells. See text for detailed explanation.

If, however, the target cell does express surface structures recognized by the cytotoxic lymphocyte the latter will be activated or triggered. The activated lymphocyte binds the target cell firmly and delivers a "lethal hit" to the target cell. The cytotoxic lymphocyte, can subsequently dissociate from the damaged target cell and mount a fresh attack against another target cell.

Part of the lethal hit consists of the cytotoxic granules discharged from the killer cell. Cytotoxic granules are directionally released by a calcium-dependent process into a small area formed by intimate contact between the killer cell and target cell membranes. Cytotoxic granules contain several proteins, including perforins (also known as cytolysins or pore-forming proteins), granule enzymes (granzymes or serine esterases), and chondroitin sulfate proteoglycans. Perforin monomers bind to the target cell membrane and polymerize to form cylindrical transmembrane pores. Polyperforin channels disrupt normal transmembrane ion gradients and can cause swelling and osmotic lysis of the target cell. Furthermore, perforin channels may allow entry of other cytotoxic components, such as the granzymes, into the target cell.

Besides inducing perforin-mediated lysis, cytotoxic lymphocytes can cause some target cells to commit suicide in a process known as apoptosis or programmed cell death. In contrast to the passive necrosis induced by perforin, apoptosis requires the active participation of the target cell. During apoptosis the target cell degrades its own DNA, fragments its nucleus, and breaks apart into small membrane-bound vesicles (apoptotic bodies) that are ingested by phagocytic cells. Killer cell-induced apoptosis is similar to the physiologic cell death that occurs normally during embryogenesis, involution of hormone-dependent tissues after hormone deprivation, and the turnover of epithelial surfaces. Some of the granzymes exocytosed with the cytotoxic granules have been implicated in apoptosis of target cells. However, induction of target cell apoptosis can occur in the absence of granule exocytosis. As discussed below, both secreted and membrane-bound proteins may deliver apoptotic signals to target cells.

THE TUMOR NECROSIS FACTOR SUPERFAMILY AND ITS RECEPTORS

Cytotoxic lymphocytes produce two related tumor necrosis factors (TNFs): cachectin (TNF-α) and lymphotoxin (TNF-β). Binding of these cytokines to specific receptors on the surface of susceptible target cells induces apoptosis of the target cell. Two distinct TNF receptors have been identified: TNF-R p55 (TNF-R1, CD120a) has an apparent molecular weight of ~55 kD and TNF-R p75 (TNF-R2, CD120b) a molecular weight of ~75 kD. TNF-α and TNF-β can bind to either of these receptors, which are widely distributed on many different cell types. TNF-α can be expressed as a membrane-bound protein on the killer cell surface or can be secreted as a soluble protein.

The genes encoding a number of other proteins are related to the genes encoding the TNF receptors. These proteins, including the nerve growth factor receptor and the CD27, CD30, CD40, and CD95 (APO-1/Fas) antigens, are said to belong to the TNF receptor superfamily. Similarly, the ligands for these receptors (nerve growth factor, TNF-α, TNF-β, CD70, CD30 ligand, CD40 ligand, and CD95 ligand) are all related to one another and belong to the TNF superfamily. These ligands and their receptors appear to be involved in the regulation of cell activation, proliferation, and death. TNF-α and TNF-β not only cause apoptosis of certain tumor cells, but can also augment the proliferation of activated T cells, B cells, and NK cells. Ligation of the CD40 antigen induces B cell proliferation and inhibits apoptosis of activated B cells. Similarly, ligation of the CD27 or CD30 antigens on activated T lymphocytes augments their proliferation. The CD95 ligand (Fas ligand) is expressed on the surface of some CTL. Interaction of CD95 ligand with the CD95 (APO-1/Fas) antigen on target cells causes apoptosis of the latter. CD95-induced apoptosis has been implicated in the target cell killing that has been observed under conditions that preclude granule exocytosis and perforin-mediated lysis. In contrast to TNF, CD95 ligand may induce programmed cell death of normal activated lymphocytes as well as of malignant tumor cells. Thus, depending on the cell type involved and its state of activation or differentiation, engagement of members of the TNF receptor superfamily can induce or enhance proliferation (mitosis) and can induce or inhibit programmed cell death (apoptosis).

INTERLEUKIN-2 AND LYMPHOKINE-ACTIVATED KILLER CELLS

Interleukin-2 (IL-2) is a cytokine that plays a prominent role in the regulation of mammalian immune responses. IL-2 promotes the proliferation and differentiation of helper T cells, cytotoxic T cells, and B cells. IL-2 also stimulates the function of NK cells and monocytes. The effects of IL-2 on lymphocytes and other cell types are mediated through specific cell surface receptors. The IL-2 receptor consists of at least three subunits: IL-2R α (or p55), IL-2R β (or p75), and IL-2R γ (or p64). Neither the β subunit nor the γ subunit appears to bind IL-2 significantly, but heterodimeric β γ complexes bind IL-2 with intermediate affinity and can mediate IL-2 internalization and signaling. The α chain by itself binds IL-2 with low affinity and does not appear to internalize the cytokine or initiate signaling. The α β γ heterotrimer appears to constitute the authentic high affinity IL-2R.

Most resting T cells do not express functional IL-2R. After activation by specific antigen, T cells express high affinity IL-2R heterotrimers as well as an excess of free α chains. Activated T cells also secrete IL-2, and thus can proliferate in an autocrine, IL-2-dependent manner. Although both CD4 and CD8 T cells can produce IL-2, CD4 cells appear to secrete larger amounts of the cytokine. Indeed, IL-2 secretion is believed to be one of the ways that CD4 T cells help activated CD8 T cells differentiate into CTL. IL-2 can also stimulate the proliferation of NK cells and activated B cells. Unlike T cells, NK cells do not appear to secrete IL-2 in quantities sufficient to support their own growth. Thus, NK cells are probably dependent on activated T cells for IL-2.

In contrast to T cells, almost all freshly isolated human NK cells express IL-2R β γ heterodimers and can respond to IL-2 in the absence of additional stimuli. After activation, the IL-2R α chain is also expressed on NK cells. Although they are spontaneously cytotoxic, unstimulated NK cells generally cannot kill freshly isolated cancer cells from patients with solid tumors. In contrast, IL-2-activated NK cells can lyse such solid tumor targets. Lysis of NK-resistant tumor cells by IL-2-activated lymphocytes is called lymphokine-activated killer (LAK) activity. IL-2 probably augments NK cell cytotoxicity in several ways. IL-2 induces the upregulation of several NK cell adhesion molecules and causes increased conjugate formation by NK cells. IL-2 also stimulates the production of cytotoxic granules and their constituents, including perforin and some of the granzymes. Moreover, IL-2-activated NK cells produce cytokines, such as TNF-α and interferon-γ, that may directly or indirectly contribute to target cell lysis. The receptors that trigger LAK activity, like those that trigger natural killing, have not been defined; it is possible that IL-2 also upregulates the expression of putative "LAK receptors" on NK cells. LAK activity is similar to NK activity in that it is not MHC restricted and does not appear to be antigen-specific. $CD56^+$ T cells and a small subset of $CD56^-$ $CD8^+$ CTL can exhibit LAK activity, but NK cells appear to be the predominant LAK effector among human blood leukocytes.

CLINICAL APPLICATIONS OF TNF, IL-2, AND LAK CELLS

TNF inhibits the *in vitro* growth of several kinds of tumor cells, including cell lines derived from breast cancer, lung cancer, colon cancer, ovarian cancer, and melanoma. Administration of TNF to tumor-bearing animals can induce hemorrhagic necrosis of the tumors in some cases. Based on these preclinical data, TNF has been used to treat advanced human cancers. TNF has been administered intravenously, intramuscularly, subcutaneously, intraperitoneally, and by direct intratumoral injection. Unfortunately, very few tumor responses have been reported in these studies. Systemic administration of TNF to humans produces a syndrome resembling the septic shock that accompanies some bacterial infections. The serious side effects produced by TNF preclude administration of the cytokine in high doses, which may explain in part the poor antitumor effects of TNF in humans. Unless methods can be developed to ameliorate the toxicity of TNF while maintaining its potential antitumor activity, it is unlikely that TNF will assume a prominent role in the systemic treatment of human cancer.

Shortly after the LAK phenomenon was described, Rosenberg and colleagues demonstrated that LAK cells could cause the regression of malignant tumors in laboratory animals. For these experiments, lymphocytes were cultured with high concentrations of IL-2 *in vitro* to generate LAK cells and then the LAK cells were injected into tumor-bearing animals. Substantial antitumor effects were seen when animals were given high doses of IL-2 *in vivo* together with the LAK cells. Based on these results, high-dose IL-2 and LAK cells were administered to patients with advanced cancers for which no effective therapy existed. These patients were given IL-2 intravenously, their peripheral blood lymphocytes were collected and stimulated for several days in cultures containing IL-2, and the resulting LAK cells were then administered intravenously together with additional doses of IL-2. Such treatment caused tumor shrinkage in 15 to 35% of patients with advanced renal cancers and ~20% of patients with metastatic melanoma. Complete disappearance of tumor occurred in ~5% of these patients. These clinical trials provided strong evidence that immune mechanisms could destroy even advanced cancers in humans. Nevertheless, high-dose IL-2 therapy was associated with many toxic side effects and only a small minority of patients appeared to benefit from the treatment. Many investigators are currently exploring ways to decrease the toxicity and increase the efficacy of IL-2-based immunotherapy. LAK cells generated by the methods described above contain a substantial proportion of cytolytically inactive cells. Techniques have been developed to purify IL-2-activated NK cells or to expand tumor-associated CTL for *in vivo* administration with IL-2. Additional cytokines, such as interferons or TNF, have been given in combination with IL-2 for cancer therapy. Monoclonal antibodies recognizing tumor-associated antigens have been used in attempts to harness ADCC that can be mediated by LAK cells. Several groups have given prolonged infusions of IL-2 in much lower doses than those used in the high-dose IL-2/LAK cell trials. Currently none of these approaches has been proved to be clearly superior to high-dose IL-2, and additional studies are ongoing. Progress in clinical immunotherapy may result from advances in basic tumor immunology, including identification of the receptors responsible for natural killing and LAK activity and demonstration of the factors underlying the poor immunogenicity of specific human cancers.

CONCLUSION

The vertebrate immune system provides a rapid, flexible, and very effective defense against many different infectious pathogens. Given this impressive defense mechanism, it is not surprising that clinical and laboratory scientists have attempted to manipulate the immune response for an antitumor effect. An extensive body of preclinical data indicates that immune responses can eliminate malignant tumors in several mammalian species. Moreover, clinical trials have shown that immunotherapy can mediate the regression of advanced cancers in humans. Major tumor responses have been the exception rather than the rule, and better approaches are clearly required. Further clinical and basic science investigation may yield new knowledge that will ultimately permit a more effective and less toxic therapy for human malignancies.

ACKNOWLEDGMENTS

This work was supported in part by NIH grants CA-41619 and CA-01730. M.J.R. was supported by the Claudia Adams Barr Program in Cancer Research.

REFERENCES

Berke, G. (1994). The binding and lysis of target cells by cytotoxic lymphocytes: Molecular and cellular aspects. Ann. Rev. Immunol. 12, 735–773.

DeVita, V.T., Hellman, S., & Rosenberg, S.A. (eds.) (1995). Biologic Therapy of Cancer, (2nd edn.). J.B. Lippincott, Philadelphia.

Doherty, P.C. (1993). Cell-mediated cytotoxicity. Cell 75, 607–612.

Herberman, R.B. (1992). Tumor immunology. JAMA 268, 2935–2939.

Houghton, A.N. (1994). Cancer antigens: Immune recognition of self and altered self. J. Exp. Med. 180, 1–4.

Paul, W.E. (Ed.) (1993). Fundamental Immunology (3rd edn.). Raven Press, New York.

Robertson, M.J. & Ritz, J. (1990). Biology and clinical relevance of human natural killer cells. Blood 76, 2421–2438.

Robertson, M.J. & Ritz, J. (1992). Role of IL-2 receptors in NK cell activation and proliferation. In: NK Cell Mediated Cytotoxicity: Receptors, Signaling, and Mechanisms (Lotzova, E. & Herberman, R.B., eds.), pp. 183–206, CRC Press, Boca Raton, FL.

Rosenberg, S.A. (1993). Principles and applications of biologic therapy. In: Cancer: Principles and Practice of Oncology, 4th edn. (DeVita, V.T., Hellman, S., Rosenberg, S.A., eds.), pp. 293–324, J.B. Lippincott, Philadelphia.

Sitkovsky, M.V. & Henkart, P.A. (eds.) (1993). Cytotoxic Cells: Recognition, Effector Function, Generation and Methods. Birkhauser, Boston.

Smith, C.A., Farrah, T., & Goodwin, R.G. (1994). The TNF receptor superfamily of cellular and viral proteins: Activation, costimulation, and death. Cell 75, 959–962.

Tartaglia, L.A. & Goeddel, D.V. (1992). Two TNF receptors. Immunol. Today 13, 151–153.

Trinchieri, G. (1989). Biology of natural killer cells. Adv. Immunol. 42, 181–377.

Chapter 11

Quantitative Analysis of Nuclear Size for Prognosis-Related Malignancy Grading

FLEMMING BRANDT SØRENSEN

Advances in Oncobiology
Volume 1, pages 221–255.

ISBN: 0-7623-0146-5

INTRODUCTION

The choice of treatment and prognostic prediction of clinical outcome for patients suffering from solid cancer are evaluated primarily by examining the type and site of the tumor, the clinical stage of the disease (i.e., the size and extent of the tumor), and the histologic grade of malignancy. The first three are settled by the collaboration of the clinician with the pathologist, while the grade of malignancy is determined solely by the pathologist. The impact on therapeutic decision, and thus on prognosis, of the malignancy grade shows great variation internationally, and it is therefore quite impossible to get an overview of the real value of histopathologic grading of malignancy. Moreover, there is no unifying concept for malignancy grading of solid cancers, because a number of different grading schemes have been elaborated. From the technical point of view, some basic problems are associated with the traditional way of morphologic grading of malignancy, in that it is a subjective discipline. The pathologist is involved in a high order of mental process in assessing often multiple histologic and cytologic variables for the determination of the malignancy grade. Among these parameters are: tumor configuration, relationship between tumor and surrounding stroma, nuclear density, nucleo-cytoplasmic ratio, nuclear polarity, cytoplasmic differentiation, nuclear size and shape, chromatin pattern, nucleolar characteristics, mitotic density, and configuration. Thus, the grade of malignancy is scored as low, medium, or high by a complex constellation of the pathologist's summarized evaluation of these many parameters. Because all the variables are assessed subjectively and are purely qualitative in nature, there is a well documented, high risk of observer variation. This fact has prompted a new development within the field of malignancy grading, namely the use of quantitative methods for estimating objectively the histologic severity of the neoplastic lesions.

This chapter provides a discussion of histopathologic malignancy grading. The definition of malignancy grading will be briefly reviewed, and various techniques of grading including their virtues and weak points are also dealt with, but the main purpose of the chapter is to present new quantitatively based methods for objective grading of malignancy.

FEATURES OF MORPHOLOGIC GRADING OF MALIGNANCY

Malignancy grading of solid neoplasms is a method for establishing the aggressiveness of the tumor based on cytologic and histologic features as seen in two-dimensional sections. Although a large number of features are evaluated, in general, they are related to the tumor tissue *per se* and/or the relationship between the tumor and the host.

The growth pattern of solid epithelial cancers vary considerably, even within the same histogenetic type of tumor originating from the same organ. The malignancy may be papillary or sessile/solid, the latter often having the most unfavorable prognosis. The mode of invasion also differs, in that some tumors invade expansively, whereas others show a tentacular or diffuse infiltrative growth at the invasive zone. These differences in the frontier between the host and the malignant neoplasm have also been linked to a differing prognostic outcome. Moreover, the immunologic defense of the host against the tumor may be visualized by lymphocytic infiltration around the tumor margin. The growth pattern, mode of invasion, and immunologic response may represent both tumor and host capabilities. With regard to the tumor itself, the degree of so-called nuclear anaplasia may be responsible for the morphology of the neoplasms, in that the nuclear content is thought to determine the metabolism and behavior of the malignant cell. Thus, the examination of structural nuclear changes associated with malignant transformation has played a central role in malignancy grading.

The nuclei of cancer cells are mostly larger than those of the cells from which they originate, and the nucleo-cytoplasmic ratio is increased in malignant cells. Clumping of chromatin and/or increased amount of DNA make the malignant nuclei hyperchromatic, and the texture of the chromatin may be drastically changed. The increased nuclear size is accompanied by abnormalities in nuclear shape. This feature may be hard to recognize during microscopic examination of two-dimensional tissue sections, but focusing through a thick (i.e., three-dimensional) plastic section of cancer tissue reveals the highly ragged and irregular shapes of malignant nuclei. Within the nucleus, at least in some cancers, the nucleolus may grow to an immense size, and, in addition, acquire a most irregular shape. The dominance of the nuclei in the tumor tissue is striking, and the nuclear density is mostly rather high not only due to the increased nucleo-cytoplasmic ratio, but also the increased proliferative activity of cancer cells. In the fixed tissue this activity may be displayed by the number of mitoses. These may sometimes show rather abnormal configurations (e.g., tripolar), may be due to an abnormal composition of the chromosomes, or due to the missing control of cell division in cancer cells. The existence of such abnormal mitoses may represent cell divisions that are never to be completed. Also, the metabolism of the malignant cell is abnormal, and the cytoplasmic differentiation may fail to appear. In malignant cells originating from glandular epithelium, the normal polarity of the cells, with the nucleus in the basal

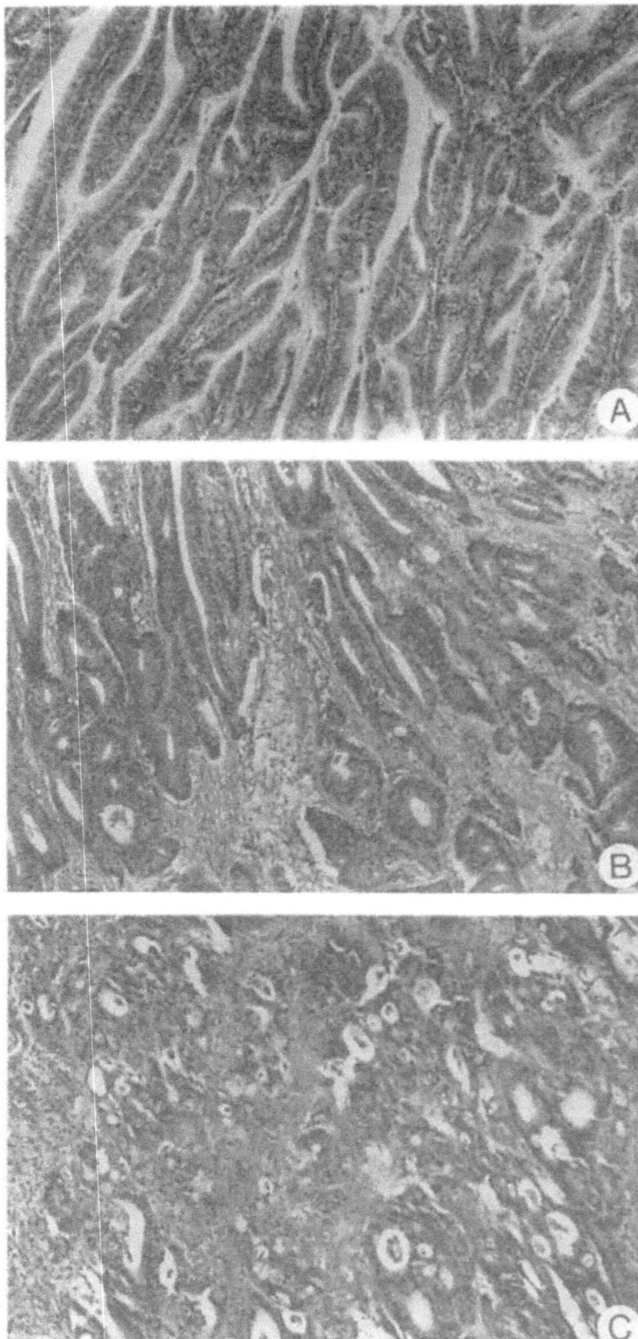
A
B
C

Figure 1. This series of histologic fields of vision has been taken from three adenocarcinomas of the large bowel. In general, the morphologic degree of malignancy or dedifferentiation is determined by the derangement of morphology in the cancer tissue, as compared to the native histology of the normal tissue, from which the cancer is derived. In the case of adenocarcinoma of the large bowel, the neoplastic epithelial cells derive from the mucosal epithelium. This is quite obvious in **A**, which is a highly differentiated adenocarcinoma of low malignancy grade. In **C** the morphologic pattern of glandular epithelium is nearly absent, and this cancer represents low differentiation with a high malignancy grade. In between these extremes is an adenocarcinoma of intermediate differentiation and medium grade of malignancy shown in **B**.

---

part and the metabolically active cytoplasm in the luminal part, is disrupted. In tumors derived from stratified epithelium, the usual maturation of the squamous cells, tends to be absent as they move toward the surface of the epithelial lining.

All these qualitative, histologic, and cytologic variables are considered for evaluating the morphologic severity of malignancy, which is customarily reported as high-, moderate-, or low-grade. An example of such scoring is that applied to adenocarcinomas of the large bowel, as shown in Figure 1. This has been the usual way of expressing the grade of malignancy ever since Broders introduced the scheme for grading malignancy in 1922. Later on more sophisticated approaches to the scoring of malignancy have of course been developed. These systems are aptly named multifactorial in that the pathologist gives each of a number of histopathologic parameters a numerical score, and these are added for an overall grade of malignancy on a semi-quantitative scale. However, both the qualitative and the semi-quantitative approach to malignancy grading suffer from the same problem, namely, an often rather poor reproducibility.

REPRODUCIBILITY OF MORPHOLOGICALLY BASED GRADING OF MALIGNANCY

The number of features examined in the process of histopathologic grading of malignancy are thus manifold. The task for the surgical pathologist is to incorporate the evaluation of each of these variables and integrate this information into an overall malignancy score or grade, which can be communicated to aid the clinician in the choice of treatment along with other characteristics of the patient and the neoplasm in question. This is a complex procedure, and the subjectivity in interpreting these qualitative parameters surely creates the problem of observer variation. Low reproducibility may, in the end, invalidate the clinical usefulness of the malignancy grade in the case of solid cancers (see Suffin et al., 1977; Ringsted et al., 1978; Græm et al., 1980; Larsen et al., 1980; Prade et al., 1980; Heenan et al., 1984; Sørensen et al., 1989; Sørensen et al., 1994). Moreover, the prognosis according to the grade of malignancy, reported in clinical trials of patients treated for solid cancers, may be more dependent on the investigator carrying out the grading than on the true nature of the given treatment for a particular type of cancer. Although it seems to be a straightforward matter to grade a neoplasm (Figure 1), the practice of malignancy grading has taught us that this is patently not the case. Also, this is not a new observation, since Willis (1948) drew attention to the problem in his book Pathology of Tumours. He succinctly writes, "Attempts at precise histological grading of malignant tumours as an index to their degrees of malignancy are very arbitrary and unscientific. Tumors show great cytological variation and the decision as to which cells shall be labeled differentiated or undifferentiated, as the case may be, greatly depends on the personal bias and preconceptions of the examiner".

A brief description of a few examples of tumors will now be given regarding the observer variation of morphologically based malignancy grading. When evaluating the reproducibility of qualitative or semi-quantitative grading schemes, one should apply so-called *Kappa* statistics (Holman et al., 1982; Svanholm et al., 1989). This statistical method takes into account the possibility that the observers might give the same grade of malignancy just by coincidence. Unfortunately, this statistical approach has only been adopted in the international medical literature this past decade.

Bladder Cancer

Papillary transitional cell carcinomas (TCC) of the urinary bladder are graded according to the WHO scheme, as described by Mostofi et al. (1973). However, in some countries, e.g., Denmark, grading of TCC malignancy is done according to the method proposed by Bergkvist et al. (1965). The two grading systems are not wholly comparable, but a translation between the two systems has been proposed (Olsen, 1984). Both systems overlap with regard to the features examined histologically, but there are differences in the evaluation and scoring. The fact that different grading schemes are available for the same type of cancer creates an obstacle for the international comparison of treatment results according to the malignancy grade of TCC. Adding to this confusion is the frequent low reproducibility within the different grading systems, as shown for TCC in a number of studies (Busch et al., 1977; Ooms et al., 1983; Colpaert et al., 1987; Abel et al., 1988; Robertson et al., 1990). Using *Kappa* statistics, a recent Japanese-Danish collaborative study has disclosed an inter-observer reproducibility of about 0.50, indicating that the investigators were only able to score the same grade of malignancy once every two TCC examined (Sørensen et al., 1994).

Squamous Cell Carcinomas

Malignancy grading of squamous cell carcinomas (SCC) dates back to 1922, when Broders (1922) published the first histologic grading method for epitheliomas. A year later, Martzloff (1923) proposed a three-tiered grading scheme based on the morphologic type of tumor cell in SCC of the uterine cervix. Also, a WHO based scheme of malignancy grading of this type of cancer is available today (Poulsen and Taylor, 1975). An alternative method for grading SCC of the uterine cervix is the multifactorial system proposed by Stendahl et al. (1980). The reproducibility of the malignancy grade reported by the WHO system and the grading scheme proposed by Stendahl et al. (1980) is, however, far from optimal (Cocker et al., 1968; Ringsted et al., 1978; Stendahl et al., 1981). A multifactorial grading scheme for SCC of the larynx (Sørensen et al., 1989), comparable to the scheme described by Stendahl et al. (1980), has been tested for observer variation using *Kappa* statistics, and in this case, the inter-observer reproducibility was at the level of chance! Thus, it does not seem likely that reproducibility increases as the grading scheme becomes more sophisticated. Moreover, multifactorial malignancy grading schemes are highly dependable on optimal quality of the biopsy specimens examined. Superficial biopsies might not include the most important features associated with the interface between tumor and host. Actually, in routine settings one has been forced to exclude about 25% of cases investigated due to poor quality of the biopsies (Bichel and Jakobsen, 1985; Sørensen et al., 1992). Studies of SCC of the oral cavity have likewise disclosed that a multifactorial grading scheme was not applicable to more than 50% of the cases (Bryne et al., 1991). Such a grading scheme is therefore not suited for routine purposes. The rather low reproducibility emphasizes that it is a waste of time to grade the malignancy by this method, although the prognostic impact of the grading has been found to be even better than that of the clinical stage of disease (Stendahl et al., 1979).

Cutaneous Malignant Melanomas

The prognostic evaluation and choice of treatment for patients with malignant melanomas of the skin are based on the examination of tumor thickness (Breslow, 1980), Clark's level of invasion (Clark et al., 1969), and histologic type of tumor (Clark et al., 1986). Reproducibility studies have shown low consistency in the evaluation of the latter two variables (Suffin et al., 1977; Larsen et al., 1980; Prade et al., 1980; Holman et al., 1982; Heenan et al., 1984), whereas the tumor thickness, measured according to the criteria described by Breslow (1980), does better in this regard. This is probably due to the fact that tumor thickness is a quantitative variable, which is reported on a continuous scale, in contrast to the two other, categorical parameters. Tumor thickness is, however, also sensitive to inter-observer variation (Breslow, 1977; Breslow, 1980; Prade, 1980; MacKie, 1982; Colloby et al., 1991) to an extent that may have clinical impact. The mitotic activity in melanomas has been associated with prognostic value (Schmoeckel et al., 1983; Tan and Baak, 1984; Søndergaard and Schou, 1985). Unfortunately, the reproducibility of mitotic counts is not optimal (Larsen et al., 1980), and, more importantly, is the varying way of reporting mitotic counts (Ellis and Whitehead, 1981) which makes it impossible to compare mitotic counts reported from different investigations. Adding to this is the possibility of variation in mitotic number due to even small departures from standardized tissue handling and conservation (Græm and Helweg-Larsen, 1979).

WHAT CAN BE DONE TO IMPROVE THE RELIABILITY OF MALIGNANCY GRADING?

After having focused on three examples of solid cancer, it is obvious that something has to be done to improve the reliability of grading methods. Because morphologic grading of malignancy often includes examination of subtle cytologic and histologic features, the problem of observer variation can be extrapolated to many other types of cancers. Moreover, the sophisticated, multifactorial, semi-quantitative grading approach does not help in this regard, because the basic method still relies on subjective evaluations. Hence, the lesson learned from experience concerning quality control in malignancy grading is that subjective, qualitative methods must be replaced by objective, quantitatively based techniques to improve the reproducibility. In essence, this is the only valid scientific approach, which can improve malignancy grading to a level where it can be used for specific stratification of therapy, and provide a sound foundation for making prognostic forecasts for the individual patient with cancer.

First, it is important to make clear the essential requirements for quantitatively based methods for malignancy grading. The methodology has to be simple, fast, efficient, internationally reproducible, and of use in all cases of specific cancers examined in the routine setting of histopathologic practice. The variables to be measured by the quantitative technique should reflect the biology in question, i.e., cancer tissue, and most importantly, be of therapeutic and prognostic value to the clinician. One may argue that it is pointless to invest much work in performing histopathologic grading of malignancy. There is, however, accumulating evidence in the literature that it is indeed worthwhile to grade the malignancy of solid cancer by quantitative and reproducible methods (Sørensen, 1992). The histopathologic grade of malignancy can in some cancers add independent prognostic information to the clinical stage of disease, which for most solid malignancies is by far the best prognostic parameter.

In defining the object for quantification, one is provided with a long list of histologic and cytologic variables, mentioned above. It is, however, both practically and methodologically impossible to measure all of these features for the definition of a mean malignancy score. In the first place, it is very difficult to describe changes in shape on an unbiased quantitative scale, and second, it would take too long to grade a single neoplasm by quantification of the mentioned parameters. It is preferable to select a few parameters that are easy to measure by methods including morphometry (Loud and Anversa, 1984; Hall and Fu, 1985; Pesce, 1987; Baak, 1991), stereology (Weibel, 1979; Gundersen et al., 1988a,b; Mattfeldt, 1990), and cytophotometry (Lovett et al., 1984; Hall and Fu, 1985). The molecular basis of cancer is, unfortunately, still unknown, and hence the pathologist must focus on measuring the structural basis of malignant neoplasia, which in cytology and histology is primarily associated with so-called nuclear anaplasia—that is, major features such as nuclear size and variation of size (Black and Speer, 1957; Meyer-Arendt and Humphreys, 1972). The three-dimensional nuclear size may be a dynamic reflection of the metabolic state of the nucleus, and may represent a physical correlate of the total content of biochemical constituents in the nucleus. Chief among the constituents are histone and non-histone proteins, inorganic materials, water, RNA, and DNA. All of these may change during malignant transformation, along with chromatin alterations and an increase in the nuclear-cytoplasmic ratio, as well as cytoplasmic, functional dedifferentiation.

Before entering upon a more detailed discussion of quantitative analysis of nuclear size for malignancy grading, it is important to realize that biology is, as always, unpredictable, and it is impossible to describe some general nuclear changes associated with all kinds of malignant neoplasia. As one example, the small cell (oat cell) anaplastic carcinoma of the lung is an extremely aggressive type of cancer, and yet this malignancy is characterized by rather small cell nuclei (Aru and Nielsen, 1989). In sharp contrast, most other solid cancers have larger cell nuclei than their native cells of origin (Sørensen, 1992). It is therefore of the utmost importance that the surgical pathologist and the cytologist be cautious when interpreting data relating to nuclear quantification. Vital information about tumor specific aggressivity, based on nuclear size measurements, can only be obtained by studying specific types of cancer separately.

In addition to estimates of mean nuclear size and nuclear size variation, other quantitative parameters are of interest within the scope of histopathologic malignancy grading. These include variables easy to measure like indices of mitotic profile activity, nuclear profile density, and volume fractions of various tissue components. The techniques required for obtaining such estimates will only be discussed briefly here, since the main objective is to focus on the estimation of nuclear mean volume and the nuclear size pleomorphy and their prognostic impact.

STEREOLOGIC ESTIMATION OF NUCLEAR MEAN VOLUME

The pathologist trying to quantify cell nuclei has to face the problem associated with the reduced information of the three-dimensional reality found in two-dimensional tissue sections. Unbiased estimates (i.e., estimates without systematic deviation from the true mean value) of nuclear mean volume cannot be obtained by model-based methods requiring unrealistic assumptions about nuclear shape and orientation. Thus, although a correlation may exist between nuclear profile diameter or profile area and the true nuclear volume, no matter how elegant the statistical model is, the true relationship is always unknown. This is the basic concept of the classical corpuscle problem, as described by Wicksell (1925), and it is related to the fact that two-dimensional samples of nuclear profiles are biased, as illustrated

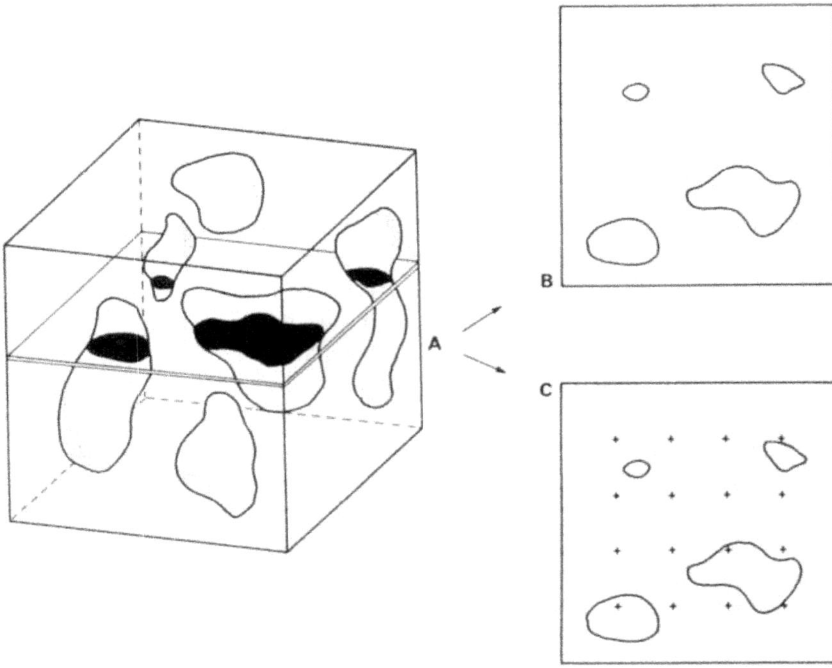

Figure 2. **A**) In this drawing of a three-dimensional tissue block, a section has been made by which four arbitrarily shaped particles are hit. **B**) Projection of this section shows the four particle profiles hit by the section plane. The particle profiles are represented on the section with a probability directly proportional to the height of the particles perpendicular to the section plane (i.e., height-weighted sampling). **C**) Supplementing the section with a test-system of zero-dimensional points, only two particles are hit by both the section plane and the points. The sampling probability is now volume-weighted in that the particles are first sampled in proportion to height by the section and then with a probability proportional to particle profile area (i.e., height times area = volume).

in Figure 2. In fact, one has to make some sort of serial tissue sectioning (Cruz-Orive, 1980; Sørensen, 1991) to obtain unbiased information about three-dimensional nuclear size. This is a rather time-consuming task, and it is not suited for the study of nuclear mean volume in the routine evaluation of malignant tumors for estimating the malignancy grade.

Recent advances in stereology have provided a completely new approach to the problem of particle number and mean volume estimation (Gundersen, 1986; Gundersen et al., 1988a,b; Cruz-Orive and Weibel, 1990). These design-based techniques are assumption-free, yield unbiased estimates, and are thus suited for the study of arbitrarily shaped particles like cancer cell nuclei. The unbiasedness of the methods is ensured by the sampling design, which relies on sampling in three-dimensional space (thin serial or thick optical tissue sections) for number estimation, and point-sampling (Figure 2C) in two-dimensional tissue sections for size estimation. Such sampling approaches are independent of shape and orientation of the sampled objects. At first glance, one may be frightened by the rather complex mathematics on which the stereology is based, but for the histopathologist both the practical aspects and use of the methods are extremely simple.

Estimation of the Number-Weighted Mean Nuclear Volume

Estimation of the true, number-weighted mean nuclear volume can be arrived at by combining two stereologic techniques, the disector (Sterio, 1984) and point-counting (Gundersen et al., 1988b). With a random tissue section, one first estimates the volume fraction of nuclei (V_V) by point-counting, that is, points on the randomly orientated test-system which hit the nuclei, are counted in relation to all points hitting the tissue section (Figure 3A). Next, the absolute number of nuclei per tissue volume (N_V) is estimated using a disector. The disector is a three-dimensional

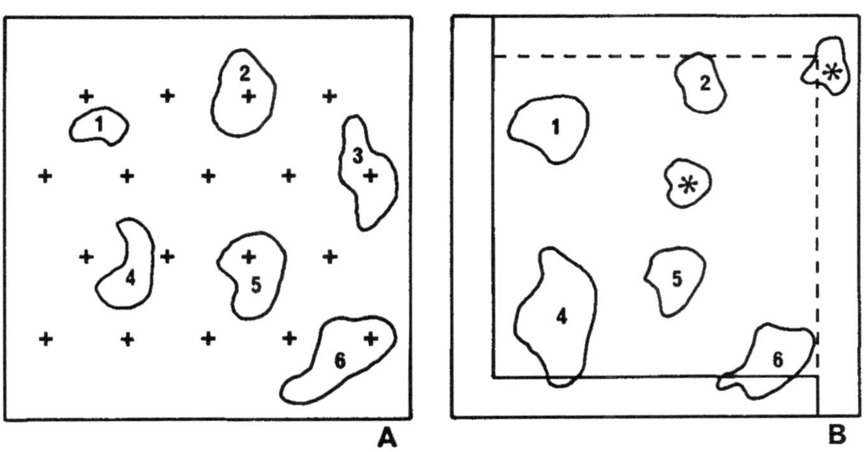

Figure 3. **A**) Diagrammatic representation of a 2 µm thick tissue section, with six nuclear profiles. A test-system with points has been superimposed onto the section, and by counting the number of points occurring in nuclei relative to the total number of points, which hit the tissue, the volume-fraction of nuclei can be estimated. **B**) The 2 µm thick tissue section immediately adjacent to the tissue section, shown in **A**, has been superimposed with a counting frame. In this section the third nuclear profile has disappeared, whereas two new nuclei are represented by their profiles (marked with *). Using an unbiased counting rule, only nuclei inside the frame or on the hatched edges are sampled, if they do not intersect with the fully drawn exclusion line (Gundersen, 1977). Thus, two nuclei are sampled in this disector, in a volume corresponding to the section thickness times the area of the counting frame.

sampling approach, involving either thick optical plastic sections, or thin, adjacent physical sections. The idea is that one counts nuclei seen in one focal plane, but which are not seen in the adjacent physical or optical focal plane a known distance from the first section plane (Figures 3A,B). By this sampling approach the nuclei are sampled uniformly according to their number, independent of shape and orientation. Combining the two methods, an estimate can be made of the number-weighted mean nuclear volume, $\bar{v}_N = V_V/N_V$. However, no information about nuclear size variability is obtained by this technique.

If information about the three-dimensional nuclear size distribution, and thus the pleomorphism of nuclear volume, is required, a much more laborious procedure has to be employed. It involves the preparation of serial tissue sections of known thickness, in which the nuclei are sampled on the first two adjacent sections, using a disector. Each sampled nucleus is hereafter followed by its profile through the stack of serial sections, and the area of each profile is estimated by point-counting

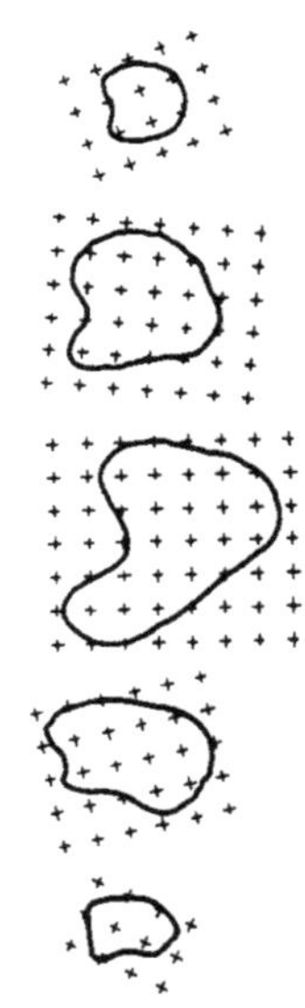

$$V(nuc) = t \cdot \sum P(nuc)$$

Figure 4. Serial sections of 2 μm thickness, t, have been prepared of a nucleus, which has been sampled by a disector. A test-system with points, P, has been thrown randomly on the nuclear profiles seen on the five sections on which the nucleus is represented. Knowing the area in μm^2 associated with a point on the test system, an unbiased estimate of the nuclear volume in μm^3 can be obtained using the given formula. This method of estimating particle size is known as the disector-Cavalieri approach.

(Figure 4). The volume of each nucleus is then calculated from $v(nuc) = t \cdot \Sigma P$, where t is the section thickness and P is the number of points (with a known associated area) which hit the profiles of the nucleus on the tissue sections. Calculating the mean of the nuclear population in question provides an unbiased estimate of the number-weighted mean nuclear volume, $\bar{v}_N$ (Jensen and Gundersen, 1987; Sørensen, 1991), and because the distribution of nuclear volumes is known by this approach, the nuclear volume variability can be readily calculated (Sørensen, 1991). Estimation of the mean nuclear volume in this way is known as the assumption-free disector-Cavalieri method, which, apart from unbiasedness, is very precise. The work load is, however, very large and time consuming.

Estimation of the Volume-Weighted Mean Nuclear Volume

There exists a much faster method for unbiased estimation of nuclear volume, which is not only more efficient than those already mentioned, but also provides for the routinely working pathologist engaged in malignancy grading an elegant and more appropriate solution to many practical limitations. This technique, which is based on the nucleator principle (Gundersen, 1988), involves the measurement of linear intercepts of point-sampled nuclear profiles on a single random, two-dimensional tissue section (Gundersen and Jensen, 1985). The nuclei are sampled with a probability directly proportional to their volume: they are first sampled by the tissue section with a probability directly proportional to nuclear height perpendicular to the section plane, and then, on the section, with a probability directly proportional to nuclear profile area (i.e., height times area = volume, Figure 2C). Because all three dimensions are utilized by this sampling strategy, estimates of the volume-weighted mean nuclear volume, $\bar{v}_V$, are independent of nuclear shape and orientation (Gundersen and Jensen, 1983, 1985). A volume distribution of nuclear volume is skewed to the right, as compared to the number distribution of nuclear volume (Figure 5). This may be an advantage in the study of cancer, in that the larger nuclei may provide the most information with regard to the feature of aggressiveness in the majority of malignant neoplasms.

After completing point-sampling, the nuclear intercepts are measured. The basic principles underlying this technique were developed by Gundersen and Jensen (1983, 1985). Two important requirements must be fulfilled in order for the estimation of nuclear $\bar{v}_V$ to be unbiased. The first is the so-called General Requirement (Gundersen, 1986) for arbitrarily shaped particles like cell nuclei; this demands that the investigator be able to unambiguously identify all profiles belonging to the same nucleus. In cancer tissue, the cell nuclei may vary in shape, but at the light microscopic level they are mostly identified as a single profile in the tissue section, due to roughly convex nuclear shape at this level of magnification. The second requirement is isotropy, which means that the nuclear orientation, or the intercepts measured within the nuclear profiles, have an equal probability of being oriented in all directions in three-dimensional space (Gundersen, 1986). To

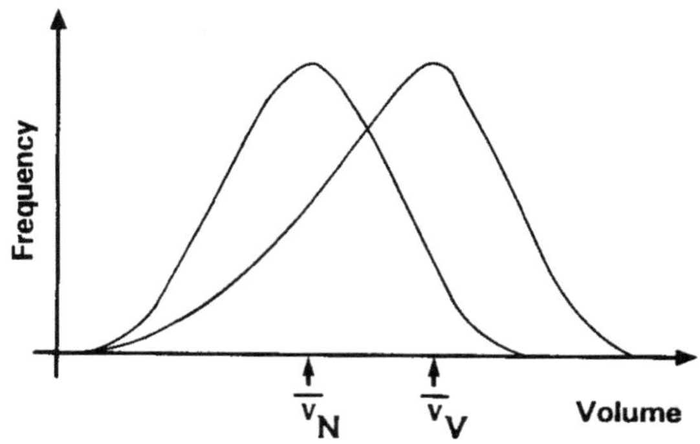

Figure 5. A population of arbitrarily shaped particles is shown in the number and volume-weighted size distributions. In the number-weighted distribution, the particles have been sampled with a uniform probability (with the same chance = one), whereas the particles in the volume-distribution have been sampled according to their individual volume. The volume-weighted distribution of particle volume is skewed to the right, as compared to the number-weighted distribution. Only in the hypothetical case of a population exclusively composed of particles with identical size is $\bar{v}_N$ equal to $\bar{v}_V$.

fulfill this requirement, the investigator can choose between two technical approaches. Either the nuclear intercepts can be measured in one arbitrary direction in globally isotropic nuclear populations, using a simple test-system (Figure 6), or they can be measured in three-dimensional isotropic directions generated in the tissue sections. It is in fact impossible to exclude anisotropy of nuclear populations, but isotropy can be achieved by the production of so-called IUR sections (isotropic, uniform, random), in which all directions of section planes are represented with uniform probability. Such tissue sections can be made by using the orientator, described by Mattfeldt et al. (1990), but this way of counterbalancing anisotropy may sometimes destroy the histologic pattern necessary for morphologic diagnosis, and is thus not always well suited for routine pathology. However, there is another approach to eliminate anisotropy which is often better because it preserves diagnostic morphology. This is based on vertical section design, described by Baddeley et al. (1986). The sampling strategy is to make vertical sections perpendicular to a fixed, but otherwise arbitrary horizontal plane of reference which can either be a natural intrinsic part of the tissue (e.g., a mucosal lining), or artificially generated. It is easy to make vertical sections in the routine practice of histopathology (Gundersen et al., 1988b), and this sampling design has been used in a large number of studies (see Baddeley et al., 1986; Cruz-Orive and Hunziker, 1986; Gundersen et al., 1988a,b; Nielsen, 1988; Sørensen, 1989b; Sørensen, 1991; Sørensen and Erlandsen, 1990). On these vertical sections it is now possible to generate three-dimensional isotropic directions (Sørensen, 1991), along which the nuclear intercepts can be measured in a simple test-system (Figure 7).

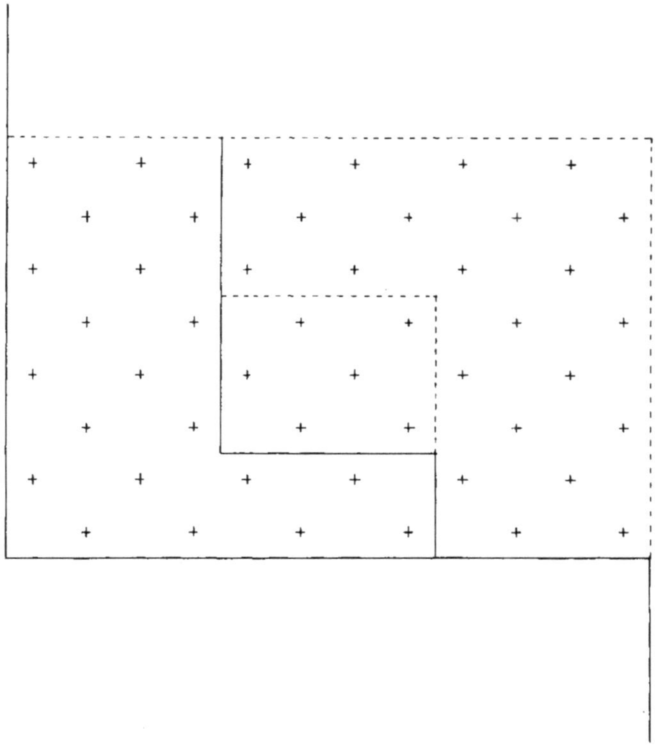

Figure 6. This so-called integrated test system contains points and a counting frame. Using point sampling of particle profiles, estimation of the volume-weighted mean particle volume can be carried out on IUR sections, while the counting frame is used for estimating two-dimensional densities, e.g., particle profile density, using the counting rule described in Figure **3B**.

After having sampled tissue sections according to the requirements, it is extremely easy to perform the measurements for obtaining estimates of nuclear $\bar{v}_V$. In the nuclear profiles of the focal plane that are hit by points (i.e., point-sampled), one need only measure the test-line length (i.e., nuclear intercept = ℓ_0, where o denotes sampling by zero-dimensional points) from nuclear boundary to nuclear boundary, through the sampling point (Figure 8). The intercepts can be measured with virtually any kind of ruler, but using a projection microscope and a crude classified ruler have proved very convenient (Sørensen, 1991). In computer-assisted systems, one can measure the absolute intercept length on the monitor screen by clicking a mouse, and letting the computer do the calculations. In most malignant neoplasms it is sufficient to measure about 75 point-sampled nuclear intercepts to obtain a robust estimate of nuclear $\bar{v}_V$. The mean of the cubed intercept

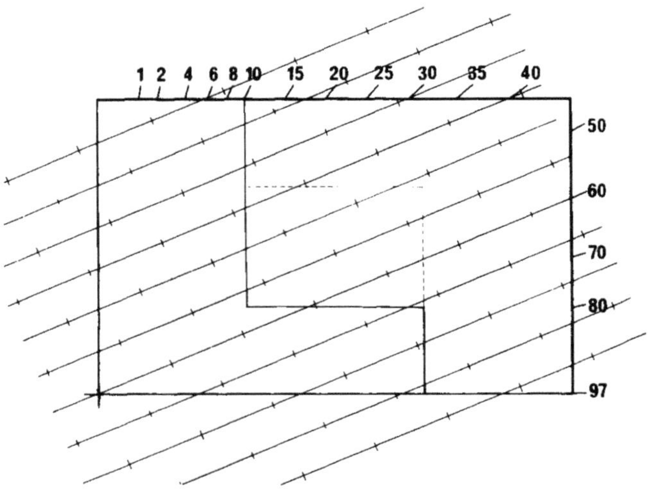

Figure 7. Test system used for the estimation of nuclear $\bar{v}_V$ from vertical sections. The short edges of the frame (i.e., the vertical axis) are aligned perpendicular to the chosen horizontal plane of reference of the system. The frame is used to orientate the linear probe and has a nonequidistant scale of numbers marked off on the edge. The density of numbers increases as the sine of the angle with the vertical axis increases (sine-weighted scale). A transparent overlay of test points positioned on parallel lines is superimposed on the orientation frame at a predetermined random orientation (in this case position 60), i.e., with one of the parallel lines passing through this number and the lower left corner of the frame. Nuclear profiles hit by points are sampled, and the nuclear intercepts are measured along the orientation lines. For the next fields of vision in the same tumor, a constant is added to the initial orientation number (e.g., 37, and the orientation numbers in the subsequent fields are 97, 37, 74, 14, 51, 88, 28, etc.). When the orientation is systematically changed the resulting intercepts are isotropically orientated in three dimensions. The orientation frame must be smaller than the field of vision to permit the measurement of intercepts in nuclei that are hit by a point inside the frame but which extend outside the frame. The integrated counting frame is used as described in Figure **3B**.

lengths, $\overline{\ell_0^3}$, is simply multiplied by $\pi/3$ (Gundersen and Jensen, 1985). An illustrated proof of this relation has been published by Cruz-Orive (1987). By supplementing the test-system with the integration of sampling frames (Figures 6, 7, and 8) it is possible, within about 15 minutes, to obtain other quantitative parameters (Table 1) along with estimates of nuclear $\bar{v}_V$ in one working process.

One must now mention the problems associated with the processing of surgical specimens, since tissue distortion as the result of shrinkage and swelling may have a profound impact (Boonstra et al., 1983) on quantitative estimation of tissue components, especially in the context of quantification of volumes. Fixation, dehydration, embedding in paraffin, and tissue sectioning need to be standardized

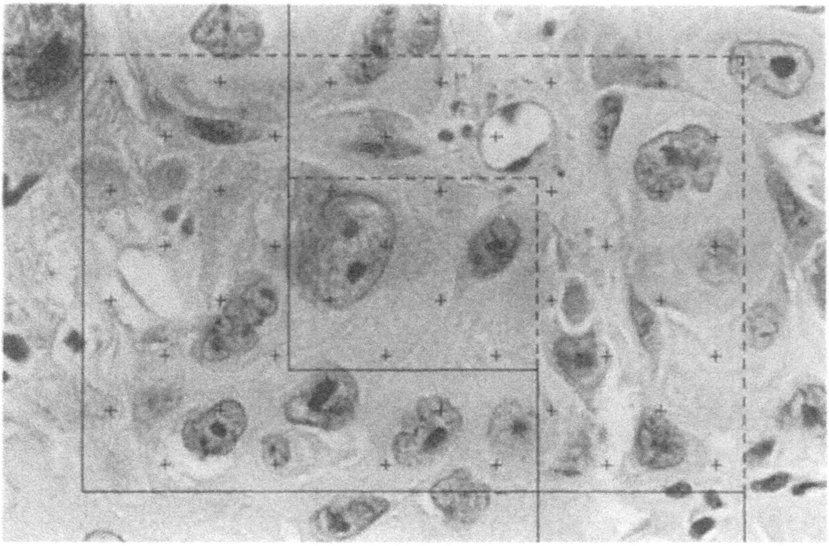

Figure 8. An IUR section from an invasive squamous cell carcinoma of the oral mucosa has been projected onto the test system shown in Figure **6** at a magnification of ×1000. Nuclear profiles in the focal plane, which are hit by points, are measured from nuclear boundary to nuclear boundary through the sampling point in one arbitrary direction for the estimation of the volume-weighted mean nuclear volume. If a nuclear profile is hit by two points it is measured twice. The two integrated counting frames can be used for estimation of nuclear profile density and/or mitotic profile density, using the counting rule described in Figure **3B**. A number of quantitative, morphometric/stereologic data can be derived using this integrated test system, as shown in Table 1.

to ensure the stability and precision of the stereologic estimators. Data from a standardized, recent study, comparing embedding in plastic with paraffin, have shown only minor differences in estimates of nuclear $\bar{v}_V$ with regard to the embedding medium (Ladekarl, 1994). Data based on two-dimensional estimates of nuclear mean profile area are already available suggesting that this parameter is rather resistant to minor variation in the tissue processing (Baak et al., 1989a). Thus, if tissue handling and processing are standardized—which is preferable even if quantitative studies are not performed—one can expect the quantitative estimates of mean nuclear size to be relatively stable. However, it is much more difficult, if not impossible, to control internal tissue variation in nuclear size within the same organ that may arise due to varying sensitivity to tissue processing of lesions of varying severity (e.g., variation in the malignancy grade). Measurements in different regions within the same tumor may therefore, hypothetically, yield highly variable estimates of nuclear mean size, and only a proper macroscopic sampling scheme can counterbalance such bias (see below).

Table 1. Formulas Used for the Calculation of Quantitative Histopathologic Parameters.

The nuclear volume fraction	$V_V(nuc/tis) = P_P(nuc/tis) = \frac{N \cdot a(p)}{n_f \cdot A}$
The nuclear profile density index	$NI = \frac{Q_1(nuc)}{n_f \cdot A_1}$
The mitotic profile index	$MI = \frac{Q(mit) \cdot A}{Q_1(nuc) \cdot A_1}$
The mean nuclear profile area	$\bar{a}_H(nuc) = \frac{a(p) \cdot N \cdot A_1}{Q_1(nuc) \cdot A}$
The volume-weighted mean nuclear volume	$\bar{v}_V(nuc) = \frac{\pi}{3} \cdot \overline{\ell_0^3}$

Note: The variables can be estimated using the test-system shown in Figures 6, 7, and 8.
$P_P(nuc/tis)$ = the fraction of points, which hit nuclei per tissue.
N = the total number of nuclei hit by points.
n_f = the number of fields of vision.
a(p) = the area associated with each point in the test-system.
$Q_1(nuc)$ = the total number of nuclear profiles counted using the small frame with area A_1.
Q(mit) = the total number of mitotic profiles counted using the large frame with area A.
The subscripts V and H indicate that the items are sampled with a probability proportional to volume and height, respectively.

Tissue sectioning may also create problems as to the robustness of estimates of nuclear $\bar{v}_V$, but embedding in plastic eliminates this source of bias. So-called lost caps or truncation, i.e., the actual or virtual disappearance of nuclear fragments near the surface of the section, has an unpredictable influence on estimates of nuclear $\bar{v}_V$ (Gundersen, 1986). Due to the reduction in point-sampling of nuclei at their periphery, this artifact tends, however, to produce an overestimation (Cruz-Orive and Weibel, 1990). Finally, the physical thickness of the tissue section may introduce overprojection, but the upper limit of overestimation of nuclear $\bar{v}_V$ is about 36%, assuming infinite section thickness and spherical nuclear shape (Gundersen and Jensen, 1985), and even negative for particles of varying size (Gundersen, 1986). Thus, standardization of tissue handling, processing, sectioning, and measurements are necessary to keep all these biasing factors to a minimum.

Estimation of Nuclear Volume Pleomorphism

Increased variability of nuclear profile area is, morphologically, a well known and characteristic feature of malignant transformation, and the degree of nuclear size pleomorphism may be of value in quantitatively grading the malignancy. It is, however, not a straightforward matter to obtain estimates of three-dimensional nuclear size pleomorphism from two-dimensional tissue sections. Until recently, it was necessary to make serial sections and apply the laborious disector-Cavalieri approach to get unbiased data of this kind (Sørensen, 1991).

The estimation of the volume-weighted mean nuclear volume is a matter of interest, because estimates of nuclear $\bar{v}_V$ combine information about three-dimensional size with size variability. There exists a simple relation between nuclear $\bar{v}_V$ and variability of nuclear volume: $\bar{v}_V = \bar{v}_N \cdot (1 + CV_N^2(v))$ (Gundersen and Jensen, 1985), where $CV_N(v)$ is the relative variation of nuclear volume in the ordinary number-weighted distribution of particle volume. The high sensitivity of nuclear $\bar{v}_V$ in malignancy grading may in part be due to this relation, since estimates of nuclear $\bar{v}_V$ increase as a function of the variability of three-dimensional nuclear size, even in the case of fixed nuclear mean size.

It is a laborious task to obtain the size-distribution of nuclear size. One can, however, take a short cut to obtain estimates of nuclear $CV_N(v)$, in that nuclear $\bar{v}_V$ and nuclear $\bar{v}_N$ can be estimated simultaneously, by the use of the so-called selector described by Cruz-Orive (1987). This technique is extremely efficient and easy to perform when using an optical disector on thick plastic embedded tissue sections (McMillan et al., 1992).

The variation associated with estimates of nuclear $\bar{v}_V$, $Var_V(v)$, is difficult to estimate because one needs to estimate the second moment of the volume-weighted distribution of volume, $\overline{v_V^2}$, because nuclear $Var_V(v) = \overline{v_V^2} - \bar{v}_V^2$ (Jensen and Gundersen, 1985). Nuclear $\bar{v}_V$ is easy to estimate, but nuclear $\overline{v_V^2}$ is much more complicated. The second moment of the volume-weighted distribution of volume is valid for IUR sections or particle distributions that are globally IUR in three-dimensional space, and is defined by: $\overline{v_V^2} = 4 \cdot \pi \cdot (\overline{a_0^2 \cdot \nabla_0})$, in which a_0 is the area of a point-sampled planar profile, and ∇_0 is the area of a random triangle generated within this planar profile (Figure 9). Obviously, the estimation of a_0 and ∇_0 calls for the use of some kind of image analyzer, especially because the precision of the mean triangle area is important here. However, in this case an easy and fast manual procedure is possible, if the investigator seeking a tremendous gain in efficiency is willing to accept a small bias in the estimation of nuclear $\overline{v_V^2}$ (Sørensen, 1991). It has been shown by computer simulation that the ratio $\overline{\nabla}_0/a_0$ is remarkably constant (~ 0.075) in light microscopic sections of neoplastic nuclear populations (Jensen and Sørensen, 1991). Thus, one has only to estimate the point-sampled nuclear profile areas, knowing that the estimator of nuclear $\overline{v_V^2}$ can be re-written to a much more simple relation: $\overline{v_V^2} = 4 \cdot \pi \cdot 0.075 \cdot \overline{a_0^3}$, where a_0 still denotes the area of a point-sampled planar profile (Jensen and Sørensen, 1991; Sørensen, 1991). It is therefore not necessary to have access to expensive image analyzers for obtaining estimates of nuclear $Var_V(v)$, and what is even more interesting is that estimates of three-dimensional nuclear size variation from simple two-dimensional measurements in tissue sections using an ordinary microscope with projection attachment—

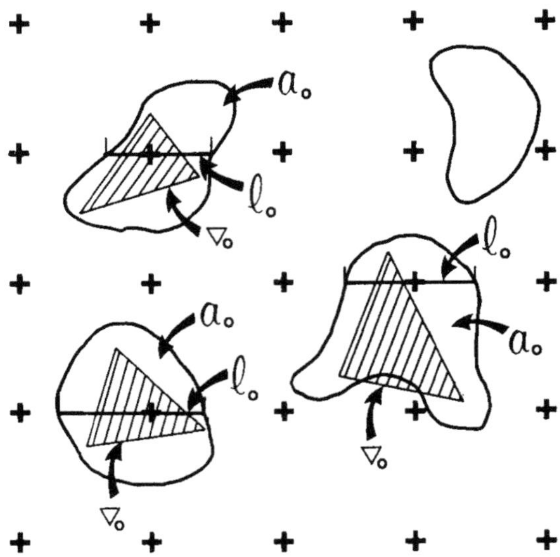

Figure 9. Four particle profiles are represented on this section, projected onto a test system with equidistant points. Only three of the particle profiles are hit by points. The point-sampled profile area, a_0 can be estimated by simple point-counting, as shown in Figure **4**. The area of the triangle inside each transect, ∇_0, is indicated by hatching, and can be estimated in the same manner. The triangles have been drawn after the random generation of three points inside each particle profile. The volume-weighted mean particle volume is estimated by measuring the point-sampled intercepts, l_0, in one arbitrary direction (here shown horizontally), assuming the particle population is isotropically orientated. Note that the general requirement of being able to identify all two-dimensional transects belonging to the same three-dimensional particle must be satisfied.

equipment that is accessible in all departments of laboratory medicine and pathology—can be obtained.

WHAT DOES NUCLEAR VOLUME EXPRESS?

Comparative studies of the relation between nuclear size and the total amount of biochemical nuclear constituents are hard to perform. Some sort of statistical analysis is the most common approach. In cancer studies, it is of special interest to estimate the changes in nuclear mean volume as a function of alterations in nuclear DNA content, i.e., the ploidy level. Such studies have been carried out but they have mainly relied on the comparison between nuclear DNA-content (expressed by the DNA index as estimated by either flow-cytometry or cytophotometry) and estimates of two-dimensional nuclear size (Helander and Tribukait, 1988; Stål and Hatschek, 1988). There are two major problems in such analyses. First, the relation between estimates of nuclear profile area and true three-dimensional nuclear size are always unknown, and second, the existence of tumor heterogeneity may create serious sampling problems at the tissue level for the two different analytical techniques involved.

The method for retrospective analysis of DNA content in paraffin embedded, routine tissue specimens, developed by Hedley et al. (1983, 1985) has, however, improved the prospects for a more realistic comparison between nuclear size and DNA content. Two adjacent sections can be analyzed separately by the simultaneous use of flow-cytometry in one (a 50 μm thick section) for the estimation of nuclear DNA-index, and stereologic estimation of realistic three-dimensional nuclear size (i.e., nuclear $\bar{v}_V$) in the other, thin section. Surprisingly, there seems to be only a weak or even nonexisting correlation between estimates of nuclear $\bar{v}_V$ and the nuclear DNA index. This has, for example, been shown in primary and metastatic malignant cutaneous melanomas (Sørensen et al., 1990), in squamous cell carcinomas of the uterine cervix (Sørensen et al., 1992), and in bladder cancer (Nielsen et al., 1989b). DNA may thus only contribute a rather small fraction to the total nuclear volume, and the cancer biologic implications are that the prognostic value of estimates of both nuclear $\bar{v}_V$ and nuclear DNA indices can only be evaluated in tumor specific studies.

SAMPLING AND EFFICIENCY

The clinical application of quantitative histopathology for objective grading of solid tumors depends on several factors, one of the most important of which is the balance between the desirability of precision and reproducibility and the costs in terms of money and manpower. In this regard estimates of nuclear $\bar{v}_V$ are recommendable, since they show a large spread among patients [a large CV (= SD/mean) within the group of patients] with the possibility to stratify differing prognoses of individual patients according to the value of nuclear $\bar{v}_V$. Based on this large biologic variation of the estimator it is unnecessary to invest a lot of work in achieving a better precision of estimates of nuclear $\bar{v}_V$. The number of measured intercepts, and the number of fields of vision, and their distribution throughout sections from the often rather heterogeneous tumors deserve more attention than that given to precision of the measurements of individual nuclear intercepts. The sampling procedure is thus the important factor, and it has been stressed that there is a need for a common approach internationally (Collan et al., 1987).

Two sampling approaches are used extensively in the quantitative grading of malignancy in solid cancers: these are selective, diagnostic sampling and systematic random sampling. The former stresses the necessity for subjectively selecting the tumor areas of relevance and nuclei to be investigated within such areas (Baak, 1984; 1987a,b), and this approach to sampling may only result in semi-quantitative malignancy grading. The latter, in contrast, provides the possibility of obtaining unbiased (and thus assumption-free) estimates of, e.g., nuclear $\bar{v}_V$, and, in addition, deserves credit for its high efficiency (Gundersen and Jensen, 1987). This is true because systematic random sampling eradicates the possible influence of tumor heterogeneity on estimates of nuclear $\bar{v}_V$. Moreover, the subjective decisions made by the pathologist in selecting measurement areas, which might be associated with personal prejudices and possible biases, are eliminated. Systematic random sampling involves sectioning of the whole tumor in a number of slabs, followed by systematic sampling of a fraction of these slabs, where the first must be taken at a random number within the sampling fraction needed. There are, however, restrictions to the use of systematic random sampling; for example, the sampling approach is not suited for most retrospective studies (i.e., usually the macroscopic surgical specimens have already been destroyed by selective sampling). On the other hand, convenient, macroscopic systematic sampling schemes for even very large surgical specimens have been developed, satisfying the basic requirements for obtaining unbiased, stereologic data at the microscopic level (Jensen, 1991).

Systematic random sampling of the macroscopic surgical specimens is, from the theoretical point of view, the only method by which it is possible to obtain internationally comparable results with respect to quantitative grading of malignancy. However, within selected sections from tumors, there seems to be only a minor difference as seen in estimates of nuclear $\bar{v}_V$ obtained by either systematic random sampling or with selective sampling within the section (Baak et al., 1994). This may not be the case with estimates of mitotic profile density in tumors, since the mitotic rate is often highest in the expansive, peripheral zone of the tumor, which is usually the most oxygenated part of solid malignant neoplasms.

Approximations to systematic random sampling, using one central section from a solid tumor, may represent a practical solution for sampling, especially with retrospective material, for such sections are often available. This approach has been used in a number of studies concerned with estimation of nuclear $\bar{v}_V$ (for a review, see Sørensen, 1992). Implementing so-called nested analysis of variance (Gundersen and Østerby, 1981) has revealed a high efficiency of estimates of nuclear $\bar{v}_V$. The contribution to overall variance by measurements of individual intercepts and by the differences between individual fields of vision is negligible, and by far the largest contributor to overall variation is the biologic differences between individual patients, i.e., placed at the highest level of sampling (Nielsen et al., 1986; 1989c; Sørensen, 1989a; Sørensen et al., 1991a,b).

THE IMPACT OF TUMOR HETEROGENEITY ON QUANTITATIVE MALIGNANCY GRADING

Primary malignant neoplasms are composed of a number of different clones of tumor cells which together constitute the solid tumor, and are thus responsible for the existence of heterogeneity at all levels of the tumor phenotype (Henson, 1982). This is also reflected by estimates of nuclear $\bar{v}_V$. In bladder cancer, blood group negative transitional cells have a significantly higher nuclear $\bar{v}_V$ than that found in blood group positive neoplastic urothelial cells (Ørntoft and Nielsen, 1989). Another example has been given by Mattfeldt et al. (1988), who found hormonal receptor negative breast cancer to have a higher nuclear $\bar{v}_V$ than that of receptor positive cancers. The clonal theory for tumor development is perhaps better illustrated when it comes to estimates of nuclear $\bar{v}_V$ by the following experiment: sampling in defined central and peripheral regions of primary skin melanomas and their associated metastases showed statistically significant differences in estimates of nuclear $\bar{v}_V$ in the two regions of the primary melanomas, but no difference in the secondary tumors (Sørensen and Erlandsen, 1990). This is in accord with the heterogeneous composition of primary tumors and the more homogeneous cell population in secondary neoplasms (Nowell, 1976; Fidler and Kripke, 1977; Poste and Fidler, 1980).

It is important when designing sampling schemes for quantitatively based grading of malignancy in solid cancer to deal rigorously with tumor heterogeneity. In this regard, it is always advisable to perform systematic random sampling in all three dimensions of the macroscopic surgical specimen.

REPRODUCIBILITY OF STEREOLOGIC ESTIMATES OF NUCLEAR VOLUME

The observer variation associated with the unbiased stereologic estimation of nuclear $\bar{v}_V$ is, like the efficiency, also of importance. From the theoretical point of view, the technique involved in estimating nuclear $\bar{v}_V$ offers superior conditions for excellent reproducibility. First, the sampling is based on a systematic random design, and second, the nuclei are point-sampled, i.e., the investigator does not, subjectively, select individual nuclei. Thus, there is no need for a highly experienced pathologist to carry out the measurements. Actually, the only possible source of observer variation is the measurement of nuclear intercepts, i.e., definition of nuclear borders in the chosen focal plane. As pointed out, the variation contributed by this lowest level of sampling is only minor.

A number of reproducibility studies have already been carried out with regard to stereologic estimation of nuclear $\bar{v}_V$ in various tumor types, such as prostatic cancer (Nielsen et al., 1989a), melanocytic skin tumors (Sørensen and Erlandsen, 1990; Sørensen and Ottosen, 1991), and squamous cell carcinomas of the uterine cervix (Sørensen et al., 1991a). An excellent intra- and inter-observer reproducibility has been found in these investigations, the correlation coefficients being above 0.85, and the slopes of correlation lines not significantly different from the ideal (= 1.00). However, if stereologic estimates of nuclear $\bar{v}_V$ are to be used on an international scale in malignancy grading of solid tumors, the reproducibility has to be analyzed, using an international study design. The first of this kind has now been completed (Sørensen et al., 1994). An excellent intra- and inter-observer reproducibility of estimates of nuclear $\bar{v}_V$ was found in this study of bladder cancer (correlation coefficients of 0.93 and 0.90, respectively, and slope of correlation lines = 1.00).

QUANTIFICATION OF NUCLEAR SIZE IN PROGNOSIS-RELATED MALIGNANCY GRADING

The stereologic technique outlined above for estimating the mean nuclear volume can be used in prognosis-related malignancy grading. The method has already been applied in many solid tumors, and it is not possible to give an extensive description of the results obtained in all performed studies. Instead, a brief review will be given of selected types of cancer, some of which have already been discussed in relation to the problems arising from traditional, morphologic grading of malignancy.

Cutaneous malignant melanoma. Two-dimensional estimates of nuclear profile areas have been widely used in the diagnostic and prognostic evaluation of melanocytic skin tumors (see for example Heenan et al., 1989; Tosi et al., 1989). However, the prognostic value of such estimates appears questionable. Estimates of nuclear mean profile area have been reported to be correlated with survival of patients treated for skin melanomas, with the large nuclear mean profile area indicating shorter survival and increased mortality (Heenan et al., 1989). Another study failed to substantiate these findings (Sørensen, 1989a). Such diverse findings may in large part be accounted for by the biases associated with two-dimensional estimates of nuclear mean profile area.

Turning to three-dimensional estimation of nuclear size, retrospectively designed studies have documented the prognostic value of estimates of nuclear $\bar{v}_V$ (Sørensen, 1989a; Sørensen et al., 1991c). The largest values of nuclear $\bar{v}_V$ are those associated with the most unfavorable prognosis, and when compared to other prognostic indicators for melanomas based on Cox analyses, estimates of nuclear $\bar{v}_V$ have an independent prognostic value (Sørensen, 1989a; Sørensen et al., 1991c). The results of retrospective investigations are, however, sensitive to changes in treatment and other factors related to secularity, and must therefore be re-investigated in contemporary, prospective studies of consecutively treated patients with malignant melanomas of the skin.

Squamous cell carcinoma of the uterine cervix. This type of cancer is an exception to the rule that cancer cells with large nuclei are more malignant than neoplastic cells with smaller nuclei. A retrospective investigation of patients treated for squamous cell carcinomas of the uterine cervix has shown that estimates of nuclear $\bar{v}_V$ correlated negatively with mortality rate (Sørensen et al., 1992). Cox analysis has indeed demonstrated the independent prognostic value of the stereologic parameter when analyzed along with information of the clinical stage of disease and DNA index, which also represent independent prognostic variables in this type of malignant neoplasm (Sørensen et al., 1992). Thus, by the use of these three reproducible parameters it is possible to construct a prognostic index for individual patients with excellent predictiveness. This clears the way to the future development of a prognostic identification pattern based on multiple, objectively assessable variables, where no single prognostic parameter contains all the pertinent prognostic information.

Transitional cell carcinomas of the urinary bladder. This type of neoplasm has been investigated intensively with the new stereologic technique for estimating nuclear $\bar{v}_V$ in a prognosis-related setting. The utility of estimates of nuclear $\bar{v}_V$ in the histopathologic, prognostic evaluation of patients with transitional cell carcinomas has been documented in a number of studies (Nielsen et al., 1986; Nielsen et al., 1989c; Sasaki et al., 1993). Of special interest is the high sensitivity of estimates of nuclear $\bar{v}_V$ in predicting the clinical course of patients with bladder tumors in clinical stage Ta. It is well documented in two retrospectively designed investigations that the estimate of nuclear $\bar{v}_V$ in the diagnostic biopsy of the initial Ta tumor of these patients is highly specific in predicting the final status of the patients after a clinical follow-up period of at least five years (Nielsen et al., 1989c; Sasaki et al., 1993). Such findings are of clinical importance, since they may influence the clinical follow-up of patients with bladder tumors in a more specified, individual fashion, and at the same time reduce costs in the health-care system.

Adenocarcinoma of the prostate gland. There seems to be a large overlap in the values of nuclear $\bar{v}_V$ obtained in incidental and clinically manifest prostate cancer (Nielsen et al., 1987; Jørgensen et al., 1988). However, one study has demonstrated the prognostic value of repeated estimations of nuclear $\bar{v}_V$ in patients requiring more than one resection for prostatic adenocarcinoma. Increased values of nuclear $\bar{v}_V$ in secondary surgical specimens are associated with shorter survival (Nielsen et al., 1989a).

Squamous cell carcinoma of the oral cavity, the larynx, and the lung. A group of 35 patients with supraglottic squamous cell carcinoma of the larynx has been investigated retrospectively with regard to the prognostic significance of estimates of nuclear $\bar{v}_V$ and various other morphometric variables (Sørensen et al., 1989). All the investigated parameters were without significant prognostic impact with regard to survival. Further studies of a larger series of patients are needed to evaluate the true prognostic value of stereologic, nuclear grading of malignancy in this type of cancer.

The prognostic value of estimates of nuclear $\bar{v}_V$ in squamous cell carcinomas of the oral mucosa has been investigated in two independent studies. In a group of 44 patients no prognostic value could be found in univariate analyses, using arbitrary cutoff points (Bryne et al., 1991). However, in the second study of 74 patients, estimates of nuclear $\bar{v}_V$ were shown to be of independent prognostic value using Cox analysis, when analyzed together with the clinical stage of the disease (Bundgaard et al., 1992). As part of a larger epidemiologic, prospective study, the latter findings have recently been corroborated regarding the clinical prognostic value of nuclear $\bar{v}_V$ estimates in squamous cell carcinomas of the oral cavity.

Recently, an investigation of 55 patients with squamous cell and adenosquamous carcinomas of the lung has shown prognostic significance of estimates of nuclear $\bar{v}_V$ (Ladekarl et al., 1995). This is quite interesting in view of the high frequency of this malignant neoplasm. Moreover, morphologic grading of squamous cell carcinomas of the lung has hitherto been lacking in histologic parameters with significant prognostic impact.

Breast cancer. Ductal carcinoma of the mammary gland has been the object of study in a large number of morphometric investigations relying on two-dimensional morphometry (see Baak et al., 1989b). Recently, the use of unbiased stereology has been evaluated in this frequent type of cancer (Ladekarl and Sørensen, 1993a). In the most frequent type of breast cancer, the ductal type, estimates of nuclear $\bar{v}_V$ have shown independent prognostic value in a Cox analysis, when tested together with the clinical stage of the disease (Ladekarl, 1995). Similar results were obtained in a study of patients with lobular carcinoma of the breast, where estimates of nuclear $\bar{v}_V$ also showed independent prognostic value (Ladekarl and Sørensen, 1993b). It thus seems that the stereologic technique for malignancy grading is of prognostic value in this very frequent type of human malignant neoplasm, and the results of these primary, retrospective studies are now being investigated in prospectively designed trials.

Cerebral glioma. In this type of tumor the pathologist is faced with serious problems relating to morphologic evaluation, because it may be extremely difficult to distinguish reactive glial cells from neoplastic cells of the same origin. From a technical point of view, estimation of nuclear $\bar{v}_V$ is optimal; the pathologist need only consider the biologic character of the relatively few glial cells hit by points of the used test system. Moreover, the shape of neoplastic glial cell nuclei is highly pleomorphic (Giangaspero et al., 1984), and estimates of nuclear $\bar{v}_V$ are therefore to be preferred, as compared to shape-dependent estimates of nuclear mean profile area. In a retrospective study of patients treated for supratentorial gliomas, it was shown that only estimates of mitotic profile index were of prognostic significance in respect of final clinical outcome after 10 years of follow-up. However, estimates of nuclear $\bar{v}_V$ could distinguish two prognostically different groups of patients in the three first postoperative years (Sørensen et al., 1991b).

Other variables may be of more prognostic relevance in the case of brain tumors, for example the macroscopic volume of the tumor. This parameter may be obtained rather easily and precisely by the use of Cavalieri's principle on CT scans for the unbiased estimation of tumor volume. It has been shown that the smallest gliomas have the worst prognosis during the first two to three-postoperative years (Sørensen et al., 1991b). This could be due to a higher growth potential in tumors giving clinical symptoms, when they are only relatively small, and may indicate that estimation of the proliferative activity by immunohistochemical markers may be of prognostic value in cerebral gliomas.

Ocular melanoma. Estimates of nucleolar $\bar{v}_V$ by Cox analyses have shown a remarkably, independent prognostic impact in two studies with long term follow-up (Sørensen et al., 1993a; 1993b). Important prognostic information may thus be obtained by quantitative evaluation of nucleoli, which has already been demonstrated in various types of human cancer by the use of the so-called AgNOR technique (Rüschoff et al., 1989). Moreover, estimates of nucleolar size variability (expressed by the standard deviation or coefficient of variation of nucleolar $\bar{v}_V$) are of prognostic value in patients with choroidal melanomas (Sørensen et al., 1993a). Stereologic estimates of three-dimensional size pleomorphism are, as compared to estimates of $\bar{v}_V$, more complicated to obtain. Cox analyses have disclosed that the prognostic value of nucleolar $\bar{v}_V$ suffices in the prediction of the clinical course of the disease in patients with choroidal melanomas. No extra prognostic value is gained by adding estimates of nucleolar three-dimensional size pleomorphism (Sørensen et al., 1993a). This fact may be due to the combined information of size and size pleomorphism contained in estimates of $\bar{v}_V$, again pointing to the importance of this parameter in histopathologic, quantitative grading of malignancy.

Other malignancies. A series of 100 patients with ovarian serous adenocarcinomas, FIGO stages 1B through IV, have been investigated for prognostic impact of estimates of nuclear $\bar{v}_V$. In this type of malignant neoplasm estimates of nuclear $\bar{v}_V$ correlated with the clinical stage of the disease and the morphologic grade of malignancy, whereas no significant correlation with survival was found in the studied group of patients (Mogensen et al., 1992).

Finally, the stereologic technique has also been tested for prognostic impact in a group of patients with plaque and tumor-stage mycosis fungoides with long term follow-up. In this malignant, cutaneous lymphoma estimates of nuclear $\bar{v}_V$ appeared to be closely correlated with clinical outcome, in that large values of nuclear $\bar{v}_V$ were suggestive of shorter survival (Brooks et al., 1994).

THE FUTURE OF QUANTITATIVE MALIGNANCY GRADING

With the new generation of unbiased stereologic estimators, which are easy to apply in routine histopathology, and are of prognostic value in a number of solid human malignant neoplasms, only a minor step has been taken towards objective and reproducible grading of malignancy. No single variable contains all pertinent prognostic information for a given type of malignant neoplasm. A number of other parameters, including clinical ones, may be of prognostic value in the evaluation of solid cancers, e.g., immunohistochemical demonstration of oncogenes and proliferation antigens, and the determination of DNA-index. Sophistication of prognostic prediction may thus be achieved by combining various objective and reproducible parameters in a multifactorial malignancy score. The construction of such indices for patients with solid malignant neoplasms may thereby advance the discipline of malignancy grading from its present state of unfit for use to a state of clinical relevance. On the other hand, the use of unbiased stereologic estimates of nuclear $\bar{v}_V$ for the purpose of classifying tumors into various types of neoplasms, or benign and malignant categories of tumors having the same origin, seems futile due to overlap between groups investigated (Sørensen, 1992; Brooks et al., 1993).

CONCLUSION

The state of the art of malignancy grading, based solely on the morphologic evaluation of tissue sections, is not in keeping with the times. The low reproducibility makes it patently obvious that new quantitatively based, objective techniques are required in histopathology to improve the clinical value of malignancy grading. Stereology and morphometry only represent two objective and reproducible methodologies of relevance in this context. They have already shown their prognostic superiority, as compared to traditional grading of malignancy, whereas due to overlap, they are mostly of no value in the classification of neoplasms. Moreover, it must be emphasized that the technique for obtaining unbiased estimates of nuclear $\bar{v}_V$ in tissue sections from solid malignant neoplasms is very simple and inexpensive. All pathology departments can adopt the technology in routine diagnostic work using simple equipment such as a projection microscope, or more conveniently, a computer-assisted system.

In selecting quantitative methods for the grading of malignancy, the modern, unbiased stereologic technique for estimating the volume-weighted mean volume of nuclei and nucleoli is recommended, because the method is easy to use, efficient and practicable in routine settings of histopathologic laboratories. In addition, it provides prognostic information for patients with various types of solid cancer. It must, however, be remembered that estimates of nuclear three-dimensional mean size can only supplement the more important prognostic parameters, such as the clinical stage of the disease. However, the use of multifactorial prognostic schemes, based on reproducible and objective variables offers hope for the development of more specific and individual prognostic evaluations of patients suffering from cancer, thus enabling standardization for comparison of international clinical cancer treatment trials and optimal choice of treatment for the individual patient.

SUMMARY

Simple measurements, carried out in two-dimensional tissue sections, can with the use of stereologic techniques form the basis for the unbiased estimation of three-dimensional nuclear mean size. The method involves point-sampling of nuclear profiles in random histologic sections, and is associated with high efficiency and reproducibility. The requirements for the estimator, including correct sampling at the macroscopic level to counterbalance the existence of heterogeneity in neoplasms, can be satisfied under routine conditions of histopathologic evaluation of human solid cancers, and no special training is necessary. Because of the point-sampling involved, the larger nuclei are sampled preferentially, and the stereologic technique thus provides an estimate of the volume-weighted mean nuclear volume, nuclear $\bar{v}_V$. This is an advantage, when the method is used in the context of solid human cancer, since one of the major characteristics of malignant transformation is an increase in nuclear size. Moreover, it is possible to estimate the nuclear three-dimensional size pleomorphism in two-dimensional tissue sections. This estimator may also be of significant prognostic value in the evaluation of patients with cancer. However, estimates of nuclear $\bar{v}_V$ may be sufficient, because they combine data of three-dimensional size with information of pleomorphism of nuclear volume. It is therefore well-suited for use in objective, histopathologic grading of malignancy. From the biological point of view, there seems to be only a weak or even nonexisting correlation between realistic estimates of three-dimensional nuclear mean size and the nuclear DNA content, expressed by the DNA index. Thus, nuclear size may represent a dynamic expression of the disturbed nuclear metabolism of cancer cells, and serve as a sensitive tool for malignancy grading.

Unbiased stereologic estimates of nuclear $\bar{v}_V$ have confirmed their prognostic value in various human malignant neoplasms, and these findings are now being re-evaluated in prospective studies of patients treated for cancer. The new, objective technology may not only provide prognostic information, but may also be useful in planning treatment and in the clinical follow-up of the patient. It is therefore hoped that stereologic grading of malignancy by simple, reproducible measurements of nuclei, can help replace the traditional, subjective technique for morphologic evaluation of malignancy, and turn histopathologic grading of malignancy into a discipline that aids the clinician in the choice of therapy for individual patients with cancer.

ACKNOWLEDGMENTS

The author thanks Anne Dalmose, Anette Larsen, and Maj-Britt Lundorf for their skillful and untiring technical assistance. Some of the studies upon which this survey are based were financially supported by Aarhus University Research Foundation; Aage Bang's Foundation; The Danish Cancer Society; The Danish Foundation for the Advancement of Medical Science; The Danish Medical Research Council; The 1870 Foundation; Valdemar & Thyra Foersom's Foundation; Johannes & Ella Fogh-Nielsen's Foundation; Frits, Georg & Marie Cecilie Glud's Foundation; Ib Henriksen's Foundation; Carl & Ellen Hertz Foundation; Hindsgaul's Foundation; The Højmosegaard Foundation; King Christian the Xth Foundation; Erik & Knudsine Leijon's Foundation; Lily B. Lund's Foundation; Jacob & Olga Madsen's Foundation; M.C. & J.K. Moltum's Foundation; Carla Cornelia S. Møller's Foundation; Leo & Karen Margrethe Nielsen's Foundation; Novo's Foundation; P. Carl Petersen's Foundation; Kathrine & Vigo Skovgaard's Foundation; August Frederik Wedell Erichsen's Foundation; Else & Mogens Wedell-Wedellsborg's Foundation; and Willumsen's Foundation.

REFERENCES

Abel, P.D., Henderson, D., Bennett, M.K., Hall, R.R., & Williams G. (1988). Differing interpretations of the pT category and grade of transitional cell cancer of the bladder by pathologists. Br. J. Urol. 62, 339–342.

Aru, A. & Nielsen, K. (1989). Stereological estimates of nuclear volume in primary lung cancer. Path. Res. Pract. 185, 735–739.

Baak, J.P.A. (1984). Basic points in and practical aspects of the application of diagnostic morphometry. Path. Res. Pract. 179, 193–199.

Baak, J.P.A. (1987a). Quantitative pathology today - A technical view. Path. Res. Pract. 182, 396–400.

Baak, J.P.A. (1987b). The principles and advances of quantitative pathology. Analyt. Quant. Cytol. Histol. 9, 89–95.

Baak, J.P.A. (1991). Manual of Quantitative Pathology in Cancer Diagnosis and Prognosis, Springer-Verlag, London.

Baak, J.P.A., Ladekarl, M., & Sørensen, F.B. (1994). Reproducibility of mean nuclear volume and correlation with mean nuclear area in breast cancer. Hum. Pathol. 25, 80–85.

Baak, J.P.A., Noteboom, E., & Koevoets, J.J.M. (1989a). The influence of fixatives and other variations in tissue processing on nuclear morphometric features. Analyt. Quant. Cytol. Histol. 11, 219–224.

Baak, J.P.A., Van Diest, P.J., Ariens, A.T., Van Beek, M.W.P.M., Bellot, S.M., Fijnheer, J., Van Gorp, L.H.M., Kwee, W.-S., Los, J., Peterse, H.C., Ruitenberg, H.M., Schapers, R.F.M., Schipper, M.E.I., & Somsen, J.G. (1989b). The multicenter morphometric mammary carcinoma project (MMMCP). A nationwide prospective study on reproducibility and prognostic power of routine quantitative assessments in the Netherlands. Path. Res. Pract. 185, 664–670.

Baddeley, A.J., Gundersen, H.J.G., & Cruz-Orive, L.-M. (1986). Estimation of surface area from vertical sections. J. Microsc. 142, 259–276.

Bergkvist, A., Ljungqvist, A., & Moberger, G. (1965). Classification of bladder tumours based on the cellular pattern. Preliminary report of a clinical-pathological study of 300 cases with a minimum follow-up of eight years. Acta. Chir. Scand. 130, 371–378.

Bichel, P. & Jakobsen, A. (1985). Histopathologic grading and prognosis of uterine cervical carcinoma. Am. J. Clin. Oncol. 8, 247–254.

Black, M.M. & Speer, F.D. (1957). Nuclear structure in cancer tissues. Surg. Gynecol. Obstet. 105, 97–102.

Boonstra, H., Oosterhuis, J.W., Oosterhuis, A.M., & Fleuren, G.J. (1983). Cervical tissue shrinkage by formaldehyde fixation, paraffin wax embedding, section cutting and mounting. Virchows. Arch. A. 402, 195–201.

Breslow, A. (1977). Problems in the measurement of tumor thickness and level of invasion in cutaneous melanoma. Hum. Pathol. 8, 1–2.

Breslow, A. (1980). Prognosis in cutaneous melanoma: Tumor thickness as a guide to treatment. Pathol. Ann. 15(1), 1–22.

Broders, A.C. (1922). Epithelioma of the genito-urinary organs. Ann. Surg. 75, 574–604.

Brooks, B., Sørensen, F.B., Olsen, S., & Holm-Nielsen, P. (1993). Classification of tubulo-papillary renal cortical tumours using estimates of nuclear volume. APMIS 101, 378–386.

Brooks, B., Sørensen, F.B., & Thestrup-Pedersen, K. (1994). Estimates of nuclear volume in plaque and tumor-stage mycosis fungoides. A new prognostic indicator. Am. J. Dermatopathol. 16, 599–606.

Bryne, M., Nielsen, K., Koppang, H.S., & Dabelsteen, E. (1991). Grading of nuclear pleomorphism in oral carcinomas. Higher prognostic value than stereological assessment of nuclear volume. APMIS 99, 333–339.

Bundgaard, T., Sørensen, F.B., Gaihede, M., Søgaard, H., & Overgaard, J. (1992). Stereologic, histopathologic, flow cytometric, and clinical parameters in the prognostic evaluation of 74 patients with intraoral squamous cell carcinomas. Cancer 70, 1–13.

Busch, C., Engberg, A., Norlen, B.J., & Stenkvist, B. (1977). Malignancy grading of epithelial bladder tumours. Scand. J. Urol. Nephrol. 11, 143–148.

Clark, W.H., Elder, D.E., & Van Horn, M. (1986). The biologic forms of malignant melanoma. Hum. Pathol. 17, 443–450.

Clark, W.H., From, L., Bernardino, E.A., & Mihm, M.C. (1969). The histogenesis and biologic behavior of primary human malignant melanomas of the skin. Cancer Res. 29, 705–726.

Cocker, J., Fox, H., & Langley, F.A. (1968). Consistency in the histological diagnosis of epithelial abnormalities of the cervix uteri. J. Clin. Pathol. 21, 67–70.

Collan, Y., Torkkeli, T., Pesonen, E., Jantunen, E., & Kosma, V.-M. (1987). Application of morphometry in tumor pathology. Analyt. Quant. Cytol. Histol. 9, 79–88.

Colloby, P.S., West, K.P., & Fletcher, A. (1991). Observer variation in the measurement of Breslow depth and Clark's level in thin cutaneous malignant melanomas. J. Pathol. 163, 245–250.

Colpaert, C., Goovaerts, G., & Buyssens, N. (1987). Factors influencing the subjective grading of bladder cancer. Virchows. Arch. A. 411, 479–484.

Cruz-Orive, L.-M. (1980). On the estimation of particle number. J. Microsc. 120, 15–27.

Cruz-Orive, L.-M. (1987). Particle number can be estimated using a disector of unknown thickness: The selector. J. Microsc. 145, 121–142.

Cruz-Orive, L.-M. & Hunziker, E.B. (1986). Stereology for anisotropic cells: Application to growth cartilage. J. Microsc. 143, 47–80.

Cruz-Orive, L.-M. & Weibel, E.R. (1990). Recent stereological methods for cell biology: A brief survey. Am. J. Physiol. 258 (Lung Cell Mol Physiol 2), L148–L156.

Ellis, P.S.J. & Whitehead, R. (1981). Mitosis counting - a need for reappraisal. Hum. Pathol. 12, 3–4.

Fidler, I.J. & Kripke, M.L. (1977). Metastasis results from preexisting variant cells within a malignant tumor. Science 197, 893–895.

Giangaspero, F., Muhlbaier, L.H., & Burger, P.C. (1984). The glioblastoma multiforme and the metastatic carcinoma: A morphometric study of nuclear size and shape. Appl. Pathol. 2, 160–167.

Græm, N. & Helweg-Larsen, K. (1979). Mitotic activity and delay in fixation of tumour tissue. The influence of delay in fixation on mitotic activity of a human osteogenic sarcoma grown in athymic nude mice. Acta. Path. Microbiol. Scand. (Sect. A). 87, 375–378.

Græm, N., Helweg-Larsen, K., & Keiding, N. (1980). Precision of histological grading of malignancy. Sources of variation in a histological scoring system for grading cancer of the larynx. Acta. Path. Microbiol. Scand. (Sect. A). 88, 307–317.

Gundersen, H.J.G. (1977). Notes on the estimation of the numerical density of arbitrary profiles: The edge effect. J. Microsc. 111, 219–223.

Gundersen, H.J.G. (1986). Stereology of arbitrary particles. A review of unbiased number and size estimators and the presentation of some new ones, in memory of William R. Thompson. J. Microsc. 143, 3–45.

Gundersen, H.J.G. (1988). The nucleator. J. Microsc. 151, 3–21.

Gundersen, H.J.G., Bagger, P., Bendtsen, T.F., Evans, S.M., Korbo, L., Marcussen, N., Møller, A., Nielsen, K., Nyengaard, J.R., Pakkenberg, B., Sørensen, F.B., Vesterby, A., & West, M.J. (1988a). The new stereological tools: Disector, fractionator, nucleator and point sampled intercepts and their use in pathological research and diagnosis. APMIS 96, 857–881.

Gundersen, H.J.G., Bendtsen, T.F., Korbo, L., Marcussen, N., Møller, A., Nielsen, K., Nyengaard, J.R., Pakkenberg, B., Sørensen, F.B., Vesterby, A., & West, M.J. (1988b). Some new, simple and efficient stereological methods and their use in pathological research and diagnosis. APMIS 96, 379–394.

Gundersen, H.J.G. & Jensen, E.B. (1983). Particle sizes and their distributions estimated from line- and point-sampled intercepts. Including graphical unfolding. J. Microsc. 131, 291–310.

Gundersen, H.J.G. & Jensen, E.B. (1985). Stereological estimation of the volume-weighted mean volume of arbitrary particles observed on random sections. J. Microsc. 138, 127–142.

Gundersen, H.J.G. & Jensen, E.B. (1987). The efficiency of systematic sampling in stereology and its prediction. J. Microsc. 147, 229–263.
Gundersen, H.J.G. & Østerby, R. (1981). Optimizing sampling efficiency of stereological studies in biology: Or 'Do more less well!' J. Microsc. 121, 65–73.
Hall, T.L. & Fu, Y.S. (1985). Applications of quantitative microscopy in tumor pathology. Lab. Invest. 53, 5–21.
Hedley, D.W., Friedlander, M.L., & Taylor, I.W. (1985). Application of DNA flow cytometry to paraffin-embedded archival material for the study of aneuploidy and its clinical significance. Cytometry. 6, 327–333.
Hedley, D.W., Friedlander, M.L., Taylor, I.W., Rugg, C.A., & Musgrove, E.A. (1983). Method for analysis of cellular DNA content of paraffin-embedded pathological material using flow cytometry. J. Histochem. Cytochem. 31, 1333–1335.
Heenan, P.J., Jarvis, L.R., De Klerk, N.H., & Armstrong, B.K. (1989). Nuclear indices and survival in cutaneous melanoma. Am. J. Dermatopathol. 11, 308–312.
Heenan, P.J., Matz, L.R., Blackwell, J.B., Kelsall, G.R.H., Singh, A., Ten Seldam, R.E.J., & Holman, C.D.J. (1984). Inter-observer variation between pathologists in the classification of cutaneous malignant melanoma in Western Australia. Histopathology. 8, 717–729.
Helander, K.G. & Tribukait, B. (1988). Modal DNA values of normal and malignant urothelial cells of the bladder in relation to nuclear size. Analyt. Quant. Cytol. Histol. 10. 127–133.
Henson, D.E. (1982). Heterogeneity in tumors. Arch. Pathol. Lab. Med. 106, 597–598.
Holman, C.D.J., James, I.R., Heenan, P.J., Matz, L.R., Blackwell, J.B., Kelsall, G.R.H., Singh, A., & Ten Seldam, R.E.J. (1982). An improved method of analysis of observer variation between pathologists. Histopathology. 6, 581–589.
Jensen, E.B. & Gundersen, H.J.G. (1985). The stereological estimation of moments of particle volume. J. Appl. Prob. 22, 82–98.
Jensen, E.B. & Gundersen, H.J.G. (1987). The corpuscle problem: Reevaluation using the disector. Acta. Stereol. 6, 105–122.
Jensen, E.B. & Sørensen, F.B. (1991). A note on stereological estimation of the volume-weighted second moment of particle volume. J. Microsc. 164, 21–27.
Jensen, O.M. (1991). Gross examination of musculoskeletal sarcomas. Estimation of tumour volume and sampling for unbiased morphometry. APMIS 99, 1031–1037.
Jørgensen, P.E., Berild, G.H., Bruun, E., Weis, N., Nielsen, K., & Gundersen, H.J.G. (1988). Stereological estimates of nuclear volume correlated with clinical stage and progression of prostatic cancer: Clinical aspects. Scand. J. Urol. Nephrol. (suppl.110), 131–135.
Ladekarl, M. (1994). The influence of tissue processing on quantitative histopathology in breast cancer. J. Microsc. 174, 93–100.
Ladekarl, M. (1995). Quantitative histopathology in ductal carcinoma of the breast. Cancer 75, 2114–2122.
Ladekarl, M., Hansen, T.B., Henrik-Nielsen, R., Mouritsen, C., Henriques, U.V., & Sørensen, F.B. (1995). Objective malignancy grading of squamous cell carcinoma of the lung. Stereologic estimates of mean nuclear size are of prognostic value, independent of clinical stage of disease. Cancer 76, 797–802.
Ladekarl, M. & Sørensen, F.B. (1993a). Quantitative histopathological variables in *in situ* and invasive ductal and lobular carcinomas of the breast. APMIS 101, 895–903.
Ladekarl, M. & Sørensen, F.B. (1993b). Prognostic quantitative histopathologic variables in lobular carcinoma of the breast. Cancer 72, 2602–2611.
Larsen, T.E., Little, J.H., Orell, S.R., & Prade, M. (1980). International pathologists congruence survey on quantitation of malignant melanoma. Pathology. 12, 245–253.
Loud, A.V. & Anversa, P. (1984). Morphometric analysis of biologic processes. Lab. Invest. 50, 250–261.
Lovett, E.J., Schnitzer, B., Keren, D.F., Flint, A., Hudson, J.L., & McClatchey, K.D. (1984). Application of flow cytometry to diagnostic pathology. Lab. Invest. 50, 115–140.

MacKie, R.M. (1982). Tumor thickness in melanoma. N. Engl. J. Med. 306, 1179.
Martzloff, K.H. (1923). Carcinoma of the cervix uteri. A pathological and clinical study with particular reference to the relative malignancy of the neoplastic process as indicated by the predominant type of cancer cell. Bull. Johns. Hopkins. Hosp. 34, 141–149.
Mattfeldt, T., Mall, G., Gharehbaghi, H., & Möller, P. (1990). Estimation of surface area and length with the orientator. J. Microsc. 159, 301–317.
Mattfeldt, T., Schürmann, G., Matthaei-Maurer, U., Feichter, G., & Möller, P. (1988). Morphometrische Äquivalente des Differenzierungsgrades bei epithelialen Tumoren und Lymphomen. Verh. Dtsch. Ges. Path. 72, 343–348.
Mattfeldt, T. Stereologische Methoden in der Pathologie, Stuttgart, Georg Thieme Verlag, 1990.
McMillan, A.-M., Sørensen, F.B., & Howard, C.V. (1992). The efficient and unbiased estimation of nuclear size variability using the 'selector'. J. Microsc. 165, 433–437.
Meyer-Arendt, J.R. & Humphreys, D.M. (1972). Quantitative morphology of cancer cells. Acta. Histochem. 44, 41–48.
Mogensen, O., Sørensen, F.B., Bichel, P., & Jakobsen, A. (1992). Nuclear volume and prognosis in ovarian cancer. Int. J. Gynecol. Cancer 2, 141–146.
Mostofi, F.K., Sobin, L.H., & Torloni, H. (1973). Histological typing of urinary bladder tumours, World Health Organization, Geneva.
Nielsen, K. (1988). Stereological estimates of nuclear volume in normal mucosa and carcinoma *in situ* of the human urinary bladder. Virchows. Arch. B. 55, 233–236.
Nielsen, K., Berild, G.H., Bruun, E., Jørgensen, P.E., & Weis, N. (1989a). Stereological estimation of mean nuclear volume in prostatic cancer, the reproducibility and the possible value of estimations on repeated biopsies in the course of disease. J. Microsc. 154, 63–69.
Nielsen, K., Colstrup, H., Nilsson, T., & Gundersen, H.J.G. (1986). Stereological estimates of nuclear volume correlated with histopathological grading and prognosis of bladder tumour. Virchows. Arch. B. 52, 41–54.
Nielsen, K., Jørgensen, P.E., Berild, G.H., Bruun, E., Weis, N., & Gundersen, H.J.G. (1987). Stereological estimates of nuclear volume correlated with clinical stage and progression of prostatic cancer. Application of a stereological method in diagnostic quantitative pathology. Acta. Stereol. 6, 59–64.
Nielsen, K., Ørntoft, T., & Wolf, H. (1989c). Stereologic estimates of nuclear volume in noninvasive bladder tumors (Ta) correlated with recurrence pattern. Cancer 64, 2269–2274.
Nielsen, K., Petersen, S.E., & Ørntoft, T. (1989b). A comparison between stereological estimates of mean nuclear volume and DNA flow cytometry in bladder tumours. APMIS 97, 949–956.
Nowell, P.C. (1976). The clonal evolution of tumor cell populations. Acquired genetic lability permits stepwise selection of variant sublines and underlies tumor progression. Science 194, 23–28.
Olsen, S. (1984). Tumors of the kidney and urinary tract, Copenhagen, Munksgaard.
Ooms, E.C.M., Anderson, W.A.D., Alons, C.L., Boon, M.E., & Velhuizen, R.W. (1983). Analysis of the performance of pathologists in the grading of bladder tumors. Hum. Pathol. 14, 140–143.
Ørntoft, T.F. & Nielsen, K. (1989). Heterogeneic expression of blood group A and H isoantigens in bladder tumors: Association with nuclear volume. J. Histochem. Cytochem. 37, 1153–1155.
Pesce, C.M. (1987). Defining and interpreting diseases through morphometry. Lab. Invest. 56, 568–575.
Poste, G. & Fidler, I.J. (1980). The pathogenesis of cancer metastasis. Nature. 283, 139–146.
Poulsen, H.E. & Taylor, C.W. (1975). Histological typing of female genital tract tumours, World Health Organization, Geneva.
Prade, M., Sancho-Garnier, H., Cesarini, J.P., & Cochran, A. (1980). Difficulties encountered in the application of Clark classification and the Breslow thickness measurement in cutaneous malignant melanoma. Int. J. Cancer 26, 159–163.
Ringsted, J., Amtrup, F., Asklund, C., Baunsgaard, P., Christensen, H.E., Hansen, L., Jakobsen, C., Jensen, N.K., Moesner, J., Rasmussen, J., Reintoft, I., Rolschau, J., Starklint, H., Thommesen, N., & Vrang, J. (1978). Reliability of histo-pathological diagnosis of squamous epithelial changes of the uterine cervix. Acta. Path. Microbiol. Scand. (Sect. A). 86, 273–278.

Robertson, A.J., Swanson Beck, J., Burnett, R.A., Howatson, S.R., Lee, F.D., Lessells, A.M., McLaren, K.M., Moss, S.M., Simpson, J.G., Smith, G.D., Tavadia, H.B., & Walker, F. (1990). Observer variability in histopathological reporting of transitional cell carcinoma and epithelial dysplasia in bladders. J. Clin. Pathol. 43, 17–21.

Rüschoff, J., Plate, K., Bittinger, A., & Thomas, C. (1989). Nucleolar organizer regions (NORs). Basic concepts and practical application in tumor pathology. Path. Res. Pract. 185, 878–885.

Sasaki, M., Sørensen, F.B., Fukuzawa, S., Yamabe, H., Olsen, S., & Yoshida, O. (1993). Quantitative histopathology in the prognostic evaluation of patients with transitional cell carcinoma of the urinary bladder. Cancer 72, 2470–2483.

Schmoeckel, C., Bockelbrink, A., Bockelbrink, H., Koutsis, J., & Braun-Falco, O. (1983). Low- and high-risk malignant melanoma - I. Evaluation of clinical and histological prognosticators in 585 cases. Eur. J. Cancer Clin. Oncol. 19, 227–235.

Søndergaard, K. & Schou, G. (1985). Survival with primary cutaneous malignant melanoma, evaluated from 2012 cases. A multivariate regression analysis. Virchows. Arch. A. 406, 179–195.

Sørensen, F.B. (1989a). Objective histopathologic grading of cutaneous malignant melanomas by stereologic estimation of nuclear volume. Prediction of survival and disease-free period. Cancer 63, 1784–1798.

Sørensen, F.B. (1989b). Stereological estimation of nuclear volume in benign melanocytic lesions and cutaneous malignant melanomas. Am. J. Dermatopathol. 11, 517–527.

Sørensen, F.B. (1991). Stereological estimation of mean and variance of nuclear volume from vertical sections. J. Microsc. 162, 203–229.

Sørensen, F.B. (1992). Quantitative analysis of nuclear size for objective malignancy grading: A review with emphasis on new unbiased stereologic methods. Lab. Invest. 66, 4–23.

Sørensen, F.B., Bennedbæk, O., Pilgaard, J., & Spaun, E. (1989). Stereological estimation of nuclear volume and other quantitative histopathological parameters in the prognostic evaluation of supraglottic laryngeal squamous cell carcinoma. APMIS 97, 987–995.

Sørensen, F.B., Bichel, P., & Jakobsen, A. (1991a). Stereological estimates of nuclear volume in squamous cell carcinoma of the uterine cervix and its precursors. Virchows. Arch. A. 418, 225–233.

Sørensen, F.B., Bichel, P., & Jakobsen, A. (1992). DNA-level and stereologic estimates of nuclear volume in squamous cell carcinomas of the uterine cervix. A comparative study with analysis of prognostic impact. Cancer 69, 187–199.

Sørensen, F.B., Brændgaard, H., Christiansen, A.O., Bojsen-Møller, M., & Reske-Nielsen, E. (1991b). Stereological estimates of nuclear volume and other quantitative variables in supratentorial brain tumors. Practical technique and use in prognostic evaluation. J. Neuro-Oncol. 10, 253–262.

Sørensen, F.B. & Erlandsen, M. (1990). Intralesional and metastatic heterogeneity in malignant melanomas demonstrated by stereologic estimates of nuclear volume. Lab. Invest. 62, 646–654.

Sørensen, F.B., Gamel, J.W., Jensen, O.A., Ladekarl, M., & McCurdy, J. (1993a). Prognostic value of nucleolar size and size pleomorphism in choroidal melanomas. APMIS 101, 358–368.

Sørensen, F.B., Gamel, J.W., & McCurdy, J. (1993b). Stereologic estimation of nucleolar volume in ocular melanoma: A comparative study of size estimators with prognosite impact. Hum. Pathol. 24, 513–518.

Sørensen, F.B., Kristensen, I.B., Grymer, F., & Jakobsen, A. (1991c). DNA-level, tumor thickness, and stereological estimates of nuclear volume in stage I cutaneous malignant melanomas. A comparative study with analysis of prognostic impact. Am. J. Dermatopathol. 13, 11–19.

Sørensen, F.B., Kristensen, I.B., Grymer, F., & Jakobsen, A. (1990). DNA-index and stereological estimation of nuclear volume in primary and metastatic malignant melanomas: A comparative study with analysis of heterogeneity. APMIS 98, 61–70.

Sørensen, F.B. & Ottosen, P.D. (1991). Stereological estimation of nuclear volume in benign and malignant melanocytic lesions of the skin. Inter- and intraobserver variability of malignancy grading. Am. J. Dermatopathol. 13, 99–107.

Sørensen, F.B., Sasaki, M., Fukuzawa, S., Yamabe, H., Olsen, S., & Yoshida, O. (1994). Qualitative and quantitative histopathology in transitional cell carcinomas of the urinary bladder: An international investigation of intra- and inter-observer reproducibility. Lab. Invest. 70, 242–254.

Stål, O. & Hatschek, T. (1988). A rapid system for static cytofluorometry enabling the simultaneous determination of nuclear size and DNA content. Path. Res. Pract. 183, 329–335.

Stendahl, U., Willén, H., & Willén R. (1979). Classification and grading of invasive squamous cell carcinoma of the uterine cervix. Acta. Radiol. (Oncology). 18, 481–496.

Stendahl, U., Willén, H., & Willén, R. (1980). Invasive suamous cell carcinoma of the uterine cervix. I. Definition of parameters in a histopathologic malignancy grading system. Acta. Radiol. (Oncology). 19, 467–480.

Stendahl, U., Willén, H., & Willén, R. (1981). Invasive squamous cell carcinoma of the uterine cervix II. Reproducibility of a histopathologic malignancy grading system. Acta. Radiol. (Oncology). 20, 65–70.

Sterio, D.C. (1984). The unbiased estimation of number and sizes of arbitrary particles using the disector. J. Microsc. 134, 127–136.

Suffin, S.C., Waisman, J., Clark, W.H., & Morton, D.L. (1977). Comparison of the classification by microscopic level (stage) of malignant melanoma by three independent groups of pathologists. Cancer 40, 3112–3114.

Svanholm, H., Starklint, H., Gundersen, H.J.G., Fabricius, J., Barlebo, H., & Olsen, S. (1989). Reproducibility of histomorphologic diagnoses with special reference to kappa statistic. APMIS 97, 689–698.

Tan, G.J.K.H. & Baak, J.P.A. (1984). Evaluation of prognostic characteristics of stage I cutaneous malignant melanoma. Analyt. Quant. Cytol. 6, 147–154.

Tosi, P., Luzi, P., Sforza, V., Santopietro, R., Vindigni, C., Miracco, C., Baak, J.P.A., Smolle, J., & Barbini, P. (1989). The nuclei in cutaneous malignant melanoma, stage I, are smaller in survivors than in non-survivors. Path. Res. Pract. 185, 625–630.

Weibel, E.R. (1979). Stereological Methods. Vol. 1: Practical methods for biological morphometry, Academic Press, New York.

Wicksell, S.D. (1925). The corpuscle problem. A mathematical study of a biometric problem. Biometrika. 17, 84–99.

Willis, R.A. (1948). Pathology of Tumours, Butterworth & Co. Publishers Ltd., London.

Chapter 12

Prostate Cancer

MARTIN GLEAVE, MARK BANDYK, and
LELAND CHUNG

Advances in Oncobiology
Volume 1, pages 257–291.

ISBN: 0-7623-0146-5

INTRODUCTION

Adenocarcinoma of the prostate is the most common malignancy afflicting men and is the second leading cause of cancer mortality in North America. Etiologic factors are poorly defined and our ability to predict the natural history of early, potentially insignificant cancers remains inaccurate. Over the past decade, substantial improvements in diagnosis and staging have been made with the combined use of digital rectal examination (DRE), prostate specific antigen (PSA), and transrectal ultrasound (TRUS). Benefits of aggressive early detection programs resulting in increased diagnosis at earlier and potentially curable stages of disease must be weighed against the risk of unnecessary treatment of clinically insignificant disease. Refinements in the treatment of localized disease have reduced the morbidity of therapy significantly. Surgical or medical castration remains the only effective therapy for advanced disease, and no significant therapeutic advances have been made in the last 50 years for patients with metastatic disease. Intensive basic and clinical research is providing important insights into the biology of prostate cancer. Future research will hopefully provide clinicians with molecular markers capable of diagnosing prostate cancer early, predicting biologic activity and risk of clinical progression, and new therapies for androgen-independent disease.

DEMOGRAPHICS AND EPIDEMIOLOGY

Current Trends

During 1994, an estimated 200,000 new cases of prostate cancer will be diagnosed (constituting 32% of all cancers in men!) and approximately 38,000 men will die of the disease in the United States (Boring et al., 1994). Figure 1 illustrates the

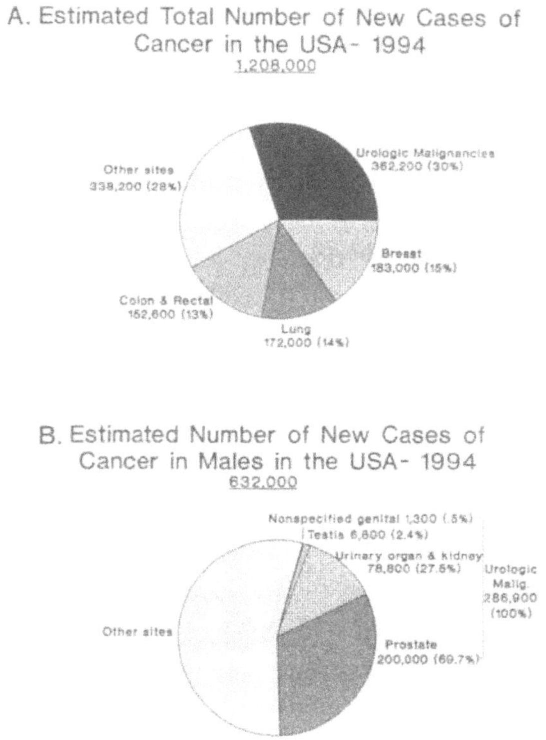

Figure 1. **A**) Urologic malignancies account for 30% of total of new cancer cases. **B**) Prostatic cancer in U.S. each year accounts for 70% of all genitourinary malignancies in men.

significant impact that prostate cancer has on the health of men in relation to other common malignancies (panel A), and the overall high incidence of male genitourinary cancers (panel B). The high incidence and mortality rates from prostate cancer translates into a diagnosis of prostate cancer every three minutes and a death from prostate cancer every 14 minutes in the U.S. alone. The incidence and age specific mortality rates have been rising gradually since 1970. The increasing incidence is due partly to a true rising incidence but also to an increase in case finding because of more aggressive early detection. Prostate cancer is a disease generally affecting older men and is rarely seen before the age of 50. One of the most dramatic observations regarding the epidemiology of prostate cancer is its marked age-related increase in incidence (Carter and Coffey, 1990). Of all cancers in humans, the incidence of prostate cancer increases most dramatically with age (Figure 2). As postwar baby boomers become older, shifts in age distribution within our society will increase the number of males older than 65 years by 64% from 1977 to the year 2000. It is estimated that over the 23 year period from 1977 to

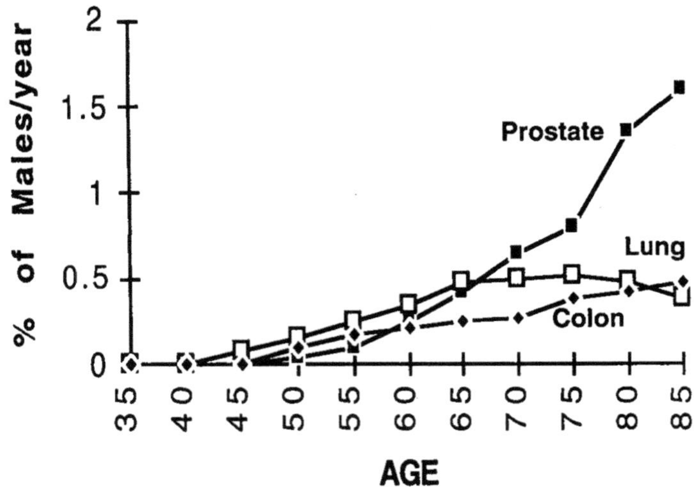

Figure 2. The incidence of prostatic cancer increases more dramatically with age than any other cancer, increasing over 20-fold from 50 to 80 years of age.

2000, the number of clinical cases of prostate cancer will increase by 90% and mortality will increase by 37%. The magnitude of the problem is apparent when one considers that 9% of 50 year old men will develop clinical prostate cancer and 3% will die from it.

Epidemiology

Autopsy studies have shown that prostate cancer is the most prevalent cancer afflicting men (Franks, 1954). Thirty percent of men over 50 years of age have histologic evidence of prostate cancer, most of which are microscopic lesions less than 1 cc in volume. However, only 1–2% of these occult cases become clinically manifest each year. The clinical incidence is therefore much less than the autopsy prevalence and clearly not all prostate cancers are a health risk that need treatment. It remains difficult to predict which occult, microscopic tumors will progress in a fashion which will pose a life threatening risk to the patient.

There is marked geographic variation in the biologic aggressiveness of prostate cancer (Carter et al., 1990). For example, although autopsy prevalence is similar throughout the world, incidence and mortality rates vary up to 30-fold between Asia and the Western world. The age-adjusted mortality rate per 100,000 men is 0.5 in Korea compared to 15 in the U.S. This difference decreases in Oriental men who immigrate to the U.S. and who have incidence rates of clinical disease intermediate between those of their country of origin and the U.S. (Carter and Coffey, 1990). The risk of progression of latent to clinically manifest prostate cancer has been proposed to be highly dependent upon the initial tumor volume in the latent cancer.

Whittmore et al. (1991) provide evidence to suggest that the rate of progression is similar for black and white men, and the differences in clinical cancers among white and black population can be best correlated and explained by the initial low volume cancers in specimens of white men.

ETIOLOGY

The most significant etiologic factors associated with the development of prostate cancer are aging and the presence of testosterone (Table 1). Of all cancers, the prevalence of prostate cancer increases most dramatically with age, tripling from the sixth to ninth decade. Testosterone has long been implicated as a possible promoter of prostatic cancer growth. The embryologic development and continued growth and function of the prostate in the adult is dependent on the presence of testicular androgens (Wilson and Griffin, 1994). Prostate cancer does not develop in eunuchs and latent prostate cancer is less frequent among cirrhotics who have elevated serum estrogen levels. Furthermore, differences in plasma testosterone have been noted between low risk (Orientals, Nigerian blacks) and high risk (African American blacks) groups. There are also marked racial differences in prostate cancer incidence, with a 44-fold increased risk in black American males compared to Oriental men living in Japan. Caucasians have an intermediate risk.

The marked geographic variation in incidence of prostate cancer, which is diminished when a man moves from a low risk to high risk region, suggests that environmental/dietary factors are important in the pathogenesis of prostate cancer. Dietary fat has been suggested as an important etiologic agent. Animal fat contributes 40% of calories in Western diet as opposed to less than 15% of calories in an Oriental diet. Furthermore, case control studies suggest that patients who regularly ingest animal fat may have twofold increased risk of developing prostate cancer (Giovannucci et al., 1993). In contrast, dietary vitamin D and A may inhibit prostate cancer cell growth and enhance cancer cell differentiation. Exposure to sunlight (which activates vitamin D) and consumption of fish oil (a rich source of vitamin A) are also associated with lower incidence of prostate cancer.

Familial associations have also been reported and the risk increases proportionately with the number of relatives with prostate cancer (Spitz et al., 1991). A man has a two- to threefold increased risk of developing prostate cancer if a first degree

Table 1. Etiology of Prostate Cancer

Genetic	2–3-fold increased risk in first degree relatives
Racial	Black:White:Oriental = 20:10:1
Environmental	Animal fats
Hormonal	Testosterone acts as promoter
Aging	Increased with age

relative has developed it. Death due to prostate cancer is three to eight times more frequent in male relatives of patients with prostate cancer than in controls. Carter et al. (1992) reported a small proportion of prostate cancer cases (9% by age 85) which follows a pattern of Mendelian inheritance. Their findings justify the search for prostate cancer-associated genes that may prove useful as diagnostic and prognostic tools in patients with a positive family history.

Numerous additional factors have been implicated as etiologic factors for prostate cancer, but the association is either tenuous or not causally related. Vasectomy has been implicated as a risk factor in some studies, but not in others. Similarly the relationship between benign prostatic hyperplasia (BPH) and cancer development is unclear. Although cancer is very frequently accompanied by BPH, there is no evidence to suggest that BPH is a precursor to cancer. No relationship exists between socioeconomic status, sexual history, or smoking and prostate cancer.

PATHOLOGY

Anatomy and Embryology

Prostatic development begins during the 12th week of life under the influence of androgenic hormones from the fetal testis. Most of the prostate is derived from the embryonic urogenital sinus, although the ejaculatory ducts and part of the central portion of the prostate gland are of Wolffian duct origin. The normal prostate is a walnut-sized gland situated at the bladder base anterior to the rectum on the urogenital diaphragm. The prostate gland surrounds the urethra like a donut and the urethra angulates upwards 35° at the verumontanum. Histologically, the prostate is a tubuloalveolar gland comprised of stromal (fibromuscular) and epithelial (glandular) tissue in roughly equal amounts. The prostate was previously thought to be comprised of various lobes; more recently, however, it has been divided into zones based on careful histologic studies (Figure 3; McNeal, 1968). The peripheral zone makes up 65% of the normal prostate and is the origin of 75% of prostatic carcinomas. This is the portion of the prostate that is felt with a digital rectal examination. The central zone is triangular and lies at the base of the prostate surrounding the ejaculatory duct and it is the site of origin of 10 to 20% of carcinomas. Periurethral glands within the transition zone are located just outside the urethra at the verumontanum and are the site of origin of BPH. The anterior zone is comprised primarily of fibromuscular tissue.

Histopathology

Prostatic carcinomas arise from epithelial cells that comprise the glandular tissue of the gland. Most cancers arise in the peripheral zone with 20–30% in the central or transition zones, and the vast majority are multifocal. Histologic grading attempts to classify the degree of differentiation of each prostate cancer and a number

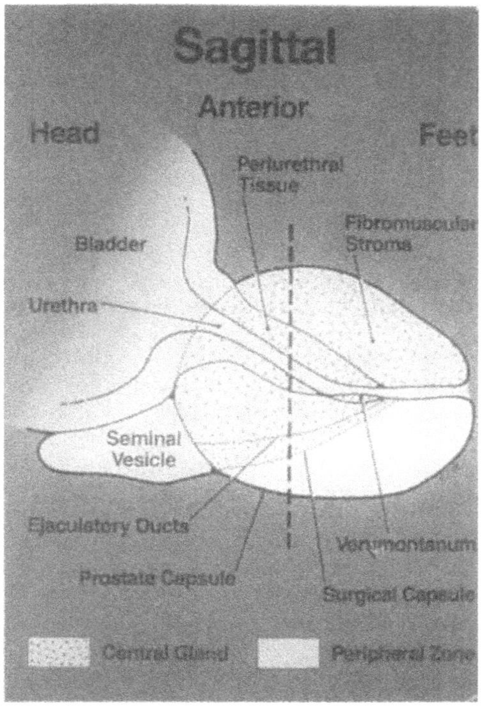

Figure 3. The prostate gland is comprised of various zones based on embryologic origin, anatomic location, and histomorphology. The peripheral zone is derived from the urogenital sinus and comprises approximately 75% of the total volume; it is the site of origin of most cancers. The cone-shaped central zone surrounds the ejaculatory ducts and is of Wolffian duct origin. The transition and periurethral zones are the site of origin of BPH, while the anterior zone is primarily comprised of fibromuscular tissue.

of grading systems have been described (Table 2). All are based on the degree of glandular differentiation, and/or the presence of nuclear abnormalities and cytologic atypia. A histologic hallmark of prostate cancer is the back-to-back arrangement of glands, which results from the loss of intervening stroma. Another characteristic is the loss of the double layer of cells lining acini of benign glands.

The Gleason grading system is the most widely employed system and describes the glandular pattern of the tumor specimen under low magnification (Gleason, 1966). The morphology of individual cells is not considered and the architectural pattern of the glands is assigned a grade from 1 to 5. Gleason grade 1 pattern has round to oval glands with no stromal invasion and grade increases with more pleomorphic glandular formation, progressing through to a Gleason grade 5 which is characterized by flat sheets of poorly organized cells with no glandular formation

Table 2. Grading Systems for Prostate Cancer

Mostofi	*Gleason*	*M.D. Anderson*
Grade I	Pattern 1	Grade I
Well differentiated glands with slight nuclear anaplasia	Single, separate uniform glands with well-defined tumor margins	75–100% of tumor forms glands (except if predominantly cribriform pattern)
Grade II	Pattern 2	Grade II
Glands with moderate nuclear anaplasia	Glands are less uniform; margins less well defined	50–75% of tumor forms glands (includes cribriform pattern)
Grade III	Pattern 3	Grade III
Undifferentiated tumor that does not form glands; marked nuclear anaplasia	Single irregular glands with poorly-defined margin	25–50% of tumor forms glands
	Pattern 4	Grade IV
	Fused glands irregularly infiltrating into surrounding stroma; cribriform, papillary or solid patterns	0–25% of tumor forms glands
	Pattern 5	
	Sheet of tumor cells irregularly infiltrating into surrounding stroma; few or no glands	

and diffuse infiltration into the normal parenchyma. The sum of the two most common patterns is termed the Gleason score; a Gleason score of 2 to 4 represents well-differentiated carcinomas, 5 to 7 represents moderately differentiated disease, and 8 to 10 is poorly differentiated (Figure 4).

The M.D. Anderson grading system relies on the overall percentage of gland formation (Brawn et al., 1982), while the Mostofi system takes into account the degree of nuclear anaplasia in addition to the glandular pattern (Mostofi, 1975). All the grading systems have comparable prognostic value. Tumor grade is one of the most clinically useful indicators of prostate cancer growth invasion and progression.

Staging

Staging systems for cancer classify patients according to anatomic extent of spread of their disease. Patients with localized disease fare much better than patients with disseminated disease. Classification of cancer by stages also permits more accurate comparison for natural history and therapy. The International Union Against Cancer (UICC) and American Urological Association (AUA) staging

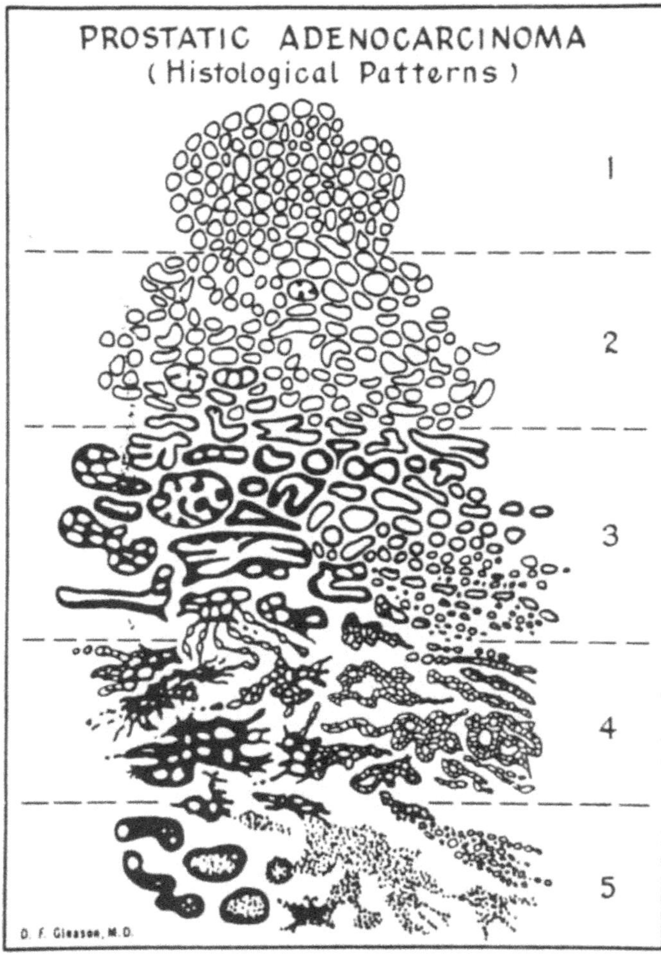

Figure 4. The Gleason system for grading of prostatic cancers classifies tumors on the basis of the glandular patterns in relation to the prostatic stroma. Well- differentiated Geason grade 1 or 2 tumors are comprised principally of round uniform acini that are sharply delineated from the surrounding stroma. Increasing variability in size and shape of glands with ragged margins are characteristic of grade 3 tumors. Poorly differentiated or anaplastic grade 4 or 5 tumors have fused glands or form sheets of cells that infiltrate widely into the surrounding stroma. The two most predominant Gleason grades are summated to give the Gleason score.

systems are compared as indicated in Table 3. Staging work-up includes the DRE, serum PSA, TRUS, bone scan, and chest X-ray. Computerized tomography (CT) scan to assess pelvic lymph nodes is not necessary prior to surgery because a staging pelvic lymph node dissection is routinely performed; CT scan is primarily used for

Table 3. Staging Systems for Prostate Cancer

AUA	*UICC*	
A	T1	Incidental finding post-TUPR
A_1	T1a	< 5% tumor volume; well differentiated
A_2	T1b	> 5% tumor volume
B	T2	Tumor clinically organ-confined
B_0	T1c	Normal DRE; elevated PSA; TRUS-guided biopsy
B_1	T2a	Tumor < 1.5 cm
B_2	T2b	Tumor > 1.5 cm or bilateral
C	T3	Extraprostatic extension
C_1	T3a	Unilateral extracapsular extension
C_2	T3b	Unilateral bilateral extracapsular extension
C_2	T3c	Seminal vesical invasion
C_2	T4	Pelvic wall fixation
D_1	N	Lymph node metastases
D_2	M	Distant metastases

simulation prior to radiation therapy. Patients with untreated prostate cancer and serum PSA levels below 10 ng/ml do not need a bone scan because the risk of osseous metastases in this population is below 1%.

NATURAL HISTORY

Natural history of prostate cancer is variable but is generally one of slow local progression with later development of regional and distant metastasis. Several clinical studies demonstrate that not all patients with prostate cancer will progress to life-threatening disease (Whitmore et al., 1991; Johansson et al., 1992). Most studies investigating the natural history of prostate cancer with "watchful waiting" have involved a select group of elderly patients (mean age 72 years) with well differentiated disease and low stage (A and B) disease. Studies of men with well differentiated, localized prostate cancer demonstrate that most have slow doubling times between two and four years; if followed for long periods of time, however, progression invariably occurs with subsequent development of local progression and/or distant metastases.

In general, the risk of progression is related to the patient's age and life expectancy; younger patients have a longer life expectancy and a higher risk of progression than older patients who may die from other competing causes of death before prostate cancer becomes symptomatic. Risk of progression is also related to tumor volume (stage) and the biologic aggressiveness of the tumor, which can be estimated by tumor grade and ploidy. Over a 10 year follow-up period, distant metastases developed in less than 10% of stage A-1 patients, 30% of stage A-2 patients, 35–80% of stage B-1 and B-2 patients, and up to 80% of stage C patients (Catalona and Scott, 1986). Approximately 50% of stage D-2 patients die within 3 years, and 80% within 5 years. Overall, more than half of patients with prostate cancer will die from their disease within 10 years, and more than two-thirds will suffer local or systemic progression despite therapy. Clinically localized prostate cancer is not a benign disease; it has a slow but steadily progressive natural history. However, molecular markers are required that more accurately identify an individual patient's risk of local and distant progression and help direct aggressive therapy towards those most likely to progress and watchful waiting for those who can be safely followed.

Patterns of Dissemination

Common sites of extraprostatic spread include direct periprostatic extension, regional metastatic spread to pelvic lymph nodes, and distant metastatic spread to the axial skeleton. Capsular penetration most frequently occurs along neurovascular branches coursing laterally through the prostatic capsule at the base and apex of the gland. Cephalad extension of the tumor may lead to bladder outlet or ureteral obstruction, both of which result in obstructive uropathy. Hydronephrosis secondary to bladder neck or ureteral obstruction occurs in up to 30–40% of patients with advanced disease. Much less frequently, caudad extension of the tumor may lead to invasion of the external urethral sphincter and urinary incontinence, while posterior extension may lead to invasion of the rectum with tenesmus and colonic obstruction.

Risk of regional lymph node metastasis is related to stage and grade of the primary tumor. The first echelon of nodal drainage are the obturator lymph nodes, followed by the hypogastric, external iliac, presacral, common iliac, and retroperitoneal lymph nodes. Incidence of pelvic lymph node metastasis varies from 5% in well differentiated stage B-1 disease to 30–40% in moderately differentiated B-2 disease and up to 80% in poorly differentiated stage C disease (Fowler and Whitmore, 1981).

Most patients dying of prostate cancer experience painful and sometimes crippling osseous metastases with up to 80% of men having bony metastases at autopsy (Jacobs, 1983). Prostate cancer selectively spreads to the cancellous bones of the spine, pelvis, femur, and ribs, where it is the only malignancy to consistently produce osteoblastic lesions (Figure 5). Why should bone be the primary metastatic site for prostate cancer, when other organs like the lungs, liver, and kidney all receive a greater percentage of the cardiac output? Several factors may be responsible for this pattern of metastases. Some investigators proposed that prostate cancer cells selectively seed the lumbar spine and pelvis via a paravertebral venous plexus (Batson's plexus) through which retrograde flow from the prostate to the spine may occur at times of increased intraabdominal pressure (Batson, 1940). However, the

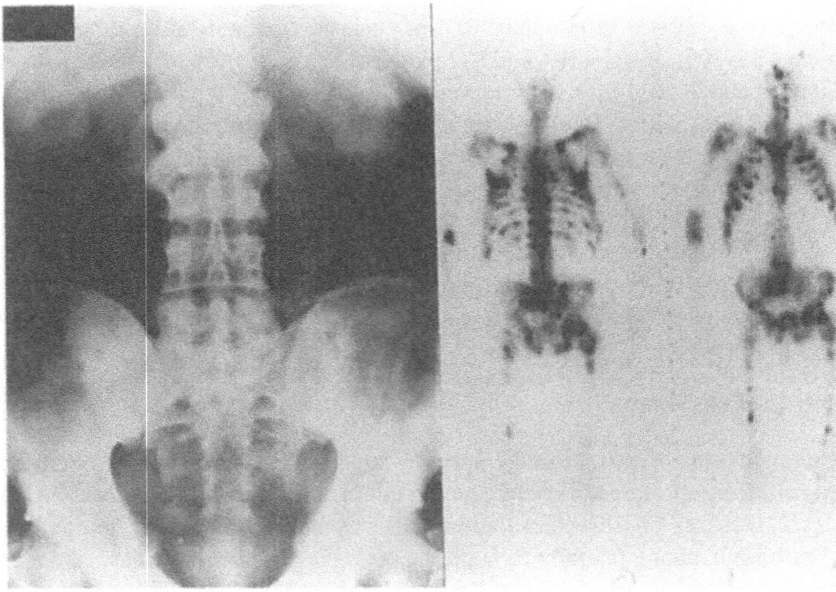

Figure 5. Prostate cancer selectively metastasizes to the axial skeleton where it is the only cancer to consistently produce osteoblastic lesions.

arrest of tumor cells in an organ does not necessarily lead to metastatic growth. Most tumor cells in the venous circulation pass through the lungs and yet the incidence of clinically apparent lung metastases in patients dying of prostate cancer is low.

The "seed and soil" hypothesis (Paget, 1889) attempts to explain the nonrandom distribution of metastases with certain tumors ("seed") that preferentially grow at particular sites ("soil"). Numerous factors likely contribute to the "fertility of the soil" of an organ, and include differential adhesion between microvessel endothelial cells or basement membrane collagen IV from one organ to another, organ-specific chemotactic factors capable of inducing migration of tumor cells from vascular to interstitial spaces, and organ-specific growth factors (or inhibitors) capable of accelerating (or preventing) the growth of tumor cell deposits. It is likely that these factors all contribute to the development of bony metastases. Experimental animal models of human prostate cancer demonstrate that bone stromal cells produce various growth factors that act in a paracrine fashion to stimulate metastatic cancer cell growth (Gleave et al., 1991).

DIAGNOSIS

The diagnosis of prostate cancer has traditionally been made following needle biopsy of an abnormal gland discovered on DRE. However, over the past decade, a number of new modalities have become available to augment the DRE including TRUS and serum PSA assays. The use of DRE and PSA with selective TRUS and biopsy, termed the diagnostic triad, has increased detection rates and has fine-tuned the investigation of suspicious or abnormal prostatic examinations (Cooner et al., 1990).

Symptoms from prostate cancer do not develop until the disease is advanced. Patients may present with obstructive urinary symptoms secondary to BPH and are subsequently found to have a nodule on digital rectal examination. Locally advanced prostate cancer may present with bladder outlet obstructive symptoms, urinary retention, hematuria, or incontinence. This sort of presentation is uncommon. Uncommonly, patients may present with bone pains and neurologic symptoms secondary to cord compression from osseous metastasis. Approximately one half of patients with prostate cancer are diagnosed with localized disease (stage A or B) at a potentially curable stage. The remaining one half of patients present with either locally advanced (stage C) or disseminated disease (stage D). Early detection programs are the only immediate method to reduce the mortality rate from prostate cancer because there is no therapeutic panacea on the horizon to cure patients with metastatic disease and because our poor understanding of its etiology precludes the use of preventive measures to decrease the rising incidence. Because only one half of patients present with localized (and potentially curable) disease, it may be possible to increase the number of patients who are diagnosed at an earlier stage through the application of early detection programs (Table 4).

Because the prevalence of occult prostate cancer is so much greater than the clinical incidence and mortality, many fear that any attempts at early detection will be too costly and result in overdetection of latent cancers that are clinically insignificant (Chodak and Schoenberg, 1989). Some clinicians believe that clinically important cancers are rarely detected when they are curable and clinically unimportant cancers do not need to be treated because patients will die with, rather than of, prostate cancer (Whitmore, 1988). However, differences between prevalence of occult cancer and incidence of clinical cancer are exaggerated when one ignores the effects of time (Scardino, 1989). A 50 year-old man has a 25 year life expectancy and a 9% chance of developing clinical cancer and a 3% chance of dying from it. Over time, one in five men with occult microscopic prostate cancer develop

Table 4. Why Do We Need Early Detection Programs for Prostate Cancer?

Rising incidence and mortality rates from prostate cancer.
Aging population.
Poor understanding of etiology and pathogenesis precludes use of preventative measures.
No therapeutic panacea for metastatic disease on the horizon.
Using DRE alone only 30–40% of patients present with localized and potentially curable disease.

a clinical cancer and one in 15 die from the disease. The most important characteristics that help to distinguish between clinically significant and clinically insignificant prostate cancer are tumor volume and grade (McNeal, 1986). Most clinically important but undetected cancers are greater than 0.5 cc in volume; these are the appropriate targets for an early detection program. In other words, the proper goal of an early detection program would not be to detect all cancers, but rather only those of larger volume that are clinically significant and potentially lethal. To increase early detection and avoid overdetection, screening tests should identify the roughly 6–9% of men older than 50 years of age with clinically important prostate cancers without detecting the 30% with clinically insignificant cancers less than 0.5 cc in volume.

Tests For Early Detection Of Prostate Cancer

Digital Rectal Examination

DRE has been the traditional method for diagnosing prostate cancer. The prostate gland is normally smooth, symmetrical, walnut or plum-sized, with a firm consistency. Malignant glands may be asymmetrical, indurated, nodular, or even benign in consistency. In eight studies of DRE screening including 13,000 patients, the detection rate for prostate cancer using DRE averaged 1.3% (Chodak and Schoenberg, 1989). The advantages of DRE are its low cost and noninvasiveness. However, DRE detects only a minority of prostate cancers at an early stage, with up to two-thirds extending beyond the prostate or being metastatic at the time of diagnosis. Furthermore, DRE is a subjective test with a great deal of variability in the accuracy of the examination as a function of the skill and commitment of the practitioner. Also, one-third to one-half of all cancers detected through DRE screening that are initially believed to be organ-confined are subsequently found to be extracapsular at the time of treatment. Because of the subjective nature of the DRE, its low sensitivity, as well as its inability to detect lesions early enough, other tools are being used to aid in the early detection of prostate cancer.

Prostate Specific Antigen

The discovery of PSA as a tumor marker for prostate cancer is a landmark in urology. PSA is a single chain glycoprotein first isolated from prostate tissue in 1979 (Wang et al., 1979). PSA is produced only by prostatic epithelial cells (e.g., normal, benign hyperplastic, and malignant prostatic cells). It is normally secreted into the lumina of the prostatic ducts and is responsible for the liquefaction of the seminal coagulum. PSA synthesis is hormonally regulated; testosterone stimulates the production and secretion of PSA, and following medical or surgical castration, PSA production decreases approximately fivefold (Gleave et al., 1992).

Serum PSA levels slowly increase with age, and therefore, normal PSA levels are age-dependent; 0 to 3.5 μg/L for men 50 to 59 years, 0 to 4.5 μg/L for men 60

Table 5. Clinical Application of PSA

Monitoring systemic therapy of metastatic disease.
Early detection of recurrence after radical prostatectomy or radiation therapy.
Immunohistochemical diagnosis of poorly differentiated carcinoma.
Diagnostic adjunct in patients with suspicious DRE.
Staging adjunct of diagnosed cases.
Early detection for men aged 50–70 years with normal DRE.

to 69 years, and 0 to 6.5 μg/L for men 70 to 79 years. There is no significant diurnal variation of PSA, and DRE has no significant effect on PSA levels (Oesterling, 1991). Prostatitis, BPH, or prostatic manipulation following cystoscopy or prostatic biopsies can all transiently increase PSA levels. Because BPH tissue variably contributes to serum PSA concentrations and has a high prevalence in men over 50 years of age, careful interpretation of an elevated PSA level is required.

Serum PSA measurements are currently recommended for monitoring systemic therapy of metastatic disease in the early detection of recurrent disease after radical prostatectomy or radiation therapy, as a staging adjunct for diagnosed cases, and as a diagnostic adjunct in patients with a suspicious DRE (Table 5). The role of PSA in the early detection of prostate cancer is currently the subject of much debate and research. Recent and accumulating data indicate that PSA is the single best test for the early detection of prostate cancer. When used as a screening tool in the appropriate population, serum PSA is the single best test for the early detection of prostate cancer, and compares favorably with screening tests for breast and cervical cancers. Because PSA is objective, reproducible, and relatively inexpensive in itself, the debate over its role in screening is primarily focusing on its cost-effectiveness and two related issues: Is PSA testing (along with DRE) specific and sensitive enough to detect organ-confined curable cancer, and will it detect an excessive number of clinically insignificant small volume cancers and lead to overtreatment?

Because BPH tissue variably contributes to serum PSA concentrations and has a high prevalence in men more than 50 years of age, careful interpretation of an elevated PSA level is required. On an individual basis, serum PSA by itself is not capable of distinguishing between patients with early organ-confined prostate cancer in an ocean of BPH. Although PSA lacks adequate specificity to be diagnostic of prostate cancer, it remains the single best test for stratifying men into groups with a high risk of having prostate cancer who should undergo definitive testing with prostatic biopsy, and those with a low risk of having prostate cancer who can be reassured and followed without additional testing (Cooner et al., 1990; Catalona et al., 1991; Brawer et al., 1992). Attempts to improve PSA sensitivity with lowering of the upper limit of normal leads to an increase in number of false positives and reduction in specificity which is most important for a screening tool. Approximately 25% of men with BPH may have elevated serum PSA levels, usually in the intermediate range between 4–10 μg/L. Although elevated serum PSA levels can occur in men with BPH, elevations occur more frequently and are higher in men with prostate cancer.

Detection of Cancer on Initial PSA Determination

Cross-sectional studies involving tens of thousands of men from several countries have been published and show very similar results regarding the ability of PSA to predict the presence or absence of cancer following TRUS-guided biopsy of the prostate. Studies involving community-based populations similarly suggest that use of serum PSA increases the sensitivity, specificity, and positive predictive value of DRE in the diagnosis of prostate cancer. Andriole and Catalona (1993) published the Washington University experience results from a screening population of 20,000 men. Overall, about 10% of screened men older than 50 years of age had a PSA greater than 4 μg/L, and one-third of these were found to have cancer on subsequent biopsy. The probability of having cancer varied with the degree of PSA elevation. PSA levels were less than 4 μg/L in 92% of men, 4–10 μg/L in 6.5%, and greater than 10 μg/L in 2%. In patients with serum PSA levels between 4–10 μg/L, 22% had cancer, and in those with PSA levels > 10 μg/L, 67% had cancer. When PSA was the only abnormal parameter, cancer was diagnosed in 20%. However, when both PSA and DRE were abnormal 31% of men had cancer, and when PSA, DRE, and TRUS were abnormal 56% had cancer (Table 6).

Brawer et al. (1992) reported similar results and concluded that use of serum PSA resulted in the best performance characteristic for the diagnosis of prostate cancer. They evaluated 1,249 men and noted a positive predictive value of 31% for PSA and a 2.6% detection rate. Labrie et al. (1992) randomly screened 1,002 men between 45 and 80 years of age in Quebec City and, using a cutoff of 3.0 μg/L, reported the sensitivity and specificity for PSA to be 80.7% and 89.6%, respectively. Based on their findings, Labrie et al. calculated the positive predictive value of PSA to be 25% at levels above 3.0 μg/L, 33% at levels above 4.0 μg/L, 51% at levels above 10 μg/L, and 90% at levels above 30 μg/L.

Table 6. PSA in Prostate Cancer Screening

Abnormal Parameter	*Percent With Cancer*
PSA> 4*	20
PSA + DRE	31
PSA + DRE + TRUS	56

Note: *PSA was 4–10 in 6.5% (22% had cancer); PSA was > 10 in 2% (67% had cancer). From Catalona et al., 1991.

Each of these cross-sectional community-based studies have their minor flaws, yet despite significant variability in the populations studied and indications for biopsy, they are remarkably consistent and indicate that serum PSA levels above 4 μg/L are associated with a 30–40% chance of finding cancer on TRUS-directed biopsies. Overall, 8–12% of screened men over the age of 50 will have an elevated serum PSA greater than 4 μg/L. The detection rate of prostate cancer using PSA in community-based populations is 2.2% to 3%, which is approximately twice that when DRE alone is used and approximately 2–3 times the detection rate of breast cancer using mammography-based screening programs (Kerlikowske et al., 1993).

Detection of Cancer in Longitudinal Studies

The evaluation of PSA sensitivity and specificity in cross-sectional studies are subject to criticisms because the true negative disease status of each study participant can never be known. For example, what proportion of men with normal PSA and DRE, or with elevated PSA and negative biopsy, at initial evaluation will subsequently develop clinically significant disease in the future? Longitudinal studies involve measuring PSA levels at the start of follow-up in stored blood from a cohort of healthy men who later are or are not diagnosed with prostate cancer. The incident cases of prostate cancer allow a less-biased estimation of test validity. Based on a retrospective, longitudinal study from the Baltimore Longitudinal Study of Aging including 54 men followed from 7 to 25 years, Carter et al. (1992) demonstrated that consistent increases in PSA of more than 0.75 μg/L identified men with prostate cancer or an average of five years before the usual diagnosis with a 72% sensitivity and 90% specificity. Differences in PSA velocity between patients with BPH and cancer were apparent as early as nine years before diagnosis and even longer in patients diagnosed with metastatic cancer.

A prospective longitudinal study evaluating the ability of PSA to predict for prostate cancer has recently been published. Gann et al. (1995) prospectively evaluated PSA testing by measuring levels in stored blood at the start of follow-up in a cohort of 366 healthy men who were subsequently diagnosed to have prostate cancer (usually without the aid of PSA) and 1,098 healthy controls. They concluded that PSA had impressively high sensitivity and specificity. Their results, which are in general agreement with retrospective longitudinal studies (Carter et al., 1992), noted that a single PSA level would have detected 80% of all aggressive cancers diagnosed within five years and about 50% of aggressive cancers as early as 9–10 years before diagnosis. Only 96 of 1,098 men who remained free of prostate cancer diagnosis over a 10-year period had an elevated PSA, which means that a false-positive test remained very low over this period of time. The mean lead-time provided by PSA testing is estimated to be 5.4 years. On the basis of their data, the authors concluded "that PSA has the highest validity of any circulating cancer screening marker discovered thus far."

PSA should not be used alone to exclude the possibility of prostate cancer. PSA is a useful adjunct to DRE in the early diagnosis of prostate cancer, and permits a rational guide to the use of TRUS and biopsy. When used in conjunction with DRE, the positive predictive value of PSA is improved. Use of PSA and DRE detects 27% more cancers than would have been detected by PSA alone, and 34% more than by DRE alone (Catalona et al., 1991; Brawer et al., 1992). For example, in a screening study of 6,630 men using PSA and DRE, 18.2% of cancers detected were in men with normal PSA levels (Catalona et al., 1994; Brawer, 1994). The positive predictive value of an abnormal DRE when PSA is normal is 10%; conversely, the positive predictive value of an abnormal PSA when DRE is normal is 20–30%.

Does PSA Predict for Clinically Significant Cancers?

The study reported by Gann et al. (1995) suggests that PSA screening will not lead to an increase in the diagnosis of clinically insignificant tumors. PSA testing may be sensitive enough to detect clinically aggressive cancers whose natural history may be altered by early detection and therapy, but not sensitive enough to identify the highly prevalent small volume indolent cancers. Use of PSA increases the proportion of cancers detected and treated at an organ-confined and potentially curable stage. More than 95% of the cancers detected in the University of Washington screening study were localized to the prostate, and about two-thirds had pathologically organ-confined disease (Andriole and Catalona, 1993).

However, it is possible that screening with serum PSA will increase the detection of clinically insignificant cancers and lead to overtreatment of some men whose quality of life or life expectancy would otherwise not have been affected by their prostate cancer. The clinical significance of a cancer is determined not only by the volume of a cancer and its grade, but also by the life-expectancy of the patient which determines the duration of time he is at risk for progression. Evidence thus far suggests that <15% of PSA-detected cancers are < 0.5 cc and probably clinically insignificant (Catalona et al., 1993; Oesterling et al., 1993b; Epstein et al., 1994). Pathologic evaluation of stage T1c cancers detected through elevations in serum PSA reveal that these nonpalpable tumors are similar to clinical stage T2 tumors in terms of their tumor volume, grade, and pathologic stage. However, as reviewed by Stamey (1995), the risk of diagnosing and treating potentially clinically insignificant disease increases to 30% when sextant biopsies contain a single positive core containing well to moderately differentiated cancer <3 mm in length. To reduce the risk of overtreatment, Stamey recommends repeat biopsies in an effort to validate the need for aggressive treatment.

Enhancing the Positive Predictive Value of PSA

Because of the substantial overlap in serum PSA between men with early prostate cancer and those with BPH, PSA lacks sufficient specificity to be considered an ideal screening test. Consequently, research efforts are focusing on the development of methods that improve the ability of PSA to predict the presence of clinically important early prostate cancers.

PSA velocity. The rate of change of PSA over time, termed PSA velocity, is one such method. Based on the retrospective, longitudinal study by Carter et al. (1992), a PSA velocity of greater than 0.75 μg/L is highly predictive of the presence of prostate cancer. The specificity for diagnosing prostate cancer increases to 90% when PSA velocity is > 0.75 μg/L in one year, compared to 60% for a cutoff of 4.0 μg/L for a single cross-sectional serum PSA determination. Differences in PSA velocity between BPH and prostate cancer patients were apparent as early as nine years before the diagnosis of cancer was made. Catalona et al. (1993) also suggested that a cutoff for PSA velocity of 0.8 μg/L per year helped distinguish between BPH and cancer patients. The mean velocity for men with cancer was 2.18 μg/L, compared to 0.48 μg/L for men without cancer. In their screening cohort of 693 men, Brawer et al. (1994) suggested that a PSA velocity of greater than 20% per year was highly suggestive of cancer and warranted further investigation. The positive predictive value in men with PSA velocity higher than 20% per year was 17.1% in the second year and 18.6% in the third year.

Serial PSA measurements over time in individual men permit longitudinal measurements of PSA which provide a more accurate reflection of pathological changes within the prostate than does a single determination, and will increase both the positive predictive value of PSA and the likelihood of diagnosing cancers while they are still organ-confined. However, one problem that has become apparent with PSA velocity results from the individual variation in serum PSA levels from one determination to another. This variability can be as high as 55–70% (Partin and Oesterling, 1994). The optimal utility of PSA velocity is not yet clearly defined and many questions regarding number of determinations, intervals between testing, and the influence of aging on interpretation of PSA velocity remain unknown. Because of the significant variation between measurements over time, it is likely that at least 3 annual PSA determinations will be necessary to establish an accurate PSA velocity for any given individual.

PSA density. Another method to improve PSA specificity is to correlate serum PSA with prostatic volume, using a parameter called PSA density, which is the quotient of serum PSA divided by prostate gland volume as determined by TRUS. PSA density is based on the concept that cancer, on a gram-for-gram basis, will increase serum PSA levels to a greater extent than will BPH or normal prostate tissue. Initial studies suggested that PSA density helped differentiate between BPH and early nonpalpable cancer. Benson et al. (1992) reported a mean PSA density of 0.04 in 41 patients with BPH compared to a PSA density of 0.58 in 20 cases of cancer, and they proposed a cutoff level for PSA density of 0.15. However, no biopsies were obtained from men with PSA densities below 0.15 in this preliminary study and therefore the true negative rate could not be calculated. Subsequent studies reported by Seamen et al. (1993) suggested that PSA density is most helpful in determining which patients with mild elevations of PSA (4–10 µg/L) and normal DRE and TRUS should have a biopsy or just be followed with serial evaluations.

Although additional investigators also found PSA density to help distinguish between BPH and cancer in men with intermediate elevations of PSA (Bazinet et al., 1994), others reported that PSA density offered no improvement over PSA alone in men undergoing prostatic biopsy (Brawer, 1994; Mettlin et al., 1994). Inconsistencies in the ability of PSA density to provide more information than PSA alone is due to difficulty in obtaining accurate and reproducible volume determinations of the prostate gland using TRUS, as well as differences in heterogenous stromal and epithelial composition between prostate glands which lead to marked variation in the amount of PSA produced per gram of prostate tissue. Use of PSA density as a strict guide to biopsy will miss some early cancers and may not be appropriate in younger men with intermediate elevations in PSA.

Age-specific PSA reference ranges. Increases in serum PSA levels with increasing age is a well-documented observation. Oesterling et al. (1993b) reported that the correlation of serum PSA and age is significant, and suggested that the performance status of PSA would be improved by selecting different age-specific cutoff levels, rather than using a single arbitrary value. Based on this data, the upper limit of normal for serum PSA in each decade of life is 2.5 µg/L during the 5th decade, 3.5 µg/L during the 6th decade, 4.5 µg/L during the 7th decade, and 6.5 µg/L during the 8th decade. The use of age-specific ranges as a guideline to biopsy will theoretically increase the detection rate of early cancers in younger men (i.e., cancers most likely to be a health threat over time) and decrease the biopsy rate and detection rate of early cancers in men over 70 years of age (i.e., cancers that may not be a health risk). Therefore, consideration of the age of the patient in interpretation of a given PSA level will help increase the sensitivity of PSA as a tumor marker in men under 60 years of age, and increase its specificity in men older than 60.

Summary

PSA is the gold standard tumor marker for prostate cancer this decade and is the single best test for the early detection of prostate cancer (Table 7). Unlike DRE and

Table 7. PSA in the Early Detection of Prostate Cancer

PSA can identify nonpalpable prostate cancers not detectable by DRE (32%–38%).
For a given population, PSA identifies approximately twice as many cancers as DRE (2.5% vs. 1.3%).
Cancers detected by PSA elevations and/or abnormal DRE are more often organ-confined (and therefore curable).
PSA cannot detect all cancers: 20% of biopsy-proven cancers have a normal PSA.

Table 8. Recommendations for the Early Detection of Prostate Cancer

Annual PSA and DRE in men between 50 and 70 years of age.
If either PSA or DRE is abnormal, then TRUS is indicated.
Directed biopsy is indicated if TRUS shows hypoechoic lesion (along with abnormal PSA or DRE).
Six quadrant biopsies are indicated when PSA > 10 but DRE and TRUS are normal.

TRUS, PSA is not a subjective evaluation; the results are reliable and reproducible and not examiner-dependent. Nevertheless, serum PSA lacks sufficient sensitivity and specificity to be used alone for diagnosing prostate cancer in an ocean of BPH, and should be used in conjunction with a DRE. When used as a screening tool in the appropriate population, serum PSA is the best tool available for the early detection of prostate cancer, and compares favorably with screening tests for breast, cervical, colon, and lung cancers. The recommendations for the early detection of prostate cancer by the American Urologic Association and the American Cancer Society are outlined in Table 8 (Mettlin et al., 1993).

Transrectal Ultrasonography

High resolution TRUS produces images of the peripheral and transitional zones of the prostate, seminal vesicles, and urethra (Lee et al., 1989). Although the sensitivity of TRUS is approximately twice that of DRE, its major limitation is that it is operator-dependent and subject to significant interobserver variability, and it can miss some larger palpable cancers. One-third of cancers detected by DRE or transurethral prostatectomy cannot be seen by TRUS. Although most tumors are hypoechoic, prostate cancer does not have a pathognomonic echopattern on TRUS which results in poor test specificity and positive predictive value which increases unnecessary biopsies, psychological stress, costs, and morbidity (Shinoharo et al., 1989). Indeed, the positive predictive value of TRUS is similar to DRE, ranging between 17 to 41%. There is currently no role for primary TRUS in screening. The development of TRUS has improved our ability to assess the internal architecture of the gland and permits accurate or systematic biopsies of the prostate in a patient with either an abnormal DRE or elevated PSA.

TREATMENT

Prostate cancer is a disease of older men that has an unpredictable and often protracted natural history which can span decades. Most incidental, clinically occult carcinomas pose no significant health risk and require no treatment, especially in elderly patients. Treatment decisions are based on both patient and cancer factors.

Patient factors include age and general health, both of which predict physiologic age and life expectancy. The life expectancy from birth of an average North American man is 75 years. Once he has reached age 65, he will likely live another 15 years. At ages 70 and 75, his life expectancy is another 12 and 9 years, respectively, which reflect a significant number of years during which he is at risk of developing complications from an untreated prostatic cancer. Cancer factors that direct treatment decisions include stage, grade, and serum PSA levels. Overall, the five year survival rate is 77% for white men and 63% for black men.

LOCALIZED DISEASE

Watchful Waiting

Observation, or watchful waiting, is the treatment of choice for most men with incidental (stage A1) carcinoma of the prostate. There is general consensus that focal (<5%), well differentiated carcinoma diagnosed following transurethral resection for BPH can be followed safely. Small volume, low grade disease has a 5 to 15% risk of symptomatic progression at eight years without treatment and an approximate 6 to 8% risk of cancer death (Cantrell et al., 1981). Overall survival is similar to age-matched controls.

Results from several studies suggest that elderly men (65 to 75 years of age) with stage A or B1 well-differentiated tumors can be observed without treatment (Whitmore et al., 1991; Johansson et al., 1992). While local progression approached 50% by five years and 70% by 10 years, the risk of metastatic progression was low (20%) and mortality (10%) at 10 years was not significantly different from treated populations. Recent studies, also from Sweden, reporting on watchful waiting with delayed hormone therapy with longer follow-up revealed a substantial increase in mortality rates in men with more than a 10 year follow-up (Hugoson et al., 1995). Prostate cancer accounted for 20 to 30% of deaths at 10 years follow-up, and 65% of deaths in men with more than 10 years follow-up. In North America, watchful waiting as a form of treatment for prostate cancer is confined to patients with stage A1 disease, and for localized well-differentiated disease in elderly men with a life expectancy of less than 10 years.

Radical Prostatectomy

Surgical procedures for the cure of prostate cancer are based on the premise that excision of all cancer cells residing within a tissue offers the greatest chance for local control and long-term survival. Only those tumors that are confined to the prostate gland are ideal candidates for cure by surgical intervention. Radical prostatectomy is indicated for patients with clinical stage A or B cancer who are less than 70 years of age and have a life expectancy of more than 10 years (Table 9).

Table 9. Indications for Radical Prostatectomy

Ideal Candidate	*Good Candidate*	*Marginal Candidate*
Age < 65 years	Age 65–70 years	Age 70–72 (good health and family history)
T1, T2 lesion	T2b, T2c lesion	T2c lesion
Gleason 4 or 5	Gleason 6 or 7	Gleason 8–10
PSA < 10	PSA 10–20	PSA > 20

An improved understanding of the anatomy of the dorsal vein complex and the neurovascular bundle has reduced the morbidity of radical prostatectomy, which may be performed via a retropubic or perineal approach. It is usually preceded by bilateral pelvic lymph node dissection to rule out regional microscopic lymph node metastases. The prostate, seminal vesicles, and surrounding fascia are removed completely with reconstruction of the bladder neck and re-anastomosis to the membranous urethra. The cavernous nerves that mediate erection travel along the inferiolateral margins on either side of the prostate close to the prostatic capsule. It is now possible to spare these nerves, although ipsilateral wide excision of the neurovascular bundle is recommended to reduce the chance of positive margins when the cancer lies within close proximity to the nerves.

Following radical prostatectomy, approximately 30 to 40% of clinical stage B2 patients are upstaged to pathologic stage C with or without extension of the tumor to the margins of resection. Upstaging to pathologic stage C is associated with an increased risk of both local and distant failure, and these patients are currently treated with either observation, postoperative adjuvant external beam radiation therapy, or hormonal therapy. The efficacy of neoadjuvant hormone therapy to reduce the rate of positive margin disease and adjuvant radiation therapy to reduce disease recurrence are both being examined in randomized trials and the results will be available within the next several years. Early studies using three months of neoadjuvant (prior to radical prostatectomy) hormonal therapy suggests that the incidence of positive margin disease can be reduced by 50% (Labrie et al., 1993). Longer follow-up is required to determine whether reductions in positive margin rates translates into a survival advantage.

Radical prostatectomy provides excellent local control of disease with 5 and 10 year disease-free survival of 80% and 70%, respectively. Overall 15 year disease-free survival rates range between 50 and 65% which are equivalent to or better than age matched controls (Walsh and Jewitt, 1980; Gibbons et al., 1989). Complications of radical prostatectomy can be either intraoperative, immediate postoperative, or long term. Intraoperative complications include blood loss requiring transfusion and injuries to associated structures such as the obturator nerve, ureter, or rectum. Rectal injuries occur in less than 1% of cases and are usually managed by primary closure. Immediate postoperative complications include deep vein thromboses, pulmonary embolism, or wound infections and the incidence of these complications remains below 3%. Long-term complications include either incontinence or impotence. Significant incontinence is seen in roughly 3 to 5% of patients and impotence may occur in 30 to 50% of patients depending on whether both or one nerve is spared during radical prostatectomy.

Radiation Therapy

Radiation therapy is indicated in the treatment of localized prostate cancer (stage A2, B, and C disease) in men with a life expectancy of longer than five years. The upper age limit for treatment varies but ranges between 75 and 80 years. The technique of radiation therapy is well described and is beyond the scope of this chapter. CT scans and a simulator are used for accurate localization of the prostate and radiation fields are shaped to provide maximum protection to normal tissues around the prostate. Patients are generally treated at a rate of 200 cGy daily for six weeks and total tumor dose ranges between 5,500 and 7,000 cGy. The standard method of radiation delivery is via external beam photon radiation. Brachytherapy techniques that implant radioactive sources into the prostate are used in some centers but, in general, have not gained popularity.

Results from radiation therapy for stage A and B disease are comparable to those of radical prostatectomy for 5 to 10 years follow-up. However, increasing local and distant recurrence rates with reduced disease-specific survival are observed for patients followed for longer than 10 years. The prognostic significance of pretreatment serum PSA levels has been shown to be the single most significant pretreatment predictor of disease outcome after radiation therapy. Using a rising PSA as an intermediate endpoint of disease relapse, Zagars and von Eschenbach (1993) identified four prognostic groups based on their pretreatment grade and PSA level. For pretreatment PSA levels <4μg/L (any grade), patients did well with a three year relapse rate below 10%. When serum PSA is between 4 and 10 μg/L, three year recurrence rates are 20% for well-differentiated tumors and 55% for moderate–poorly differentiated tumors. When pretreatment PSA levels are between 10 and 30 μg/L, recurrence rates are about 50% for well–moderately differentiated tumors but 90% for moderate–poorly differentiated disease. Finally, any serum PSA > 30 μg/L, regardless of grade or local T-stage, predicted for a 90% relapse rate at three years. These pretreatment prognostic variables can therefore identify patients at high risk of disease recurrence who may benefit from multimodality therapy such as neoadjuvant androgen withdrawal therapy.

These observations suggest that radiation therapy may be less effective than previously thought in eradicating prostate cancer. However, defining therapeutic failure as a rising PSA may not be a clinically important endpoint in elderly patients with short life expectancies because it may not translate into clinical treatment failure. Conversely, young patients with rising PSA levels following radiation therapy are at risk of developing local and distant progression and may require further therapy. There is little data, however, to support a rational decision between observation, salvage prostatectomy, and early hormonal therapy in patients with recurrent disease after radiation therapy.

Radiation therapy is generally well tolerated. Most patients develop a mild radiation cystitis and proctitis manifested by urinary frequency, urgency, dysuria, hematuria, tenesmus, and diarrhea. Late complications are uncommon and approximately 5% of patients are bothered by chronic radiation cystitis or enteritis.

Management of Localized Disease: A Summary

It is impossible to directly compare radiation and radical prostatectomy series because of staging errors, nonrandomized and selection errors, and lack of prospective randomized trials. In the absence of clear data showing relative benefit of one treatment over the other, patients should be presented with either treatment option. Given the data indicating a high likelihood of rising PSA (and eventual clinical progression) in irradiated patients, it seems reasonable to recommend radical prostatectomy for younger patients with a mean life expectancy of 15 years or more. This includes healthy men under the age of 65. The younger the patient, the greater the potential survival benefit of radical prostatectomy over radiation therapy. Conversely, patients older than 70 years of age are less likely to benefit from radical prostatectomy compared to radiation therapy. Radical prostatectomy and radiation therapy are both effective forms of treatment with comparable 10 year survival rates. Radical prostatectomy can provide 15 year disease-free survival rates in appropriately selected patients equivalent to that of age matched controls.

METASTATIC DISEASE

Stage D prostate cancer includes any patient with metastatic disease and is subdivided on the basis of both the anatomic extent of metastasis and hormonal responsiveness of the tumor. Stage D-O refers to any patient with an elevated enzymatic prostatic acid phosphatase and no radiologic evidence of metastasis. Stage D-1 disease includes patients with pelvic lymph node metastasis while stage D-2 includes patients with extrapelvic metastasis to either the retroperitoneal lymph nodes or bone, and Stage D-3 refers to patients with androgen-independent disease progression.

Endocrine Therapy

Since the Nobel Prize winning reports by Dr. Charles Huggins in 1941 that documented the androgen-dependent growth of prostate cancer cells, endocrine manipulation has remained the cornerstone of management for advanced prostate cancer. Testicular Leydig cells produce 90% of the body's testosterone and are under homeostatic control by luteinizing hormone (LH) from the pituitary gland and LH releasing hormone (LHRH) from the hypothalamus (Figure 6). The adrenal

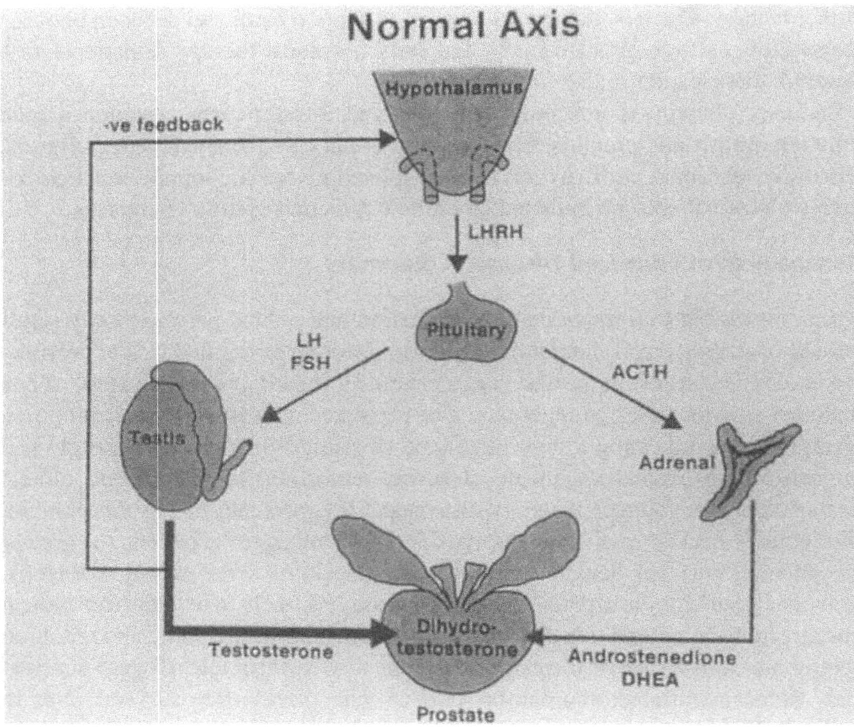

Figure 6. Hypothalamic-pituitary-gonadal axis and the sites of action of various endocrine therapies for prostatic carcinoma.

glands account for the remaining 10% of daily testosterone production. Testosterone is converted to dihydrotestosterone (DHT) in target organs through the action of 5-alpha reductase and then binds to the androgen receptor to exert its hormonal effect. Since DHT is the most potent androgen that is responsible for simulation of prostatic growth and maintenance of its secretory function, a specific type 2 5α-reductase inhibitor, finasteride, has been developed and effectively inhibits the conversion of testosterone to DHT. Finasteride is being tested in a prospective trial to determine if prolonged administration and inhibition of 5α-reductase reduces the incidence of prostate cancer. The mechanism of action of androgen and a rational basis for interference of androgen action is illustrated in Figure 7.

Choices of Androgen Ablation

Surgical Castration. Bilateral scrotal orchiectomy eliminates the primary source of testosterone and results in prompt reduction of circulating serum testosterone and DHT levels. Bilateral orchiectomy offers the advantages of being a

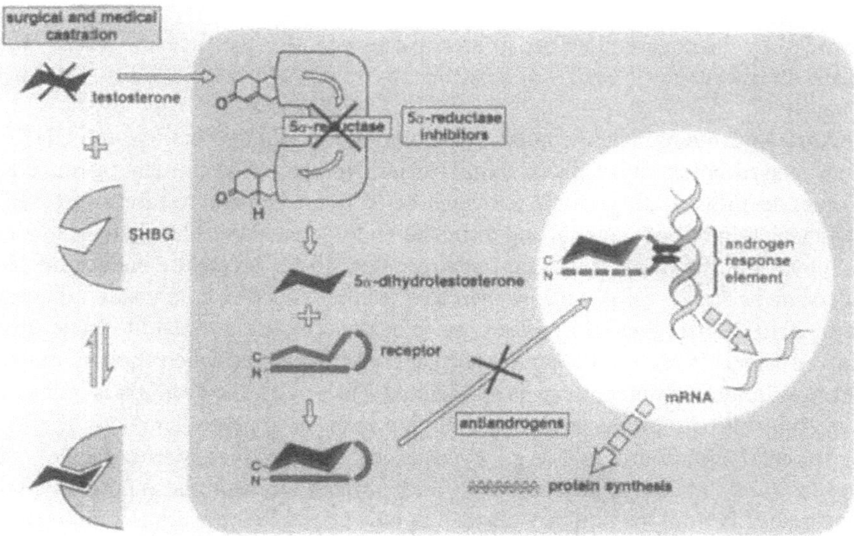

Figure 7. Mechanism of androgen action on prostate cancer cells. Following passive diffusion into prostate cancer cells, testosterone is converted to the more potent dihydrotestosterone (DHT) by 5α-reductase. DHT then binds to the androgen receptor and the activated complex then binds to specific sequences in the 5′ upstream region of androgen-dependent genes to activate transcription. Certain therapies can inhibit these processes, as illustrated. (Courtesy of Dr. N. Bruchovsky.)

simple, one-time, cost effective procedure with 100% compliance. However, its main disadvantage is the psychological impact of surgical castration.

Medical Castration. Various forms of medical castration achieve castrate levels of testosterone through the suppression of LH production by the pituitary gland. Diethylstilbestrol (DES) is the most commonly used estrogen for carcinoma of the prostate. DES is an inexpensive form of therapy with the obvious advantage of eliminating a surgical procedure and the associated surgical and psychological problems. Its main drawback is a significant increase in cardiovascular and thromboembolic complications (10%). Other problems include a decreased libido and impotence, hot flashes, and gynecomastia. The latter can be avoided by breast irradiation prior to DES therapy.

LHRH agonists produce castrate levels of testosterone and DHT by down-regulating LHRH receptors in the hypothalamus. The adrenal production of androgens is not affected. Compliance is improved by monthly Depo injections and there are no cardiovascular complications with this therapy, but hot flashes and loss of libido do occur frequently. Additionally, a "flare" phenomenon exists in the first two weeks of therapy, at which time a surge of LH release by the pituitary can cause a transient increase in testosterone and increased cancer growth. The "flare" can be avoided by the concomitant use of an anti-androgen for the first few weeks. The main disadvantage of LHRH agonist therapy is its high cost ($400 to $500 per month).

Anti-Androgens. These compounds inhibit binding of testosterone and DHT to androgen receptors in peripheral target tissues. Non-steroidal anti-androgens (e.g., flutamide, nilutamide) result in an elevation of serum LH and testosterone levels, but the action of both adrenal and testicular androgens on peripheral target tissues is blocked at the level of the androgen receptor. These agents are not considered adequate as single agents and are therefore recommended in combination therapy with LHRH analogues or orchiectomy to achieve total androgen blockade (discussed later). There are no cardiovascular side effects and when used as monotherapy, libido and potency can be maintained. Their main disadvantage is high cost and an inability to produce reliable androgen ablation as monotherapy.

Steroidal anti-androgens (e.g., cyproterone acetate, megesterol acetate) are agents of mixed action, blocking both the androgen receptor and suppressing the release of LH from the pituitary gland. The advantages of these drugs include their dual site of action, the absence of hot flashes because of their progestational activity, and intermediate cost. Their main disadvantage is an "escape" phenomenon characterized by a return in pituitary LH release and subsequent testicular testosterone production.

Inhibitors of Steroid Synthesis. Ketoconazole is a synthetic imidazole analogue developed originally as an antifungal agent that inhibits cytochrome P-450-dependent enzymes essential in the production of sex steroids and glucocorticoids. It inhibits both testicular and adrenal androgen production and has a very rapid onset of action (within hours) which can be useful in emergency situations. However, its disadvantages include a short half-life necessitating t.i.d. dosage (200–400 mg every 8 hours), as well as GI upset and potential hepatotoxicity. It is usually reserved for the treatment of androgen-independent progression.

In summary, orchiectomy remains the gold standard for hormone ablation therapy for metastatic prostate cancer. However, various forms of medical castration are equally efficacious, but are in general more expensive. Two commonly used forms of medical castration include monthly Depo LHRH agonists with or without an anti-androgen like flutamide, or a combination of cyproterone acetate (CPA) (50 mg twice daily) with low-dose (0.1 mg once daily) DES. The choice depends on the philosophy of the physician and on patient preferences. Responses can be anticipated in 80% of patients. Ketoconazole is usually reserved for the treatment of urgent situations and for the treatment of androgen-independent progression.

Timing of Hormone Therapy

Patients with metastatic prostate cancer who are symptomatic should be treated without delay. The debate on immediate versus delayed hormonal therapy centers on patients who are asymptomatic at the time of presentation. The central controversial issue is whether endocrine therapy at diagnosis (early) provides prolonged tumor control and superior patient survival rates when compared with therapy started when the patient becomes symptomatic (delayed). Early studies from the 1940s suggested that early endocrine therapy provided a survival advantage over no treatment. However, these studies utilized control groups from an earlier era that did not have the advantage of antibiotics. Subsequent randomized studies failed to demonstrate a significant survival with early endocrine therapy compared to delayed therapy. Prostate cancers are heterogeneous tumors composed of a spectrum of androgen-dependent, androgen-sensitive, and androgen-independent cancer cell subpopulations. Those who argue for delayed therapy believe that early endocrine therapy will lead to unrestricted growth of hormone-resistant clones of tumor cells. However, early hormone therapy is supported by experimental evidence from animal models. As tumor volume increases, development of androgen-independent and anaplastic subpopulations are more likely through a process of genetic drift. It is generally believed by most urologists today that early endocrine therapy is superior to delayed treatment for postponing the onset of tumor progression in most patients. Therefore, hormonal therapy should be started at the time of diagnosis of metastatic prostate cancer, except in those asymptomatic patients who wish to remain sexually active.

Total Androgen Blockade

Ninety percent of all testosterone produced in the male is testicular in origin, and the remaining 10% is of adrenal origin. Since traditional forms of hormone ablation therapy suppress only the production of testicular androgens, adrenal androgens are not affected. The possibility that adrenal androgens may contribute to the progression of conventionally treated patients with metastatic cancer has been studied for decades. Over the past decade, several multi-institutional randomized trials were conducted in the U.S., Canada, and Europe and although conflicting results exist, there is a general consensus that total androgen blockade may provide a slight increase in time to progression and survival (Crawford and Nabors, 1991). It is important to stress, however, that hormone therapy is only palliative and that eventually most patients will die of their disease unless additional non-hormonal forms of systemic therapy are developed.

Androgen-Independent Prostate Cancer Progression

The median survival of patients progressing with prostate cancer while on hormone therapy is only eight months and 85% will develop osseous metastases before death (Eisenberger et al., 1987; Kozlowski et al., 1991). Therapy at this stage is aimed at improving quality of life by alleviating symptoms without causing significant treatment-related toxicity. Palliative measures include continuation or additional hormonal manipulation such as the addition of ketoconazole or flutamide to monotherapy regimens. Radiation can relieve bone pain or local symptomatology. Prednisone can help improve energy levels, appetite, and mood. Unfortunately, there is no therapeutic panacea on the horizon for stage D-3 prostate cancer. The inability to fully characterize the mechanisms mediating androgen-independent progression has compromised the development of new therapeutic modalities. Described in this section are some options for treating stage D-3 disease, as well as experimental agents on the horizon.

Cytotoxic Therapy

Although cytotoxic therapy has been the mainstay of management, response rates in clinical trials with either single agents or combination therapies have been disappointing (10–20%). Active single agents include mitomycin C, doxorubicin, and vinblastine (Eisenberger et al., 1987). A well tolerated out-patient regimen consisting of ketoconazole (400 mg p.o. t.i.d.) and weekly Adriamycin is currently being utilized at the University of Texas M.D. Anderson Cancer Center. Estramustine is an estrogen linked to an alkylating agent with weak cytotoxic activity; its mechanism of action is presumably through interference with microtubular formation. Response rates (partial) of 20–35% have been reported in stage D-3 disease. As an oral medication (14 mg/kg/day in three divided doses for six weeks) with few side effects, estramustine may be a reasonable option.

The poor response to cytotoxic chemotherapy in patients with prostate cancer is likely due to both tumor and patient factors. Prostate cancer has a long doubling time which suggests a cell kinetic basis for resistance to chemotherapy. Also, tumor burden is advanced in patients with stage D-3 disease which tends to limit the success of most chemotherapeutic protocols. Furthermore, patients are generally old with a poor performance status which compromises their chemotherapeutic schedules. Newer agents that kill nondividing cells will be required in order to make a significant contribution to the treatment of patients with stage D-3 disease.

Tumor-Stromal Cell Interactions

Interactions between stromal and epithelial compartments are essential for normal development and function of the prostate gland. Also, both benign hyperplastic and malignant prostate cell growth are stimulated by factors produced by prostatic fibroblasts. Experimental data exist to verify the presence of bidirectional stimulatory pathways between prostate cancer cells and their surrounding fibroblasts (Gleave et al., 1991; Chung et al., 1991). This interaction is mediated through the release of soluble peptide growth factors and extracellular matrix. The interactions between prostate cancer cells and their surrounding stroma is tissue-specific, and the production of soluble growth factors by bone stromal cells may play a role in site-specific metastasis characteristic of prostate cancer. Growth factors may act in endocrine (i.e., produced in one organ and transported via the serum to act on a distant target organ), paracrine (i.e., produced by one cell type but acts on a different neighboring cell type), and autocrine (i.e., produced by one cell type and acts on a

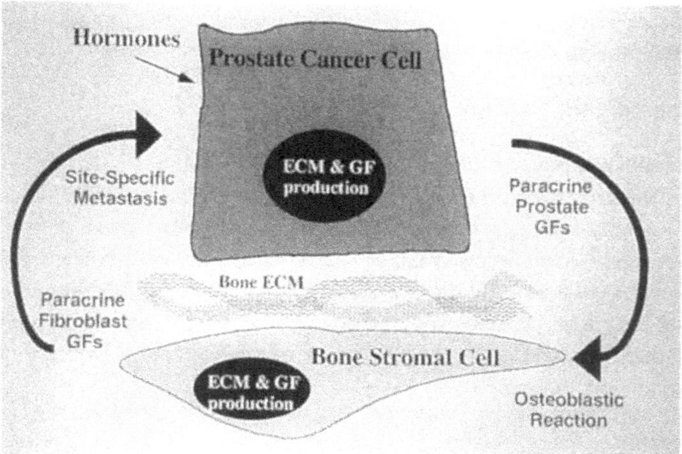

Figure 8. Growth regulatory signals can affect prostate cancer cells via autocrine, paracrine, and endocrine pathways. Both endocrine (via blood stream) and paracrine (local tissue) factors play critical roles in the growth and progression of prostate cancer. Androgens like testosterone stimulate the growth rate of prostate tumor cells, and paracrine GFs produced by bone fibroblasts and prostate tumor cells help produce the site-specific osteoblastic metastases characteristic of prostatic carcinoma. (ECM, extracellular matrix; GF, growth factor.)

similar neighboring cell type), as illustrated in Figure 8. Recent work has provided some evidence that prostate cancer cell growth may be under autocrine influences involving androgen-mediated up-regulation of transforming growth factor alpha, epidermal growth factor receptor, or basic fibroblastic growth factor (bFGF). Also, paracrine growth factors (e.g., bFGF) released by the stromal compartment also play an important role. In addition, two growth factors related to the bFGF family, keratinocyte growth factor (KGF), and the androgen-induced growth factor (AIGF), as well as insulin-like growth factors (IGFs) and their binding proteins and receptors have all been implicated in the regulation of prostate cancer growth. An improved understanding of the underlying mechanisms that mediate tumor-stromal cell interactions will provide a means to block this interaction with various therapeutic agents. Suramin, a synthetic polyanionic compound has been evaluated in the treatment of stage D-3 prostate cancer. Suramin is a heparin analogue and a nonspecific extracellular growth factor antagonist which functions to block the biological activities of several growth factors. Approximately 20–30% of patients will achieve significant responses with this therapy; however, responses are not durable and last for less than three months. More recently, monoclonal antibody technology has provided a means to produce pure clones of antibodies that block growth factor receptors, and these are currently being tested in both *in vitro* and *in vivo* animal models.

Immunoconjugate Therapy

Monoclonal antibody technology has also provided a means to produce "magic bullets": biologic compounds that are fused with cytotoxic or radioactive compounds. For example, pseudomonas exotoxin has been fused with antibodies directed against various growth factor receptors that are overexpressed in various malignant tissues to serve as site-specific cytotoxic therapy. Additionally, antibodies directed against prostate specific antigen make for an ideal magic bullet; however, their efficacy is impaired because of neutralization by circulating PSA molecules.

Gene Therapy

Based upon the understanding of both the intrinsic properties of the tumor cells and their interactions with the host microenviroment, a number of strategies are being developed to target the tumor cells, their communication with neighboring cells, or the host immune system by delivering appropriately engineered molecules to affect prostate tumor growth and metastases in experimental animal models. For example, the retroviral-based granulocyte macrophage stimulating factor (GM-CSF) was found to "cure" a small fraction of Dunning rat prostate tumors when delivered *in vivo*, a feat that has not been previously possible by conventional chemotherapy or radiation therapy (Sanda et al., 1994). Increasing the expression of tumor suppressor genes or decreasing the expression of oncogenes delivered by efficient delivery systems (e.g., adenoviral or herpes) are also being evaluated in many laboratories around the world. In particular, it may be possible to specifically target prostate cancer cells with recombinant vectors expressed under tight regulation by specific androgen-responsive promoters.

REFERENCES

Andriole, G.L.L. & Catalona W.J. (1993). Using PSA to screen for prostate cancer. The Washington University experience. Urol. Clin. N. Amer. 20, 647–652.

Batson, O.V. (1940). The function of the vertebral veins and their role in the spread of metastasis. Ann. Surg., 112, 138–149.

Bazinet, M., Meshref, A.W., Trudel, C., et al. (1994). Prospective evaluation of prostate specific antigen density and systematic biopsies for early detection of prostatic carcinoma. Urology 43, 44,19651.

Benson, M.C., Whang, I.S., Olsson, C.A., et al. (1992). The use of prostate-specific antigen density to enhance the predictive value of intermediate levels of serum prostate-specific antigen. J. Urol. 147, 817–821.

Boring, C.C., Squires, T.S., Tong, T., & Montgomery, S. (1994). Cancer statistics. Cancer J. Clinicians. 44, 7–26.

Brawer, M.K., Chetner, M.P., Beatie, J., et al. (1992). Screening for prostatic carcinoma with prostate-specific antigen. J. Urol. 147, 841–845.

Brawer, M.K. (1994). Prostate specific antigen: Critical issues. Urology 44, 917–933.

Brawn, P.N., Ayala, A.G., & von Eschenbach, A.C. (1982). Histologic grading study of prostate adenocarcinoma: The development of a new and comparison with other methods - a preliminary study. Cancer 49, 525–533.

Cantrell, B.B., De Klerk, D.P., Eggleston, J.C., Boitnott, J.K., & Walsh, P.C. (1981). Pathologic factors that influence prognosis in stage A prostatic cancer: The influence of extent vs grade. J. Urol. 125, 516–521.

Carter, H.B. & Coffey, D.S. (1990) The prostate: An increasing medical problem. Prostate 16, 39–48.

Carter, B.S., Carter, H.B., & Isaacs, J.T. (1990). Epidemiologic evidence regarding predisposing factors to prostate cancer. Prostate 16, 187–197.

Carter, H.B., Pearson, J.D., Metter, E.J., et al. (1992). Longitudinal evaluation of prostate-specific antigen levels in men with and without prostate disease. JAMA 267, 2215–2220.

Catalona, W.J., Smith, D.S., Ratliff, T.L., et al. (1991). Measurement of prostate-specific antigen in serum as a screening test for prostate cancer. N. Engl. J. Med. 324, 1156–1161.

Catalona, W.J. & Scott, W.W. (1986). Carcinoma of the prostate. In: Campbell's Urology (Walsh, P.C., Gittes R.E., Perlmutter, A.D., & Stamey, T.A., eds.), pp. 1463–1534. W.B. Saunders, Philadelphia.

Catalona, W.J., Smith, D.S., Ratliff, T.L., & Basler J.W. (1993) Detection of organ-confined prostate cancer is increased through prostate-specific antigen screening. JAMA 270, 948.

Catalona, W.J., Richie, J.P., Ahmann, F.R., et al. (1994). Comparison of digital rectal examination and prostate specific antigen in the early detection of prostate cancer: Results of a multicenter clinical trial of 6,630 men. J. Urol. 151, 1283–1290.

Chodak, G.W. & Schoenberg, H.W. (1989). Progress and problems in screening for carcinoma of the prostate. World J. Surg. 13, 60–64.

Chung, L.W.K., Gleave, M.E., Hsieh, J.T., Hong, S.J., & Zhau, H.E. (1991). Reciprocal mesenchymal-epithelial interaction affecting prostate cancer growth and hormonal responsiveness. Cancer Surveys 11, 91–121.

Cooner, W.H., Mosely, B.R., Rutherford, C., et al., (1990). Prostate cancer detection in a clinical urological practice by ultrasonography, digital rectal examination and prostate specific antigen. J. Urol. 143, 1146–1154.

Crawford, E.D. & Nabors, W.L. (1991). Total Androgen Blockade: The American experience. Urol. Clinics North Am. 18, 55–64.

Epstein, J.I., Walsh, P.C., Carmichael, M., & Brendler, C.B. (1994). Pathologic and clinical findings to predict tumor extent of nonpalpable (stage T1c) prostate cancer. JAMA 271, 368–374.

Eisenberger, M.A., Bezerdjian, L., & Kalash, S. (1987). A critical assessment of the role of chemotherapy for endocrine-resistant prostatic carcinoma. Urol. Clin. North Am. 14, 695.

Fowler, J.E. & Whitmore, W.F. (1981). The incidence and extent of pelvic lymph node metastases in apparently localized prostate cancer. Cancer 47, 2941–2945.

Franks, L.M. (1954). Latent carcinoma of the prostate. J. Pathol. Bacteriol. 68, 603–616.

Gann, P.H., Hennekens, C.H., & Stampfer, M.J. (1995). A prospective evaluation of plasma prostate-specific antigen for the detection of prostate cancer. JAMA 273, 289–294.

Gibbons, R.P., Correa, R.J., Brannen, G.E., & Weissman, R.M. (1989). Total prostatectomy for clinically localized prostatic cancer: Long-term results. J. Urol. 141, 564–569.

Giovannucci, E., Rimm, E.B., Colditz, G.A., et al. (1993). A prospective study of fat and risk of prostate cancer J. Natl. Cancer Inst. 85, 1571–1579.

Gleason, D.F. (1966). Classification of prostatic carcinomas. Cancer Chemother. Rep. 50, 125–129.

Gleave, M.E., Hsieh, J.T., Gao, C.A., von Eschenbach, A.C., & Chung, L.W.K. (1991). Acceleration of human prostate cancer growth in vivo by factors produced by prostate and bone fibroblasts. Cancer Res. 51, 3753–3761.

Gleave, M.E., Hsich, J.T., Wu, H.C., von Eschenbach, A.C., & Chung, L.W.K. (1992). Serum prostate specific antigen levels in mice bearing human prostate LNCaP tumors are determined by tumor volume and endocrine and growth factors. Cancer Res. 52, 1598–1605.

Hugosson, J., Aus, G., Bergadahl, C., & Bergadahl, S. (1995). Prostate cancer mortality in patients surviving more than 10 years after diagnosis. J. Urol. 154, 2115–2117.

Jacobs, S.C. (1983). Spread of prostatic carcinoma to bone. Urology, 21, 337–344.

Johansson, J.-E., Adami, H.-O., Anderson, S.-O., et al. (1992). High 10-year survival rate in patients with early, untreated prostatic cancer. JAMA 267, 2191–2196.

Kerlikowski, K., Grady, D., Barclay, J., Sickles, E.A., Eaton, A., & Ernster, V. (1993). Positive predictive value of screening mammography by age and family history of breast cancer. JAMA 270, 2444–2450.

Kozlowski, J.M., Ellis, W.J., & Gray hack, J.T. (1991). Advanced prostatic carcinoma. Early versus late endocrine therapy. Urol. Clin. North Am. 18, 15–24.

Labrie, A., Dupont, J.L., Gomez, Y., et al. (1995). Beneficial effect of combination therapy administered prior to radical prostatectomy. Urology 44, 29–33.

Labrie, F., Dupont, A., Suburu, R., et al. (1990). Serum prostate specific antigen as pre-screening test for prostate cancer. J. Urol. 147, 846–852.

Lee, F., Torp-Pedersen, S.T., Littrup, P.J., et al. (1989). Hypoechoic lesions of the prostate: Clinical relevance of tumor size, digital rectal examination, and prostate specific antigen. Radiology 170, 29–32.

McNeal, J.E. (1968). Regional morphology and pathology of the prostate. Am. J. Clin. Pathol. 49, 347–357.

McNeal, J.E., Kindrachuk, R.A., Freiha, F.S., et al. (1986). Patterns of progression in prostate cancer. Lancet 1, 160–164.

Mettlin, C., Jones, G., Averette, H., et al. (1993). Defining and updating the American Cancer Society guidelines for the cancer-related checkup: prostate and endometrial cancers. CA Cancer J. Clin. 43, 42–46.

Mettlin, C., Littrup, P.J., Kane, R.A., et al. (1994). Relative sensitivity and specificity of serum prostate specific antigen (PSA) level compared with age-referenced PSA, PSA density, and PSA change. Cancer 74, 1615–1620.

Mostofi, F.K. (1975). Grading of prostatic carcinoma. Cancer Chemother. Rep. 59, 11.

Oesterling, J.E. (1991). Prostate specific antigen: A critical assessment of the most useful tumor marker for adenocarcinoma of the prostate. J. Urol. 145, 907–923.

Oesterling, J.E., Jacobson, S.J., Chute, C.G., Guess, H.A., Girman, C.J., Panser, L.A., & Lieber, M.M. (1993a). Serum prostate-specific antigen in a community-based population of healthy men. Establishment of age-specific reference ranges. JAMA 270, 860.

Oesterling, J.E., Suman, V.J., Zincke, H, & Bostwick, D.G. (1993b). PSA-detected (clinical stage T1c or B0) prostate cancer. Pathologically significant tumors. Urol. Clin. N. Amer. 20, 687.

Paget, S. (1889). The distribution of secondary growths in cancer of the breast. Lancet, 1, 571–573.

Partin, A.W. & Oesterling, J.E. (1994). The clinical usefulness of prostate specific antigen: Update 1994. J. Urol. 152, 1358.

Sanda, M.G., Ayyagari, S.R., Jaffee, E.M., et al. (1994). Demonstration of a rational strategy for human prostate cancer gene therapy. J. Urol. 151, 622–628.

Scardino, P.T. (1989). Early detection of prostate cancer. Urol. Clin. North Am. 16, 635–655.

Seaman, E., Katz, A., Cooner, W.H., et al. (1993). An algorithm for prostate cancer detection based on PSA, TRUS, DRE, PSA density, and PSA velocity. J. Urol. (Part 2) 149, 414A.

Shinoharo. K., Wheeler, T., & Scardino, P.T. (1989). The appearance of prostate cancer on transrectal ultrasonography: Correlation of imaging and pathological examinations. J. Urol. 142, 76–82.

Spitz, M.R., Currier, R.D., Fueger, J.J., Babaian, R.J., & Newell, G.R. (1991). Familial patterns of prostate cancer: A case-control analysis. J. Urol. 146, 1305–1307.

Stamey, T.A. (1995). Making the most out of six systematic sextant biopsies. Urology 45, 2–12.

Walsh, P.C. & Jewitt, H.J. (1980). Radical surgery for prostatic cancer. Cancer, 45, 1906–1910.

Wang, M.C., Valenzuela, L.A., Murphy, G.P., & Chu, T.M. (1979). Purification of a human prostate specific antigen. Invest. Urol. 17, 159–163.

Whitmore, W.F., Jr. (1988). Overview: Historical and contemporary. Natl. Cancer Inst. Monogr. 7, 7–11.
Whitmore, W.F., Warner, J.A., & Thompson, I.M. (1991). Expectant management of localized prostatic cancer. Cancer 67, 1091–1096.
Whittemore, A.S., Keller, J.B., & Betensky, R. (1991). Low-grade, latent prostate cancer volume: Predictor of clinical cancer incidence? J. Nat. Cancer Inst. 83, 1231–1235.
Wilson, J.D. & Griffin, J.E. (1994). Disorders of sexual differentiation. In: Harrison's Principles of Internal Medicine. 13th ed. (Isselbacher, K.J., Braunwald, E., Wilson, J.D., Martin, J.B., Fauci, A.S. & Kasper, D.L., eds.), pp. 2039–2051. McGraw-Hill, Inc., New York.
Zagars, G.K. & von Eschenbach, A.C. (1993). Prostate-specific antigen. An important marker for prostate cancer treated by external beam radiotherapy. Cancer 72, 538–548.

Chapter 13

The Biology of Human Melanoma

S.A. LYNCH, P.M. DOSKOCH,

S. VIJAYASARADHI, and A.N. HOUGHTON

INTRODUCTION

Melanoma is a malignant disease of the melanocyte, a pigment-producing cell found primarily in the epidermis. Although melanoma was once a relatively rare disease, since the second World War its incidence in many parts of the world has

Advances in Oncobiology
Volume 1, pages 293–302.

ISBN: 0-7623-0146-5

risen much faster than that of any other malignancy (Grin-Jorgensen et al., 1992). While the cutaneous location and dark pigmentation of most lesions make early self-detection possible, melanoma remains a frequently fatal disease because of its aggressive metastatic behavior and resistance to conventional therapies.

Although the origins of melanoma are multifaceted, exposure to ultraviolet (UV) radiation is considered a prime factor. Within a given population, the disease is most prevalent at latitudes where solar UV intensity is the greatest, and individuals who develop melanoma commonly have histories of severe, intermittent sunburns. Other commonly cited risk factors include fair skin, blond or red hair, the tendency to burn rather than tan, and the presence of numerous benign nevi. In addition, there is compelling evidence that melanoma has a heritable component. Cases of families in which a high proportion of members develop melanoma suggest that some genetic factor strongly predisposes such individuals to develop this malignancy. Although familial melanomas are identified in only about 10% of the cases of cutaneous disease (Greene and Fraumeni, 1979), genetic analysis of these families is likely to provide crucial information regarding the pathogenesis of melanoma.

Whether inherited or the result of environmental factors, genetic alterations clearly play a pivotal role in transforming a normal melanocyte into a melanoma cell. The aim of this chapter is to compare and contrast the melanocyte with its malignant counterpart, examine the etiology of transformation, and discuss some strategies for disease intervention.

BIOLOGY OF MELANOCYTES AND MELANOMA CELLS

Found primarily in the basal layer of the epidermis, but also in the eye and other epithelia, melanocytes are most distinctive for their polydendritic morphology and their ability to synthesize the pigment melanin. Melanocytes package melanin in lineage-specific organelles called melanosomes, which are transferred through the melanocyte dendrites to the surrounding keratinocytes. The putative melanocytic precursor, the melanoblast, has not been identified. In the developing embryo, migrating neural crest cells eventually give rise to several cell types (Weston, 1970), including neurons, adrenal medulla, connective tissue, and melanocytes. The melanocyte progenitor's complex journey from the neural crest to the epidermis requires proteases, such as neutral serine protease, to facilitate migration through tissue and extracellular matrix (LeDouarin, 1984). This process resembles the metastatic behavior of a melanoma cell as it invades surrounding tissue during tumor progression, underscoring how malignant cells can recapitulate properties of normal precursor cells (albeit in an inappropriate context).

With the addition of exogenous growth factors, melanocytes can be cultured as adherent dendritic cells (Eisinger and Mark, 1982), although their *in vitro* life span is finite. Melanocytes derived from fetal and newborn skin are lightly pigmented and show a bipolar morphology; over time they become polydendritic and darkly pigmented, resembling melanocytes derived from adult skin. In contrast, melanoma

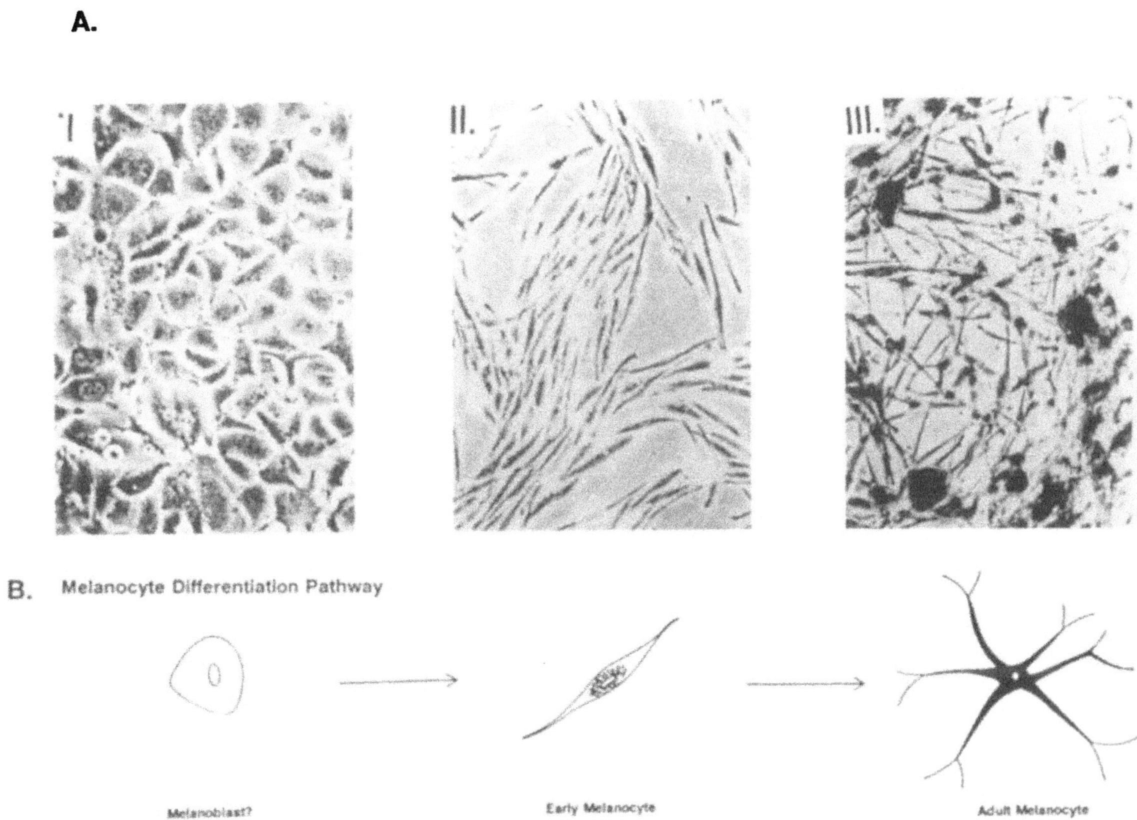

Figure 1. Phenotypic heterogeneity of melanocytic cells. **A)** Morphology of melanoma cells displaying an I. early, II. intermediate, or III. late phenotypes. **B)** Proposed pathway of melanocyte differentiation extrapolated from melanoma cell morphology and cell surface markers. (Adapted with permission from Houghton et al. (1982). J. Exp. Med., 156, 1755–1766.)

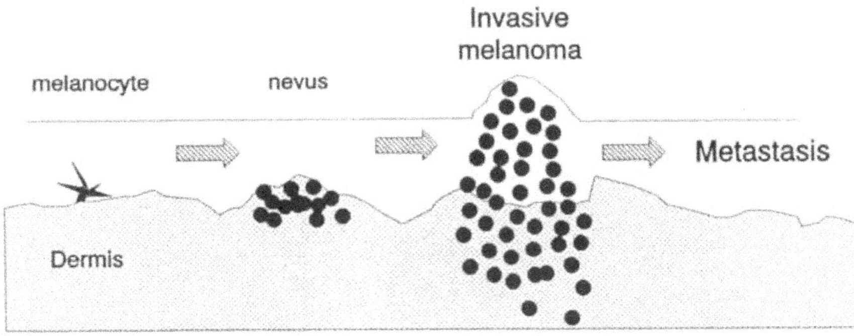

Figure 2. Schematic for tumor progression of a melanocyte to melanoma. Sequential outline of events taking place as a normal skin melanocyte transforms into melanoma cell capable of rapid expansion and metastatic behavior.

cells proliferate indefinitely in culture, are able to grow in an anchorage-independent fashion, and can often be cultured in the absence of exogenous growth factors. Although mature melanocytes are uniform in appearance, melanoma cells are heterogenous (Figure 1A); even those within the same tumor may vary in morphology, pigmentation, and antigen expression (Houghton et al., 1987).

Primary melanoma lesions *in vivo* commonly exhibit a biphasic growth pattern (Clark, 1967) (Figure 2). The initial radial growth phase is characterized by superficial, intraepidermal expansion. These melanomas are generally not capable of invasion or metastasis. During the subsequent vertical growth phase, the tumor infiltrates the lower dermis and subcutaneous fat, forming an expansion nodule capable of metastasis. Metastatic potential (and patient mortality) increases directly with tumor thickness (Balch et al., 1992). Some lesions, known as nodular melanomas, appear to bypass the radial growth phase (Clark, 1967). These observations suggest that multiple events occur as a melanocyte changes to an invasive and metastatic melanoma cell.

MELANOMA ANTIGENS

The diverse range of antigens expressed by melanoma cells can be classified into two groups: (a) antigens present on mature melanocytes and on some or all melanomas, and (b) antigens absent from mature melanocytes. The expression of antigens in the first group simply reflects the cells' melanocytic origins. Markers in the second group may be upregulated during malignant transformation. Alternatively, these antigens in the second group are expressed in early stages of normal melanocytic development but are lost as a melanoblast differentiates. The idea that normal cell development can be deduced from the phenotype of malignant cells is a familiar theme; much of the current understanding of the hematopoietic differen-

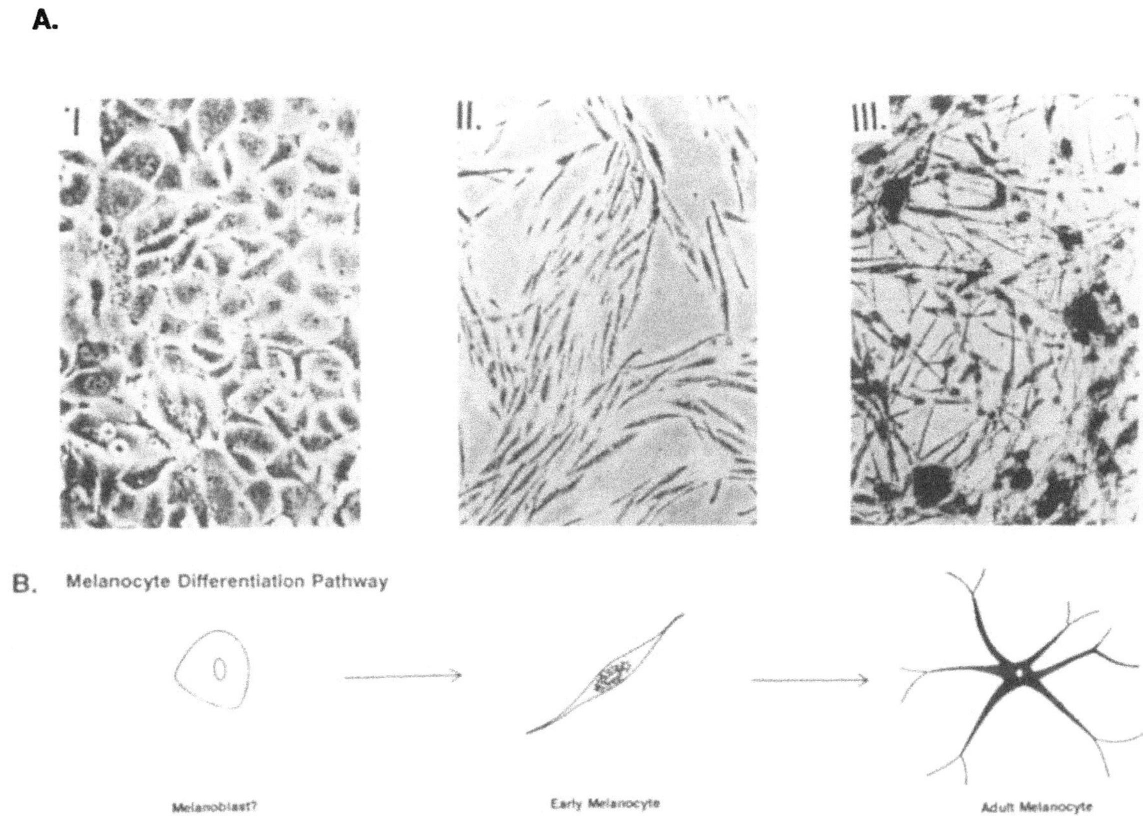

Figure 1. Phenotypic heterogeneity of melanocytic cells. **A)** Morphology of melanoma cells displaying an I. early, II. intermediate, or III. late phenotypes. **B)** Proposed pathway of melanocyte differentiation extrapolated from melanoma cell morphology and cell surface markers. (Adapted with permission from Houghton et al. (1982). J. Exp. Med., 156, 1755–1766.)

al., 1988) as well as in both primary and metastatic lesions taken from the same patient (Quinn et al., 1977, Balaban et al., 1986), implicating alteration of this chromosome in tumor initiation (Fountain et al., 1990). Most melanomas, for example, contain subchromosomal alterations in chromosome region 9p21 (Fountain et al., 1992), which appears to contain a melanoma suppression gene.

The Mendelian inheritance pattern of familial melanoma suggests that a single autosomal dominant gene is responsible. Some studies (Bale et al., 1989) have implicated a locus on chromosome 1, but other investigators (Cannon-Albright et al., 1990) have failed to confirm any linkage between familial disease and known markers. Chromosome 9p contains a gene that is closely linked to familial melanomas, and this suppressor gene appears to account for at least a subset of hereditary melanomas (Fountain et al., 1992).

Although activated *ras* oncogenes are only found in about 10% of melanomas (Albino et al., 1984), transforming melanocytes *in vitro* with dominant-acting oncogenes provides a system for studying the progression from a normal to a malignant phenotype. Melanocytes transfected with the viral oncogene *v*-Ha-*ras* (Albino et al., 1986), for example, no longer exhibit contact inhibition, are capable of anchorage-independent growth in soft agar, and upregulate GD3, a ganglioside detected at elevated levels on melanoma cells. Although *v*-Ha-*ras* transformed melanocytes are initially dependent on exogenous growth factors, they eventually become growth factor independent. Concurrently, the cells produce basic fibroblast growth factor (bFGF), an autocrine growth stimulator of melanoma (and a paracrine factor for normal melanocytes). Transformed melanocytes lose expression of a 120 kD cell surface glycoprotein known as adenosine deaminase binding protein (ADAbp). This molecule, expressed by cultured epidermal melanocytes but not melanoma cells, is a cell surface ectopeptidase (Morrison et al., 1993) that could possibly degrade crucial growth factors (such as bFGF), thereby rendering cells growth factor dependent.

These studies serve as a reminder that tumor pathogenesis need not involve novel or aberrant proteins. The activation of an unaltered developmental gene at an inappropriate time or an unaccustomed tissue site may play a significant role in malignant transformation and tumor progression. For example, the proteases produced by neural crest cells during migration can facilitate invasion and metastasis in transformed cells. Differentiation and transformation are not entirely unrelated processes.

MELANOMA: RECOGNITION BY THE IMMUNE SYSTEM

The immune response to melanoma is the best studied of any human malignancy. T lymphocytes, although rarely detected in metastatic lesions, frequently infiltrate regressing primary lesions. *In vitro*, melanoma cells can stimulate autologous cytotoxic lymphocytes (CTLs), which recognize and respond to antigenic peptide fragments bound to major histocompatibility (MHC) molecules on the surface of the melanoma cell (Wolfel et al., 1989). Moreover, antigens that can potentially elicit a humoral or cellular immune response have been identified. It is unclear what role these antigens play in initiating an immune response or whether these responses can lead to effective tumor destruction. Nonetheless, at least one-third of melanoma patients produce antibodies against these antigens.

The best characterized antigens are gangliosides, acidic glycolipids expressed on the cell membrane. Although the function of gangliosides is not understood, they may facilitate cell-cell and cell-substrate interactions. Melanoma patients who produce antibodies against gangliosides appear to have a more favorable prognosis (Livingston et al., 1987). More than 70% of metastatic melanoma patients immunized with the ganglioside GM2 plus immune adjuvant generate a specific antibody response against GM2 (Livingston et al., 1989).

The first tumor antigen recognized by human T cells has been defined on melanoma. The antigen, designated MAGE1 (for Melanoma AntiGen E), was discovered using CTL recognition of melanomas (van der Bruggen et al., 1991). DNA from a melanoma cell line which had elicited a strong autologous CTL response was transfected into a melanoma variant that had lost the antigen recognized by the CTLs. These transfectants were used to isolate the gene encoding the antigenic determinant responsible for CTL stimulation. The MAGE1 gene, which is presented to the immune system by MHC class I molecule HLA-A1, is expressed by approximately 40% of melanomas, as well as several other cancers. MAGE1 expression is not observed on most normal tissues but is present in normal testes. Tyrosinase peptide is also recognized by CTLs. The antigen is presented by the MHC HLA-A2 molecule (Brichard et al., 1993).

FUTURE THERAPEUTIC STRATEGIES

While surgery is indicated for primary or isolated metastatic melanomas, widespread metastatic disease calls for a more systemic approach. Because few patients respond to chemotherapeutic agents, and radiotherapy is useful only under particular conditions (Peters et al., 1992), a variety of alternative strategies are being examined.

Early studies utilizing **monoclonal antibodies** have shown that these molecules can target tumor cells and produce occasional clinical responses (Parkinson et al., 1992). Investigators are exploring the efficacy of antibodies linked to toxins and of chimeric molecules which combine the variable region of murine antibodies to the constant region of human immunoglobulins.

Another therapeutic strategy involves enhancing the natural immune response. Here **cytokines** may prove clinically valuable. All interferons increase expression of class I MHC molecules, and some elevate MHC class II expression as well. Interleukin-2 (IL-2) can promote growth of activated T cells, increase natural killer cell activity, and induce production of other cytokines. Although single agent clinical trials utilizing cytokines have produced occasional regressions, complete long-term responses have been uncommon. Current studies are investigating novel, more efficient delivery mechanisms for these molecules, but their greatest value may lie when used in combination with other agents, such as monoclonal antibodies.

By inducing expression of therapeutic molecules on tumor cells, **gene therapy** might also be used to boost the immune response of melanoma patients. For example, since cytokines appear to be the most effective over short distances, melanoma cells engineered to express these molecules locally might effectively stimulate a T cell response. Mice can reject injected tumor cells transduced with an IL-2 gene (Gansbacher et al., 1990); preliminary studies are underway in human subjects.

The development of a melanoma **vaccine** remains an area of active investigation. Vaccines have been derived from whole melanoma cells (Morton, 1986), shed antigens in tissue culture supernatant (Bystryn et al., 1986), purified melanoma antigens (Livingston et al., 1987), and recombinant viruses encoding such antigens (e.g., p97) (Estin et al., 1988). Another potentially valuable approach utilizes anti-idiotypic antibodies, created by immunization with an antibody against the desired antigen. Anti-idiotypic antibodies in effect mimic the original antigenic epitope, but are potentially more immunogenic than purified antigen (Chapman and Houghton, 1992).

While a clinically effective vaccine could be used to stimulate any melanoma patient's immune response, particular applications may prove especially valuable. After surgical removal of a primary tumor, for example, vaccination might ablate the development of residual micrometastases. Eventually, healthy but susceptible patients (e.g. those with multiple and recurrent precancerous nevi, or with a strong history of familial disease) could be candidates for vaccination.

SUMMARY

Incidence of melanoma, a malignant disease of pigment producing melanocytes, has climbed dramatically in recent years. Ultraviolet radiation probably plays a role in disease pathogenesis. The process of melanoma progression is better understood than that of most human cancers. It is generally accompanied by specific chromosomal alterations and changes in the expression of growth factors, their receptors and other cell surface molecules. While surgery can cure the majority of patients with melanoma, current therapies for metastatic disease are not widely effective. However, since melanoma patients often exhibit a potential weak immune response to the disease, an understanding of melanocyte biology, particularly antigen expression, is leading to new therapeutic strategies. Several promising immunotherapies, including the use of cytokines, anti-melanoma antibodies, and vaccines, are currently under investigation.

REFERENCES

Albino, A.P., Houghton, A.N., Eisinger, M., Lee, J.S., Kantor, R.R.S., Oliff, A.I., & Old, L.J. (1986). Class II histocompatibility antigen expression in human melanocytes transformed by Harvey murine sarcoma virus (Ha-MSV) and Kirsten MSV retroviruses. J. Exp. Med. 164, 1710–1722.

Albino, A.P., Le Strange, R., Oliff, A.I., Furth, M.E., & Old, L.J. (1984). Transforming *ras* genes from human melanoma: A manifestation of tumour heterogeneity? Nature 308, 69–71.

Balaban, G.B., Herlyn, M., Clark, W.H., & Nowell, P.C. (1986). Karyotypic evolution in human malignant melanoma. Cancer Genet. Cytogenet. 19, 113–122.

Balch, C.M., Soong, S., Shaw, H.M., Urist, M.M., & McCarthy, W.H. (1992). In: Cutaneous Melanoma (Balch, C.M., Houghton, A.N., Milton, G.W., Sober, A.J., & Soong, S., eds.), pp. 165–187, J.B. Lippincott Company, Philadelphia.

Bale, S.J., Dracopoli, N.C., M.A.T., Clark, W.H., Fraser, M.C., B.Z. S., Green, P., Donis-Keller, H., Houseman, D.E., & Greene, M.H. (1989). Mapping the gene for hereditary cutaneous melanoma-dysplastic nevus to chromosome 1p. N. E. J. Medicine 320, 1367–1372.

Bouchard, B., Fuller, B.B., Vijayasaradhi, S., & Houghton, A.N. (1989). Induction of pigmentation in mouse fibroblasts by expression of human tyrosinase cDNA. J. Exp. Med. 169, 2029–2042.

Brichard, V., Van Pel, A., Wolfel, T., Wolfel C., De Plaen E., Lethe, B., Coulie, P., & Boon, T. (1993). The tyrosinase gene codes for an antigen recognized by autologous cytolytic T lymphocytes on HLA-A2 melanomas. J. Exp. Med. 178, 489–495.

Bystryn, J.-C., Jacobsen, S., Harris, M., Roses, D., Speyer, J., & Levin, M. (1986). Preparation and characterization of a polyvalent human melanoma antigen vaccine. J. Biol. Response Mod. 5, 211–224.

Cannon-Albright, L.A., Goldgar, D.E., Wright, E.C., Turco, A., Jost, M., Meyer, L.J., Piepkorn, M., Zone, J.J., & Skolnick, M.H. (1990). Evidence against the reported linkage of the cutaneous melanoma-dysplastic nevus syndrome locus to chromosome 1p36. Am. J. Human Genetics 46, 912–918.

Chapman, P.B. & Houghton, A.N. (1992). In: Biologic Therapy of Cancer (DeVita, V.T., Hellman, S., & Rosenberg, S.A., eds.), pp. 1–9, J.B. Lippincott, Philadelphia.

Clark, W.H. (1967) In: Advances in Biology of the Skin (Pergamon, Oxford).

Cowan, J.M., Halaban, R., & Francke, U. (1988). Cytogenetic analysis of melanocytes from premalignant nevi and melanomas. J. Natl. Can. Inst. 80, 1159–1164.

Eisinger, M. & Mark, O. (1982). Selective proliferation of normal human melanocytes *in vitro* in the presence of phorbol ester and cholera toxin. Proc. Natl. Acad. Sci. USA 79, 2018–2022.

Estin, C.D., Stevenson, U.S., Plowman, G.D., Hu S.L., Sridhar, P., Hellstrom, I. et al. (1988). Recombinant vaccinia virus vaccine against the human melanoma antigen p97 for use in immunotherapy. Proc. Natl. Acad. Sci. USA 85, 1052–1056.

Fountain, J.W., Bale, S.J., Houseman, D.E., & Dracopoli, N.C. (1990). Genetics of melanoma. Cancer Surveys 9(4), 645–671.

Fountain, J.W., Karayiorgou, M., Ernstoff, E.S., Kirkwood, J.M., Vlock, D.R., Titus-Ernstoff, L., Bouchard, B., Vijayasaradhi, S., Houghton, A.N., Lahti, J., Kidd, V.J., Housman, D.E., & Dracopoli, N.C. (1992). Homozygous deletions of human chromosome band 9p21 in melanoma. Proc. Natl. Acad. Sci. USA 89, 10557–10561.

Gansbacher, B., Zier, K., Daniels, B., Cronin, K., Bannerji, R., & Gilboa, E. (1990). Interleukin 2 gene transfer into tumor cells abrogates tumorigenicity and induces protective immunity. J. Exp. Med. 172, 1217–1224.

Greene, M. & Fraumeni, J. (1979). In: Human Malignant Melanoma (Clark W., Goldman, L., & Mastrangelo, M., eds.), pp. 139–166, Grune & Stratton, New York.

Grin-Jorgensen, C.M., Rigel, D.S., & Friedman, R.J. (1992). In: Cutaneous Melanoma (Balch, C.M., Houghton, A.N., Milton, G.W., Sober, A.J., & Seng-jaw, S., eds.), pp. 27–39, J.B. Lippincott Company, Philadelphia.

Houghton, A.N., Eisinger, M., Albino, A.P., Cairncross, J.G., & Old, L.J. (1982). Surface antigens of melanocytes and melanomas. J. Exp. Med. 156, 1755–1766.

Houghton, A.N., Real, F., Davis, L., Cordon-Cardo, C., & Old, L. (1987). Phenotypic heterogeneity of melanoma. Relation to the differentiation program of melanoma cells. J. Exp. Medicine 164, 812–829.

Kwon, B.S., Haq, A.K., Pomerantz, S.H., & Halaban, R. (1987). Isolation and weauence fo a cDNA clone for tyrosinase that maps at the mouse *c*-albino locus. Proc. Natl. Acad. Sci. USA 84, 7473.

LeDouarin, N. (1984). Cell migrations in embryos. Cell 38, 353–360.

Livingston, P., Natoli, E.J., Calves, M.J., Oettgen, H.F., & Old, L.J. (1987). Vaccines containing purified GM2 ganglioside elicit GM2 antibodies in melanoma patients. Proc. Natl. Acad. Sci. USA 84, 2911–2915.

Livingston, P.O., Ritter, G., Srivastava, P., Padavan, M., Calves, M.J., Oettgen, H.F., & Old, L.J. (1989). Characterization of IgG and IgM antibodies induced in melanoma patients by immunization with purified GM2 ganglioside. Cancer Res. 49, 7045–7050.

Morrison, M.E., Vijayasaradhi, S., Engelstein, D., Albino, A.P., & Houghton, A.N. (1993). A marker for neoplastic progression of human melanocytes is a cell surface ectopeptidase. J. Exp. Med. 177, 1135–1143.

Morton, D.L. (1986). Active immunotherapy against cancer: Present status. Semin. Oncol. 13, 180–185.

Parkinson, D.R., Houghton, A.N., Hersey, P., & Borden, E.C. (1992). In: Cutaneous Melanoma (Balch, C.M., Houghton, A.N., Milton, G.W., Sober, A.J., & Soong S.-J., eds.), pp. 522–541, J.B. Lippincott, Philadelphia.

Peters, L.J., Byers, R.M., & Ang, K.K. (1992). In: Cutaneous Melanoma (Balch, C.M., Houghton, A.N., Milton, G.W., Sober, A.J. & Soong, S., eds.), pp. 509–521, J.B. Lippincott Company, Philadelphia.

Quinn, L.A., Woods, L.K., Merrick, S.B., Arabasz, N.M., & Moore, G. E. (1977). Cytogenetic analysis of twelve human malignant melanoma cell lines. J. Natl. Can. Inst. 59, 301–305.

Real, F., Rettig, W., Garin-Chesa, P., Melamed, M., Old, L., & Mendelsohn, J. (1986). Expression of epidermal growth factor receptor in human cultured cells and tissues:relationships to cell lineage and stage of differentiation. Cancer Res. 46, 4726–4731.

van der Bruggen, P., Traversari, C., Chomez, P., Lurquin, E., Van den Eynde, B., Knuth, A., & Boon, T. (1991). A gene encoding an antigen recognized by cytolytic T lymphocytes on a human melanoma. Science 254, 1643–1647.

Vijayasaradhi, S., Bouchard, B., & Houghton, A.N. (1990). The melanoma antigen gp75 is the human homologue of the mouse *b* (*BROWN*) locus gene product. J. Exp. Med. 171, 1375–1380.

Weston, J. (1970). The migration and differentiation of neural crest cells. Adv. Morphogen 8, 41–114.

Wolfel, T., Klehmann, E., Muller, C., Meyer Zum Buschenfelde, K.-H., & Knuth, A. (1989). Lysis of human melanoma cells by autologous cytolytic T cell clones. J. Exp. Med. 170, 797–810.

Chapter 14

Aspects of the Treatment of B Cell Malignancies

A.Z.S. ROHATINER, J.S., MALPAS, and R.K. GANJOO

Advances in Oncobiology
Volume 1, pages 303–316.

ISBN: 0-7623-0146-5

INTRODUCTION

Two B cell malignancies, myeloma and follicular lymphoma, have been chosen to illustrate the use of different treatment modalities. Both are currently incurable with conventional therapy, both have a well-documented natural history and are currently the subject of much debate and interest in terms of new drugs, the use of biological therapy, and high-dose therapy. This section will focus on principles of management and on new approaches that are currently being evaluated, rather than describing drug dosages and schedules in detail.

MULTIPLE MYELOMA

Multiple myeloma (MM) is a cancer characterized by a clonal proliferation of abnormal plasma cells which arise in the bone marrow and disseminate to other tissues and organs. These plasma cells produce paraproteins, the light chain fraction of immunoglobulin, interleukin-1 (IL-1), tumor necrosis factor (TNF), and possibly other products yet to be defined, which, for example, activate osteoclasts (resulting in osteoporosis, lytic lesions in bones, and pathological fractures), impair hemopoiesis (leading to anemia, leucopenia, and thrombocytopenia), depress normal immunoglobulin production (resulting in infection), and damage the kidney (with consequent renal failure).

MM can have a variable course. Occasionally it is a devastating disease, with death occurring within a few months of diagnosis, but rarely a very prolonged course has been described. Treatment will be influenced by this feature, and by others such as MM's occurrence in an elderly population subject to a variety of other pathologies which may result in less satisfactory toleration of specific chemotherapy. It is also important to differentiate overt, progressive MM from such conditions as monoclonal gammopathy of undetermined significance (MGUS), indolent and smouldering MM, solitary plasmacytoma of bone (SPB) and extramedullary plasmacytosis (EMP). Some of the distinguishing features of these conditions are outlined in Table 1.

The diagnosis of MM depends on the careful assessment of the presenting features, and major and minor criteria for diagnosis (Table 2). For example, lytic lesions in the bone are considered to be minor criteria because they may be produced by other pathologies. Marrow infiltration by plasma cells, similarly, is a minor criterion because plasmacytosis may be seen as a response to chronic infection.

Table 1. Some Plasma Cell Dyscrasias—Differentiating Features

Feature	*SPB*	*EMP*	*MGUS*	*Smouldering MM*	*Indolent MM*	*Overt MM*
Symptoms	±	±	—	—	—	±
Bone lesions	solitary	—	—	—	1 or 2+	+++ or –
Anemia	—	—	—	—	±	±
Hypercalcemia	—	—	—	—	—	±
Renal failure	—	—	—	—	—	±
Marrow infiltration with plasma cells	0– <5%	—	<10%	>10%,<30%	>20%,<30%	usually >30%
"M" component						
IgG (g/1)	35		35	>35,<70	>35,<70	>35
IgA (g/1)	20		20	>20,<50	>20,<50	>20
BJ protein (g/24 hr)	1.0		1.0	>1.0	>1.0	>1.0
Labeling index			<1%	<1%	<1%	>1%

Significant amounts of paraprotein in the serum (Table 1), urinary light chain excretion, and a biopsy showing massive infiltration with malignant plasma cells are the major diagnostic criteria. Various schemes for diagnosis have been adopted, and are outlined in Table 2.

Table 2. Diagnostic Criteria for Multiple Myeloma

A Symptomatic progressive disease
B One major and one minor criterion
or
Three major criteria including a and b below
Major Criteria
Plasmacytoma
Bone marrow plasmacytosis (>30% plasma cells)
"M" component: IgG >35 g/1
IgA >20 g/1
Light chain (K or L) excretion >1.0 gm/24 hr
Minor Criteria
a Bone marrow plasmacytosis (>10% to 30%)
b "M" component present but less than noted for major criteria
c Lytic bone lesions present
d So-called immunoparesis with low normal immunoglobulins
Normal ranges: IgG 8.0–16 mg/ml
IgA 1.4–4.0 mg/ml
IgM 0.5–2.0 mg/ml

PRETREATMENT EVALUATION

MM's variable course makes it important to assess prognosis when the disease first presents. Salmon and Durie demonstrated a correlation between the tumor burden of myeloma and survival, and produced a clinical staging scheme which has been in general use. The clinical features associated with low, intermediate, and high tumor burdens are shown in Table 3.

When applied to a large series of patients, these stages have been effective in discriminating between good and bad risk patients. For example, on average, Stage I patients have a median survival of five years, Stage II patients three years, and Stage III patients only nine months. More recently, other independent prognostic factors have been evaluated. These include beta-2 microglobulin, C-reactive protein, serum albumin and labeling index. All of these have been shown to be independent variables on multivariate analysis. One of the most useful is beta-2 microglobulin, where a level of <6 mg/1 in the serum is associated with a good prognosis, and one of >16 mg/1 is indicative of a poor outlook.

Table 3. Myeloma Staging System of Salmon and Durie

Stage I:	Low myeloma cell mass ($<0.6 \times 10^{12}$ cells/m^2)
	For this stage all the following must be present:
	Hemoglobin >10 g/dl
	Serum calcium <3 mmol/1
	Solitary lesions in bone or normal skeletal survey
	"M" component IgG <50 g/1
	IgA <30 g/1
	Light chain excretion <4 g/24 hrs
Stage II:	Indeterminate myeloma cell mass ($0.6–1.2 \times 10^{12}$ cells/m^2)
	Fitting neither I nor III
Stage III:	High myeloma cell mass ($>1.2 \times 10^{12}$ cells/m^2)
	Any of the following criteria should be present:
	Hemoglobin <8.5 g/dl
	Serum calcium >3 mmol/1
	Multiple lytic lesions on skeletal survey
	"M" component IgG >70 g/1
	IgA >50 g/1
	Light chain excretion >12 g/24 hrs
Suffix A	Relatively normal renal function (serum creatinine <0.1 mmol/1)
Suffix B	Renal dysfunction (serum creatinine >0.1 mmol/1)

POSTTREATMENT EVALUATION

Response has been defined as a fall of 50% or 75% in the quantity of paraprotein measured at presentation. Sometimes patients subsequently achieve a low but steady level of "M" component, and are said to have achieved a plateau phase. This is associated with relief of symptoms and signs, independence of blood transfusion requirement, and may last for a variable period. Eventually, relapse inevitably occurs. Pathological evidence of persistent myelomatous infiltration can always be found, and so, as in any other malignancy, the possibility of cure cannot be entertained.

With the chance that new therapies might successfully eliminate residual tumor, criteria for complete remission have been defined, and have brought MM into line with those terms employed in other B cell malignancies.

Complete remission (CR) should now indicate disappearance of the "M" band on conventional laboratory analysis, together with any evidence of light chain excretion. The bone marrow should be normal with <2% of normal plasma cells present. Partial response (PR) is defined as >50% reduction in the quantity of paraprotein. Stable disease (SD) or plateau is when the patient is stable clinically and the "M" band shows no tendency to rise. Relapse is associated with a >10% increase in the paraprotein level.

MANAGEMENT

The first question to be asked when a patient is seen with cancer is: "Does the patient need treatment, and is chemotherapy appropriate?" The variants of MM which may resemble this highly malignant condition have already been discussed, and some of their features outlined in Table 1. MGUS and indolent and smouldering myeloma do not require chemotherapy. There is no evidence that this increases survival, and patients may be done a disservice if chemotherapy is used too soon. Drug treatment should only be used when these patients become symptomatic and there is evidence of progression.

Overt MM commonly presents with symptoms related to skeletal damage, pain in the back or limbs, symptoms of anemia, leukopenia or thrombocytopenia giving rise to (for example) breathlessness, repeated infections, or bleeding. Neurological complications of cord compression causing paraplegia or nerve root compression with pain and (very occasionally) peripheral paraesthesia are also seen, and all these symptoms are indications for therapy. Occasionally when the lesion is localized, such as in spinal cord compression by an intraspinal mass of myeloma, radio-therapy is the treatment of choice, in combination with dexamethasone. Rapid relief of pain in spinal and limb deposits may also be achieved with radiotherapy.

Chemotherapy must be planned with care. It is a paradox that the patient may need intensive supportive maneuvers, with blood transfusion, antibiotics, fluid replacement and measures to lower serum calcium and uric acid before chemotherapy can safely be given. This general attention is part of the management required for most B cell neoplasms.

THE DRUG TREATMENT OF MM

Alkylating Agents and Early Therapy

Untreated, overt progressive MM has a median survival of about nine months. In the early 1950s the first chemotherapeutic agents capable of modifying the course of this disease became available. The first was sarcolysin, an isomer of phenylalanine mustard, which was shown by the Russians to relieve symptoms and produce objective responses in MM. In the West, phenylalanine mustard (melphalan, or L-PAM) became the preferred agent, and together with another alkylating agent, cyclophosphamide, has remained the basis of most MM treatment programs. In randomized trials, melphalan and cyclophosphamide were shown to be equally effective in producing a response, and in improving survival. Cyclophosphamide was more easily managed in the presence of marrow suppression, and melphalan was contraindicated in the presence of renal failure. With either agent, a response was seen in about 50% of patients treated, and the median survival was improved from nine months to two to three years. Details of these alkylating agents are given in Chapter 15.

Steroid Hormones

Prednisolone, methyl prednisolone and dexamethasone have been used in MM. These can be shown to affect plasma cell proliferation, and to reduce calcium loss produced by osteoclast activation. Melphalan and prednisolone have become the most frequently used combination in MM. It has been used in short courses over five to seven days every three to four weeks. Between 50 and 60% of patients show a response, and have a median survival of about 24 months, although a range of 19–54 months has been recorded in various studies.

The use of dexamethasone in large doses as a single agent has been reported, and a good response has been seen, even in patients who are refractory or relapsed after chemotherapy. The usual complications of steroid therapy have been seen, but in the case of elderly people treated with high-dose dexamethasone, mental disturbance has been considerable.

Prednisolone has been added to multi-alkylating drug regimens such as VBAP, VCAP, VMCP (Table 4). Regimens such as VAD and VAMP, where dexamethasone and methyl prednisolone are used and alkylating agents are not included, have used larger steroid doses. Steroids were included in 17 out of 18 combinations reported in a recent survey of multidrug chemotherapy in MM.

Few randomized studies have been carried out to determine whether steroids add benefit to alkylating or other agents. In a Medical Research Council trial of

Table 4. Some Combination Chemotherapy Regimens for Multiple Myelomas

Regimen		*Drugs/Doses*	*Schedule*	*Interval (Days)*
VBMCP	V	0.03 mg/kg i.v.	Day 1	
	B	1 mg/kg i.v.	Day 1	
	M	0.1 mg/kg i.v.	Days 1–7	28
	C	10 mg/kg i.v.	Day 1	
	P	1 mg/kg p.o.	Days 1–7	
VMCP	V	0.03 mg/kg i.v.	Day 1	
	M	5 mg/m^2 p.o.		
	C	100 mg/m^2 p.o.	Days 1–7	28
	P	60 mg/m^2 p.o.		
Alternating with				
VBAP	V	0.03 mg/kg i.v.		
	B	30 mg/m^2 i.v.	Day 1	28
	A	30 mg/m^2 i.v.		
	P	60 mg/m^2 p.o.	Days 1–7	
VAD	V	0.4 mg/24 hr	By continuous infusion	
	A	9 mg/m^2/24 hr	over 4 days	28
	D	40 mg p.o.	Days 1–4 repeat Day 9 and 17	
VAMP	V	0.4 mg/24 hr	By continuous infusion	
	A	9 mg/m^2/24 hr	over 4 days	28
	MP	1.5 g/m^2 i.v./p.o.	Days 1–4	
C-VAMP	C	560 mg i.v.	Day 1, 8, 15	
	V			
	A	as for VAMP		28
	MP			

Note: V, vincristine; B, BCNU (carmustine); M, melphalan; C, cyclophosphamide; A, adriamycin (doxorubicin); D, dexamethasone.

multidrug chemotherapy with or without prednisolone, there was no benefit in the prednisolone arm of the study. Although there is still debate about the benefit of conventional doses of steroids, there seems little doubt that high doses of dexamethasone or methyl prednisolone are effective, although the benefit is usually transient.

The Vinca Alkaloids

Vincristine, one of the vinca alkaloids, has been used for many years in combination with other drugs in MM. The evidence for any benefit is scanty. In one early study, slight and transient benefit was seen in two patients. Other small studies were unable to confirm this, but it is interesting to note that vincristine is used in 14 out of 18 combinations recommended for MM. Vincristine may produce neurotoxicity and other unpleasant side-effects, so the rationale for its continued use is debatable.

After many years of randomized trials comparing these regimens with melphalan and prednisolone, the consensus view is that, while multidrug regimens probably show an increase in response rate, there is no increase in survival, and certainly none of the regimens is curative.

Other Agents Employed in the Treatment of MM

The Nitrosoureas

BCNU (carmustine) has the property of an alkylating agent, and has been shown to be non-cross-resistant with melphalan and cyclophosphamide. Because of this it has been used in the VBAP regimen, which was effective without intolerable side-effects.

The Anthracyclines

Doxorubicin (adriamycin) was used early on in combination with BCNU, and was noted to be effective in relapsed or resistant MM. It is a constituent of the VBAP and VCAP regimens, and also the non-alkylating agent regimens VAD and VAMP. It has been given as a continuous low-dose infusion, to lessen the risk of cardiotoxicity.

The Epipodophyllotoxins

Etoposide and teniposide are podophyllin derivatives, active against B cell malignancy such as leukemia and lymphoma. In a few studies, etoposide has been shown to be relatively inactive in MM, although it was tried in previously-treated patients, and its assessment needs to be continued. Teniposide has shown more promise, but awaits further investigation.

The Interferons

The introduction of interferons produced by recombinant techniques has allowed their use in MM. Earlier studies of leukocyte interferon showed that objective responses could be obtained in MM by the use of this agent on a long-term basis.

$\alpha 2\beta$-Interferon and γ-interferon have been investigated. In IgA, IgG, and light-chain MM, $\alpha 2\beta$-interferon has produced responses in 17, 29, and 7% of patients, respectively.

When combined in induction therapy, results have been equivocal. In regular low doses, interferon is well tolerated, and has significantly improved the duration of remission in a randomized study; in patients who are in CR, it appears to increase survival. Neither the optimal dose for remission maintenance, nor duration of maintenance therapy, has been determined.

Antimetabolites

These have been singularly ineffective in MM, and this is difficult to explain, when they have been so useful in other B cell malignancies. This applies equally to new agents such as fludarabine and pentastatin.

NEW DEVELOPMENTS

So far, no treatment for MM has been curative, although very long survivals have been noted. It is possible that the few long-term survivors have owed this more to the biology of their MM than to the treatment they received. Although still controversial, it is probable that in a drug-sensitive tumor, intensification of therapy would lead to an increase in tumor cell death, and therefore to a better response which, in turn, would result in long-term survival. Although this can be shown not always to be true in MM, the use of high-dose melphalan or busulfan supported by bone marrow transplantation was begun in the 1980s. More CRs were certainly seen, up to half the patients achieving CR in some studies. Unfortunately, many patients relapsed 18–24 months later, but recent reviews of long-term follow-up show that a number of patients who did not relapse are still alive 8–10 years later.

Where an allogeneic donor has been available, patients with MM have had high-dose chemotherapy with or without total body irradiation. In 90 patients entered on the European Bone Marrow Transplant Registry, actuarial survival has revealed that 40% of these patients are alive at 76 months. Nevertheless, caution must be advised in the interpretation of all these studies, since MM is a slowly-proliferative tumor, and relapse could still occur after a considerable period.

In an attempt to achieve more durable remissions, interferon has been given as continuation (maintenance) therapy, and in a small randomized study has been shown significantly to improve remission duration and survival. These findings await confirmation.

The role of cytokines needs further exploration. It has been shown that IL-6 promotes plasma cell proliferation, and it is possible to intervene in the paracrine loop, either by using specific monoclonal antibodies directed at IL-6, or by blocking IL-6 receptors where the MM cells are IL-6 dependent.

FOLLICULAR LYMPHOMA

Non-Hodgkin's lymphoma are a heterogeneous group of disorders characterized by lymph node enlargement, with or without extranodal involvement. Broadly speaking, they can be divided on the basis of histological subtype into those of low or high grade (Kiel classification) and the subtype correlates with prognosis.

Although the low grade lymphomas are generally responsive to chemotherapy, they remain incurable, whereas a proportion of patients with high grade lymphoma may be cured with intensive combination chemotherapy.

Follicular lymphoma is the commonest low grade non-Hodgkin's lymphoma in the Western world. The typical presentation is that of a person aged about 50, with lymphadenopathy at several sites. The diagnosis is made on the basis of a lymph node biopsy. The bone marrow is frequently involved and characteristically shows paratrabecular infiltration with clumps of centrocytes.

Histopathology

Follicular lymphoma is characterized by the presence of nodules of lymphoma cells easily seen on the cut surface of the lymph gland. The nodular architecture is outlined by a reticulin network between the follicles which are themselves composed of two cell types: small cleaved cells (centrocytes) and larger cells (centroblasts). Depending on the relative proportions of these, three subtypes of follicular lymphoma have been described, viz. (a) predominantly single cleaved cell; (b) mixed small cleaved cell and large-cell; and (c) predominantly large-cell.

The nodules are composed of monoclonal B cells expressing B cell antigens such as CD19, CD20, CD22, and CD24; they also express the common ALL antigen (CD10). Immunoglobulin gene rearrangement studies have confirmed the monoclonal nature of the disease. The majority of patients with follicular lymphoma die as a consequence of the disease, usually following histological transformation to high grade histology. The incidence of transformation has probably been underestimated in the past, but if serial lymph node biopsies are performed at each recurrence it becomes apparent that transformation occurs in the majority of patients.

Molecular Aspects

Follicular lymphoma is associated with a non-random translocation between the long arms of chromosomes 14 and 18. The breakpoint on chromosome 14 is located in most cases within or immediately adjacent to the joining region (J_H) of the immunoglobulin heavy chain gene. The breakpoint on chromosome 18 is located on a region called BCL-2. In cells containing the t(14;18) translocation, the BCL-2 gene is juxtaposed with the IgH gene resulting in deregulated expression of the BCL-2 gene. The precise mechanisms by which the BCL-2 protein exerts its oncogenic effects are unknown but it has been demonstrated that BCL-2 protein inhibits red cell death (apoptosis) of the lymphocytes, thus favoring a "malignant" process. Transformation of follicular lymphoma to high grade histology is probably associated with the occurrence of other gene rearrangements.

MANAGEMENT

Investigation

The diagnosis having been confirmed, it is important to establish the extent and distribution of disease. Staging has historically been based upon modifications of the Ann Arbor classification, originally put forward for Hodgkin's disease. Careful physical examination is supplemented by radiography of the chest, together with computed axial tomography (CAT) scans to assess nodal involvement. A bone marrow aspirate and trephine biopsy are essential.

Treatment

Radiotherapy is very effective and is regarded as the treatment of choice for the small proportion of patients presenting with localized disease. However, the majority of patients present with more widespread disease, particularly bone marrow involvement, thus systemic therapy is indicated. However, some patients may not require any treatment at the time of diagnosis. Certainly, there are no data to suggest that survival is any better if patients who are asymptomatic are treated sooner, rather than later. It should, however, be remembered that such patients comprise the minority.

Since 1955 standard treatment has been the alkylating agent chlorambucil. Given orally as intermittent cycles of treatment, remissions are seen in approximately 75% of patients at presentation, and indeed subsequently, at first and second recurrence. The fact that such remissions are usually incomplete and that within about two years recurrence almost always supervenes, led to the use of combination chemotherapy such as cyclophosphamide, vincristine, and prednisolone. However, a series of randomized comparisons failed to confirm any advantage for the combination in terms of overall survival, although the complete response rate is higher. Similarly, the use of more prolonged treatment, so-called maintenance chemotherapy, again, prolongs time to progression but does not affect survival. More recently, the use of adriamycin-containing combinations has been evaluated and again, there appears to be no benefit in terms of survival.

Thus for the average patient with follicular lymphoma, treatment has altered the natural history of the disease, in that in comparison with the observations of Gall and Mallory in 1942, when without treatment, the median survival was five years, this has now been extended to between 9 and 10 years but this can hardly be considered acceptable for patients with an illness the median age of which at presentation is 55 years.

New Treatment Modalities

In the last 10 years, four new treatments have been evaluated. These are: (a) interferon; (b) antibody and antibody targeted therapy; (c) purine analogues; and (d) myelo-ablative therapy with autologous bone marrow transplantation.

Interferon

Interferons are naturally occurring substances which have antiviral, immunoregulatory and antiproliferative characteristics. The rationale for investigating interferon in lymphoma is based on its efficacy in both preventing the evolution of and treating L1210 leukemia and AKR lymphoma in mice. Early studies by Ion Gresser were followed by small studies in patients with low and high grade lymphoma in whom conventional treatment had failed. It became very clear from the outset that, whereas interferon did not appear to be effective in high grade lymphoma, in low grade lymphoma, and follicular lymphoma in particular, it might have a role in treatment. The earliest studies were conducted with small quantities of leukocyte interferon obtained from buffy coats; subsequently, larger studies using interferons derived from recombinant DNA technology confirmed the impression that regression of lymphadenopathy occurred in patients with low grade lymphoma. Recent trials have focused on answering two questions: does the addition of interferon to conventional chemotherapy confer any advantage, and does the use of "maintenance" interferon following conventional therapy prolong duration of remission and hence survival?

The results of these studies are somewhat difficult to interpret; there is consensus about the fact that the use of interferon for a period of time after a response has been achieved confers a modest advantage in terms of freedom from progression, and one report has indicated that the combination of an adriamycin-containing regimen, given for 18 months with interferon, increases response rate, duration of remission and overall survival. At the present time, however, its precise role remains to be defined. The treatment is not without drawbacks, i.e., it has to be given as a subcutaneous injection, the first few injections are usually associated with flu-like symptoms and patients complain of feeling tired, although most are able to continue normal activity. There is much controversy about the mechanism of action, antiproliferative effects being invoked, as also have been immunologically-mediated mechanisms such as induction of natural killer cells and killer cells.

Antibody Therapy, Targeted Therapy

Follicular lymphoma is a B cell disease and the antigenic determinants are well defined. It is therefore an ideal target for an immunologically directed approach. Monoclonal antibodies such as the "humanized" rat antibody CAMPATH- 1H have been used and responses reported in follicular lymphoma. Attempts have also been made to exploit the specific idiotype of an individual lymphoma by raising anti-idiotype antibodies to each person's specific tumor. Again, responses have been seen but the logistics of this approach are complicated. More recently, monoclonal antibodies have been conjugated to toxins and to radioactive isotopes. Recently reported studies are very encouraging, particularly those where radioactive iodine has been conjugated to anti-CD20. The results are preliminary but this represents a highly promising area of research.

Purine Analogues

Most of the data in follicular lymphoma relate to fludarabine. It has become clear that this interesting new drug can induce responses in patients in whom other treatments have failed. The drug has the advantage of being devoid of the side-effects normally associated with cytotoxic chemotherapy, i.e., patients do not lose their hair, and nausea and vomiting are very rare. There is, however, a major problem with myelosuppression, together with the additional problem of infection due to T cell dysfunction. However, response rates of approximately 50% are consistently being seen in patients with recurrent or resistant disease and studies are currently in progress to better define the true role of this group of compounds.

Myelo-Ablative Therapy With Autologous Bone Marrow Transplantation

It has now become clear that very intensive treatment comprising chemotherapy, or chemotherapy followed by whole body irradiation supported by autologous bone marrow transplantation, can cure a proportion of patients with high grade lymphoma in whom conventional treatment had failed, provided the disease remains responsive to conventional treatment. There is a dose-response relationship for both cyclophosphamide and for radiotherapy in lymphoid malignancies, hence the rationale for using such treatment in patients with follicular lymphoma. The number of patients throughout the world who have received such treatment is, however, relatively small—between two and three hundred—and the follow-up relatively short, in view of the long natural history of follicular lymphoma. However, some preliminary conclusions can be drawn. Duration of remission is better than that of patients receiving conventional treatment, but at present there is no improvement in survival. The treatment has a potential, albeit small (i.e., less than 5%), mortality and an appreciable morbidity. The majority of people remain in hospital for between four and six weeks while the blood count recovers and may not be able to return to work for between three and six months afterwards. Such treatment makes both men and women permanently infertile, which is obviously a consideration in younger patients.

Much attention has focused recently on the prognostic significance of rendering the autologous marrow negative for t(14;18) containing cells, minimal residual disease being demonstrated by means of the polymerase chain reaction. Longer follow-up is required to determine whether such treatment will be curative for a proportion of people. Meanwhile the studies continue. It would certainly be wrong to be complacent.

SUMMARY

B cell malignancies have become a major topic of interest and a focus both for clinical and laboratory research. The fact that they are incurable at present, with a slow evolution and a variable response to different treatment modalities, has suggested a pattern for more common solid tumors. This, coupled with the greater understanding of the role of cytokines in (for example) myeloma, where IL-6 has been shown to be of importance in controlling myeloma cell division, and the discovery not only of the chromosome abnormality of 14;18 translocation in follicular lymphoma, but also the way in which this inhibits apoptosis or programmed cell death in this condition, have made both of these conditions the target for experimental treatment programs. For example, interferon, antibody targeted therapy, and high-dose therapy with either autologous bone marrow or peripheral blood stem cell support are now being explored. The experience gained from high-dose therapy in both myeloma and follicular lymphoma is being translated into new treatment strategies for breast and ovarian cancer.

REFERENCES

Alexanian, R. & Dimopoulos, M. (1994). The treatment of multiple myeloma. N. Engl. J. Med. 330, 484–489.

Berlogie, B. (1992). In: Multiple myeloma. Hematology/Oncology Clin. North. Amer. Vol. 6, W.B. Saunders, Philadelphia.

Dunbar, C.E. & Nienhuis, A.W. (1993). Multiple myeloma—new approaches in therapy. JAMA 269, 2412–2416.

Gregory, W.M., Richards, M.A., & Malpas, J.S. (1992). Combination chemotherapy versus melphalan and prednisolone in the treatment of multiple myeloma—an overview of published trials. J. Clin. Onc. 10, 334–342.

Greipp, P.R. (1992). Advances in the diagnosis and management of myeloma. Sem. Hemat. 29, 24–25.

Malpas, J.S., Bergsagel, D.E., & Kyle, R.A. (1995). Myeloma Biology and Management. Oxford University Press, Oxford.

Chapter 15

Principles of Cancer Chemotherapy

J.S. MALPAS and A. ROHATINER

Advances in Oncobiology
Volume 1, pages 317–350.

ISBN: 0-7623-0146-5

SECTION I: HISTORICAL PERSPECTIVE

Although the chemotherapy of cancer is a relatively recent development, it must not be forgotten that even before the 19th century efforts were made, using various metals including zinc, silver, and mercury, in an attempt to treat cancer, and some success was reported in folk medicine from the use of either metals or plant extracts such as colchicine. It was not until 1865 that Lissauer reported the beneficial effect of potassium arsenite in chronic leukemia. Although the use of metals did not return for more than 100 years, when the platinum compounds were successfully introduced into cancer chemotherapy, a concept was born which was strengthened by the successful use of chemotherapeutic agents, first against protozoa and later against bacteria.

The modern era of chemotherapy begins with the introduction of nitrogen mustard by Wilkinson in Great Britain and Gilman and Goodman in the United States. Significant clinical responses were obtained in patients with Hodgkin's disease, when for the first time it was shown that a chemotherapeutic agent could affect a malignancy which had become widely disseminated. Until then, surgery or radiotherapy were the only available treatments to localize disease, and once cancer had spread, the patient inevitably died. It must be said that *pari passu* with chemotherapy, the use of hormones was shown to be effective. Dodds synthesized stilbestrol and used it in disseminated prostatic cancer, and Hickman and Kendall introduced cortisone for lymphoid malignancies in 1949. New derivatives of nitrogen mustard were synthesized by Ross at the Chester Beattie Institute in London, and melphalan, chlorambucil, and myeloran were introduced to clinical practice by Haddow, Galton, and others.

There still seemed no possibility of treating acute leukemia until the synthesis of the folic antagonist aminopterin by Seeger and the demonstration by Farber and his colleagues, in 1948, that it could produce remission in children with acute leukemia. These new substances, called antimetabolites, relied on an increasingly sophisticated knowledge of cell metabolism, and an ability to synthesize analogues of purines and pyrimidines. The work of Hitchings and Elion ushered in a golden age of chemotherapeutic development. Among the many antimetabolites produced, 5-fluorouracil (synthesized by Heidelberger) stands out as a drug that was specifically designed to treat carcinoma, and today remains one of the most effective agents available.

Advances in the knowledge of biochemistry led to some interesting attempts to exploit the biochemistry of the tumor cell. It was thought that the essential amino acid phenylalanine might enable a drug attached to it to gain easier entry to the cell. This was the reason for the synthesis of melphalan, and although the principle did not work, nevertheless a very useful chemotherapeutic agent was produced. The same was true of the compound cyclophosphamide, which is split by phosphatases present in high quantity in tumors. The high local phosphatase content was supposed to liberate the cyclophosphamide locally, and avoid damage to local tissues. Unfortunately this hypothesis was not borne out in practice, but nevertheless cyclophosphamide has remained an important alkylating agent. A number of other drugs were introduced by serendipity: plant, bacterial, or fungal molds became a source of a wide variety of important compounds, many effective antibacterial agents. Drugs such as actinomycin D, daunorubicin, and doxorubicin, were developed by an increasingly sophisticated pharmaceutical industry which was aware of the potential of these compounds as anticancer agents. Serendipity also came to the aid of the chemist, when (for example) extracts of the Madagascar periwinkle were being examined as a possible antidiabetic agent, and were shown to reduce the white cell count in rabbits. Inhibition of growth of tumor cells was noted, and the two compounds vinblastine and vincristine were extracted. These remain two of the most potent and widely used anticancer agents.

Along with the development of new agents came the realization of the best mode of their employment. Higher rates of response and more durable remissions could be obtained by using drugs in combination, particularly if the agents had different specific toxicities. Thus, while the antitumor effect summated, the toxic side-effects (which were the main disadvantage of chemotherapy) were limited. The highly successful MOPP regimen introduced by DeVita and his colleagues for the treatment of Hodgkin's disease, combinations of anthracyclines and alkylating agents for the treatment of childhood solid tumors, and the combination of platinum compounds with bleomycin and etoposide, resulting in the cure of testicular cancer, are some examples of the successful use of curative combinations of drugs.

Although the history of chemotherapy is relatively short, major strides have been made towards the control of many previously fatal malignant conditions.

PHARMACOLOGY OF ANTICANCER DRUGS

Pharmacokinetic Principles

Before considering the main classes of anticancer drugs, it is necessary to review the principles by which these drugs achieve their effects, and the factors which govern their absorption, distribution, and excretion: their pharmacokinetics. Anticancer drugs, when administered orally, may be wholly or partially absorbed. Absorption may be influenced by a number of factors. The blood level of the drug then rapidly rises. During its passage through the liver it may undergo metabolic changes—so-called first pass metabolism—and various metabolites may start to circulate, being removed either locally or by excretion in the urine. In general, the blood level achieved will give an indication of the exposure of the tumor to the anticancer agent. The effective exposure will be a function of concentration

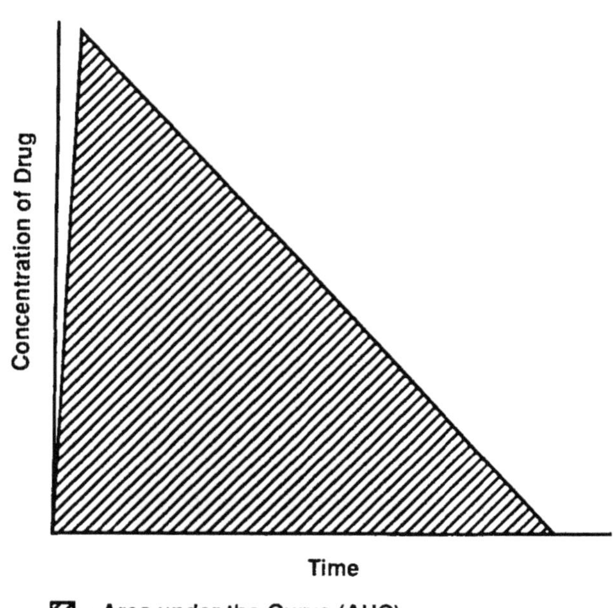

Figure 1. Curve showing variation of concentration of drug with time and derivation of the area under the curve (AUC).

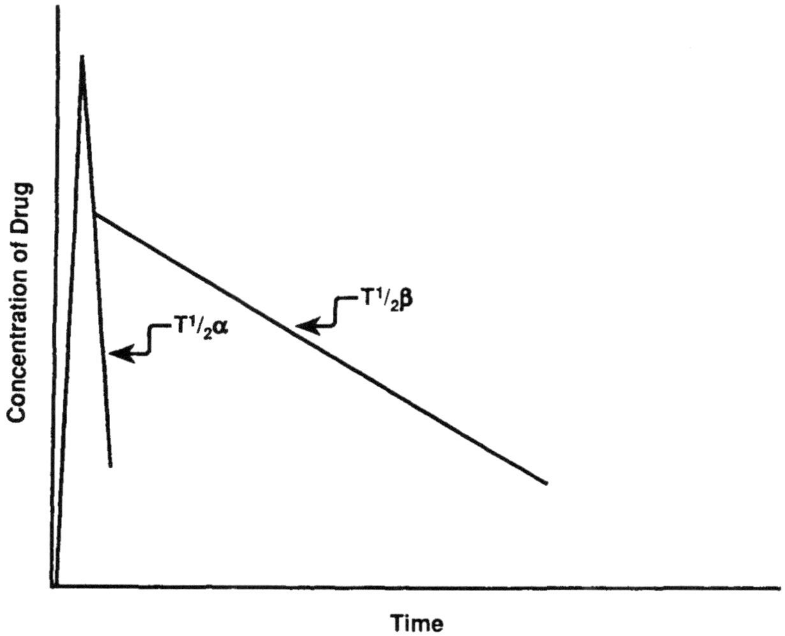

Figure 2. Curve showing variation of concentration of drug with time when drug enters two separate compartments.

multiplied by time. A typical exposure of a drug rapidly absorbed, uniformly distributed, and completely excreted, showing the curve for concentration and time, is given in Figure 1. The exposure of the tumor to the drug is measured by the area under the curve (AUC), and is shown in the shaded area in the figure. If the drug is distributed between two compartments, the curve of concentration in the blood will be modified (Figure 2). The rate of initial excretion is described as the time for half the drug to be excreted ($t^{1/2}\alpha$), and the time for the slower rate of excretion is $t^{1/2}\beta$.

Bioavailability of a drug is a frequently used term. This can be measured by assessing the AUC for the intravenously administered dose of the drug, and dividing this into the AUC for the same oral dose of the drug.

$$\text{Bioavailability} = \frac{\text{AUC for oral dose}}{\text{AUC for i.v. dose of drug}}$$

It can be shown that in the case of some drugs such as melphalan, an alkylating agent, there can be a wide variation in bioavailability, ranging from 10 to 50%, even within an individual. The same is true for 6-mercaptopurine and this may be of importance when long-term oral mercaptopurine therapy is being used for the treatment of childhood acute lymphoblastic leukemia.

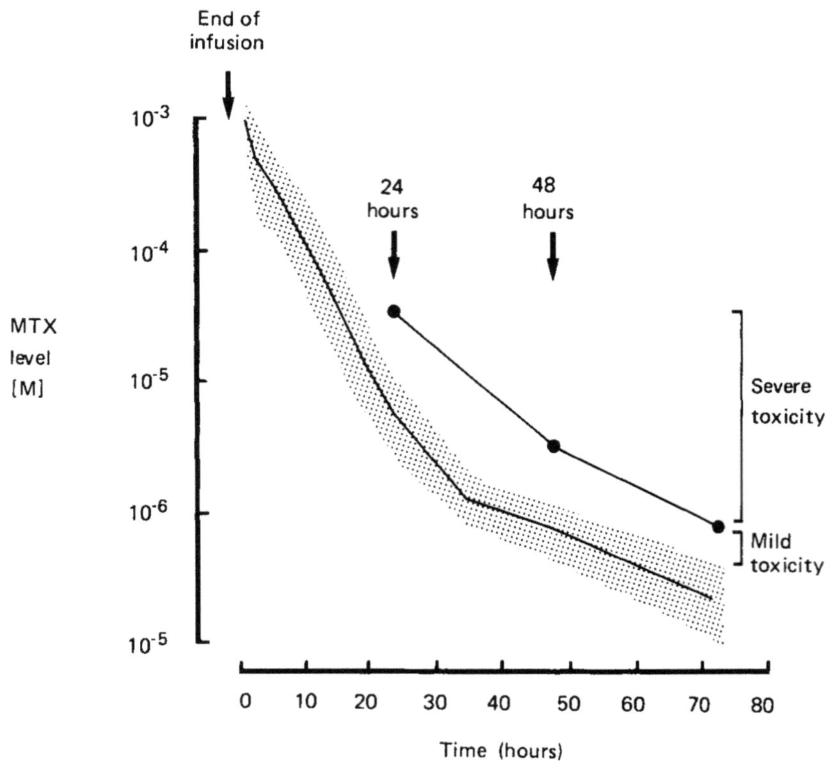

Figure 3. Curve showing poor excretion of methotrexate in a patient compared to the normal range.

As well as giving an indication of the exposure of the tumor to the drug, the AUC is a measure of the exposure of normal tissues to the drug. Many chemotherapeutic agents are toxic to the bone marrow, and it can be shown that there is a close correlation between the increasing AUC and lowered platelet and white blood cell counts.

For a drug that is mainly eliminated by renal excretion, deterioration in renal function may lead to an increased AUC. Methotrexate, for example, is largely excreted via the kidneys, and a practical example of how delayed excretion could raise the AUC and give rise to possible life-threatening toxicity is shown in Figure 3. Excretion and metabolic degradation of cytotoxic drugs may be influenced by many other factors, including the co-administration of other anticancer drugs, nonspecific medication used to control patients' symptoms, etc. Even the sequence in which certain anticancer drugs are used may inhibit or enhance the action of a specific compound.

PHARMACOLOGY OF ALKYLATING AGENTS

Alkylating agents were historically the first group of anticancer drugs to show consistent beneficial effect. From the beginning they have maintained their importance in the armamentarium available, and have shown their versatility by their adoption in high-dose chemotherapy schedules (dealt with later).

Mechanisms of Alkylation

The structure of mechlorethamine (nitrogen mustard) is shown in Figure 4. The ethylinamine group CH_2CH_2Cl is a highly reactive component which reacts with so-called nucleophilic elements in a wide variety of biological molecules. A relatively simple example is shown in the reaction with the amino acid alanine (Figure 5).

Monofunctional alkylating agents are those with a single CH_2CH_2Cl grouping able to perform reactions such as those shown in Figure 5. Bifunctional alkylating agents are those with two ethylinamine groups, which, in addition to causing damage and breaks in DNA, can bridge and form cross-links in the double helix. It can be shown that one of the commonest bonds attached is that of guanine, and that it is specifically alkylated at the seventh nitrogen (Figure 6). The bond formed between two guanine molecules is shown in Figure 7.

Because the parent substance mechlorethamine was a sclerosant, it was not possible to give it orally. Patients had to be admitted for therapy at great inconvenience, and were often subject to severe and intractable vomiting. A search was therefore made for oral preparations which would be better tolerated. This led to the introduction of chlorambucil and melphalan (Figure 8). The addition of the modified ring structure to the nitrogen mustard moiety leads to greater stability. These drugs can be administered orally, although their bioavailability may vary considerably. Besides giving stability, it was thought that in some cases the addition of an essential amino acid such as phenylalanine would result in the agent being more avidly taken up by the tumor cells. In practice this was not the case, but nevertheless melphalan remains a very important alkylating agent. Another interesting idea was that the cleavage of cyclophosphamide at the site of the phosphorus atom by the high tumor content of phosphatase and phosphorylase might work in

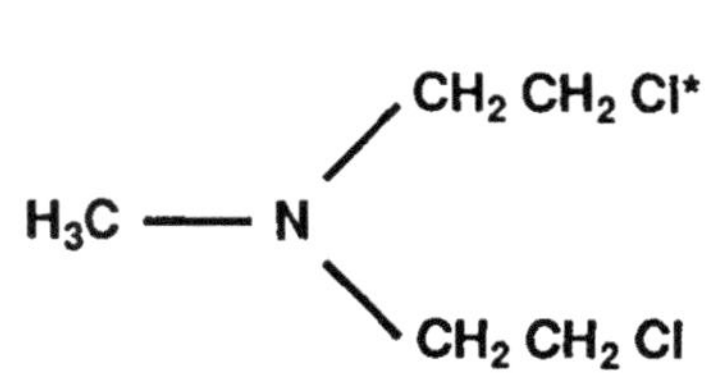

Figure 4. Methlorethamine (nitrogen mustard).

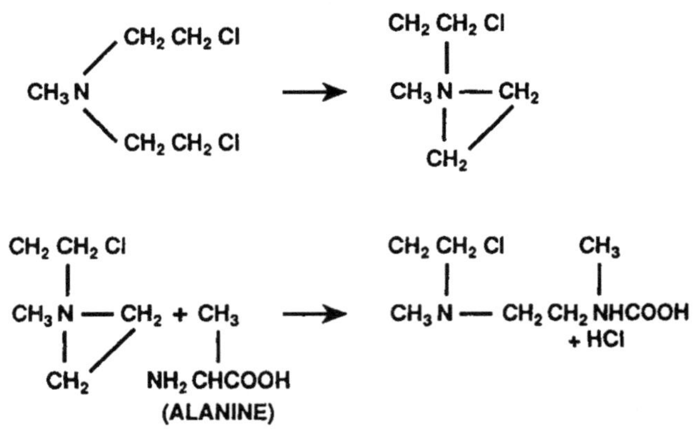

Figure 5. Alkylation of an amino acid.

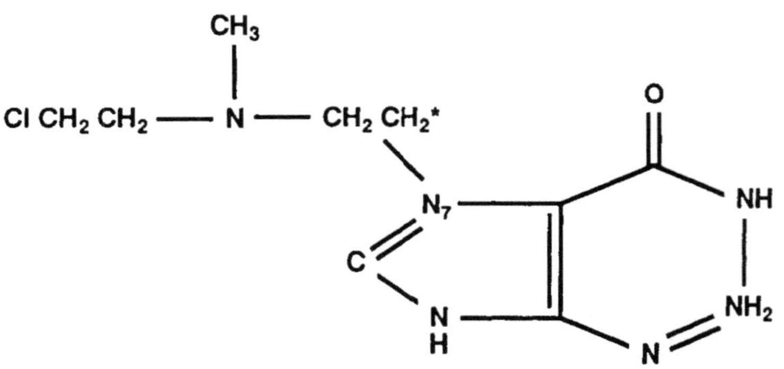

Figure 6. Alkylation of guanine at the N7 position.

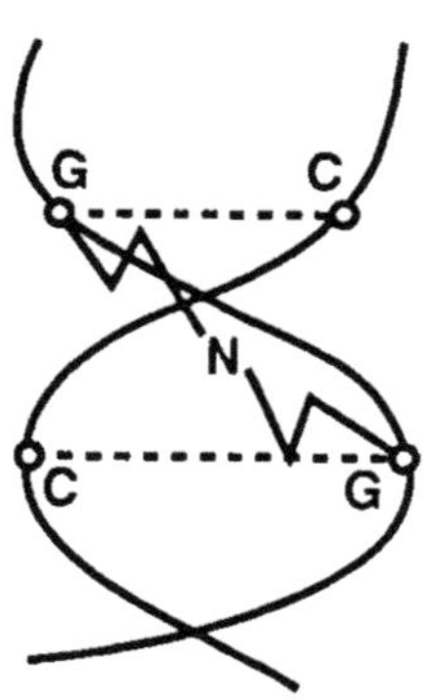

Figure 7. Alkylation with a bond formed between the two guanine molecules.

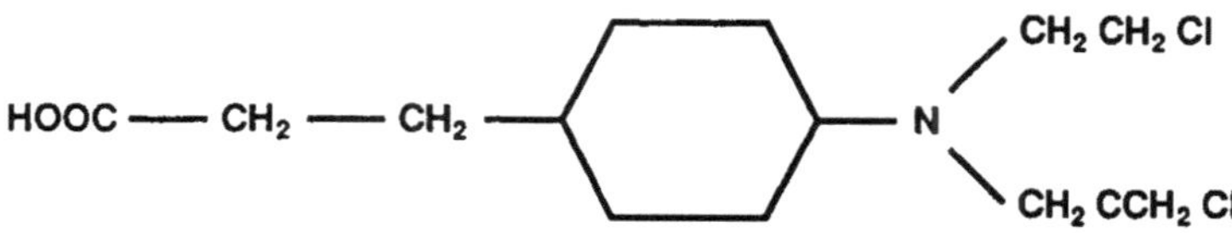

CHLORAMBUCIL

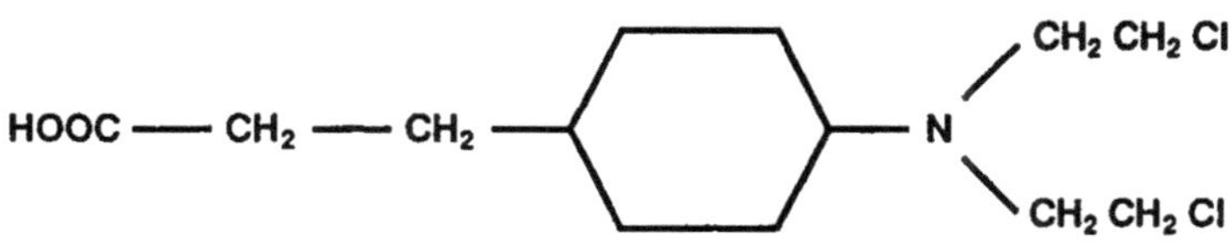

MELPHALAN

Figure 8. Structure of chlorambucil and melphalan which can be absorbed orally, compared with mustine hydro-chloride, which cannot.

practice to produce high local concentration of the antitumor agent (Figure 9). It is now known, of course, that cyclophosphamide is inert, and only becomes active when it is converted to 4-hydroxycyclophosphamide on first-pass metabolism (Figure 10).

Although the original concept of the action of cyclophosphamide is faulty, nevertheless this drug, too, remains an extremely effective alkylating agent. It is important to note that not only do some of the metabolic products (such as phosphoramide mustard) have anticancer effects, but other metabolites such as acrolein produce toxic effects on the lining of the bladder, giving rise to hemorrhagic cystitis (and possibly, ultimately, carcinoma) which is a feature of the use of this agent. 4-hydroxycyclophosphamide has been found to be active *in vitro*, and

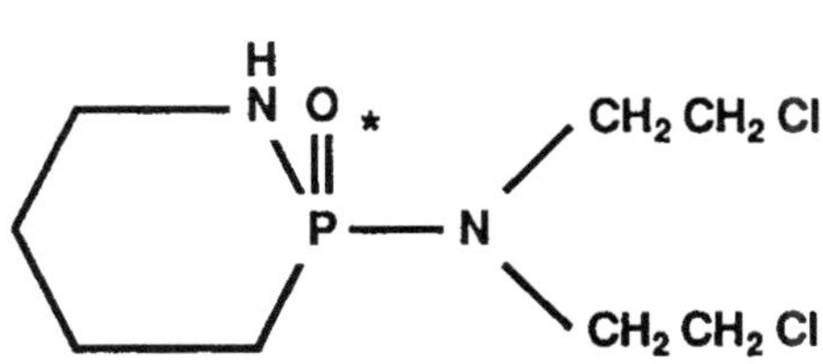

* SITE OF ACTION OF PHOSPHATASE

Figure 9. Site of action of phosphatase.

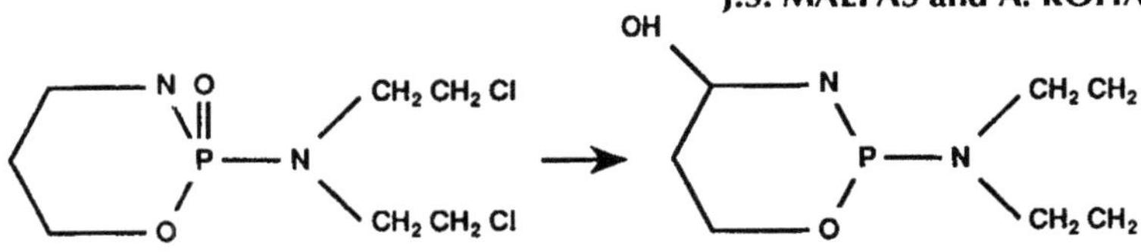

Figure 10. Formation of 4-hydrocyclophosphamide, the active derivative of cyclophosphamide.

has made a useful contribution to the removal of malignant cells in bone marrow being prepared for autologous transfusion. It has been found to be between three and four times more toxic to tumor cells than to surrounding myeloid precursor cells, and consequently purging or cleaning of bone marrow is possible.

Many other agents are known to have alkylating properties. Some of these, such as busulphan, have established a powerful role in the management of specific conditions such as chronic myeloid leukemia, while others such as the nitrosoureas demonstrate wide-ranging activity against a variety of solid tumors, including lymphomas, colon cancer, and brain tumors. After a quiescent period, alkylating agents are again becoming of interest because of the application of new techniques of structural analysis which allow the design of more effective agents and their development for clinical use. The production of drugs targeted on defined nucleotide sequences may open the way for more effective agents. New information about the repair mechanisms underlying the resistance that is acquired by tumor cells treated with alkylating agents may also be helpful. Specific methods of inhibiting these repair processes are now being developed.

PHARMACOLOGY OF ANTIMETABOLIC DRUGS

Antimetabolic drugs aim at disruption of essential metabolic pathways concerned in the manufacture of DNA or RNA. They work by competitive inhibition of key enzymes in these metabolic pathways. The most important drugs in this group include methotrexate, 5-fluorouracil, 6-mercaptopurine, and cytosine arabinoside.

Methotrexate

Methotrexate was the most successful of a number of analogues. They were designed to inhibit the enzyme dihydrofolate reductase (DHFR). The essential role of vitamin B_{12} and folic acid in cell metabolism had been established earlier, and in looking for an antagonist to folic acid, methotrexate was produced. This was shown by Farber et al. in 1948 to produce dramatic but transient remission in childhood acute lymphoblastic leukemia. Extremely tight, irreversible binding of methotrexate to DHFR inhibits the pathway (Figure 11). The pool of reduced folate is required for the production of thymines and eventually DNA. Although metho-

Figure 11. Block on the production of tetrahydrofolate production by methotrexate.

trexate is exceedingly effective at blocking the activity of DHFR, it only requires a few molecules of the enzyme to return to activity to enable this metabolic block to be overcome. The administration of purines or a source of reduced folate such as 5-formyltetrahydrofolate (leucovorin or folinic acid) will overcome the metabolic block and rescue normal cells that have been exposed to methotrexate.

Methotrexate is remarkable in that there is good bioavailability and it can be given safely intravenously, intramuscularly, and intrathecally. Care needs to be taken if there are large additional compartments in which the drug can be taken up and sequestrated, and then slowly released. Examples of this are the presence of ascites or large pleural effusions. The drug enters this third compartment and is released slowly. If rescue with leucovorin is stopped prematurely while this slow release is occurring, severe toxicity and even death may result.

Methotrexate was the first drug to cure experimental leukemia in mice, to induce remission in childhood acute leukemia, to cure the rare solid tumor choriocarcinoma (when used alone), and to form part of the curative program for acute leukemia, used as continuation or maintenance therapy. It has a wide spectrum of activity and is relatively well tolerated. As has been noted above, it is excreted through the kidneys. Renal function therefore must be good if severe toxic side effects are to be avoided. Its short-term toxicity includes mucositis and bone marrow suppression, which is of fairly rapid onset and short duration. In more intense dosages, hepatotoxicity, pulmonary toxicity, and adverse effects on the central nervous system may occur.

5-Fluorouracil

5-Fluorouracil (5FU) is notable as an antimetabolite which was designed and synthesized to produce a particular biochemical effect to inhibit cell growth in tumor cells which were noted to have avidity for uracil. Up to the time of its production, most antimetabolites that had been produced were effective against leukemia or lymphoma. An antimetabolic agent with efficacy against solid tumors (particularly adenocarcinoma) was sought, and a team led by Heidelberger set out to produce such an agent. 5FU remains a drug of great interest and versatility. It

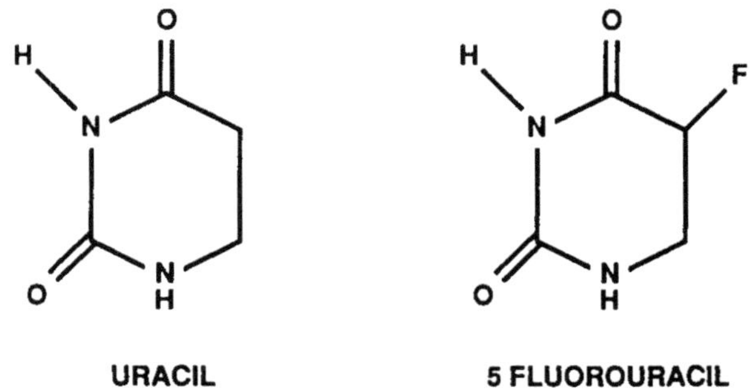

Figure 12. The similar structures of uracil and 5-fluorouracil (5FU).

achieves antitumor activity against breast cancer and colon adenocarcinoma. New knowledge of biochemistry has allowed its antitumor effect to be enhanced by using other agents in conjunction with it, as well as with radiotherapy.

The structures of uracil and 5-FU are shown in Figure 12 and the mechanism of action of 5FU is thought to work in two main ways. It is first converted to the nucleoside 5-fluorouridine diphosphate and triphosphate (FUDP) (Figure 13). These are incorporated into RNA by the action of RNA polymerase. These specious compounds then interfere with RNA function. The other pathway is the conversion to fluorodeoxyuridine monophosphate (FdUMP) which competitively inhibits the enzyme thymidylate synthetase, thus preventing the production of thymidylate, essential for DNA synthesis.

It has subsequently been found that sequential use of methotrexate enhances the toxicity of 5FU, and as might be expected, the addition of 5-formyltetra-hydrofolate (leucovorin) which enhances the reduced folate available, improves the antitumor activity of 5FU.

5FU is catabolized to dihydrofluorouracil in the liver. Some 90% is eliminated by metabolism, and some 5% is excreted unchanged in the urine. It is also poorly absorbed, and is best given intravenously. Because of detoxification at first pass in the liver, continuous infusion is more effective and less likely to lead to toxic effects than bolus administration. However, prolonged infusion gives rise to the major toxic effect of myelosuppression and mucous membrane ulceration. Skin rashes, conjunctivitis, ataxia due to cerebellar damage, cardiotoxicity and peeling of the

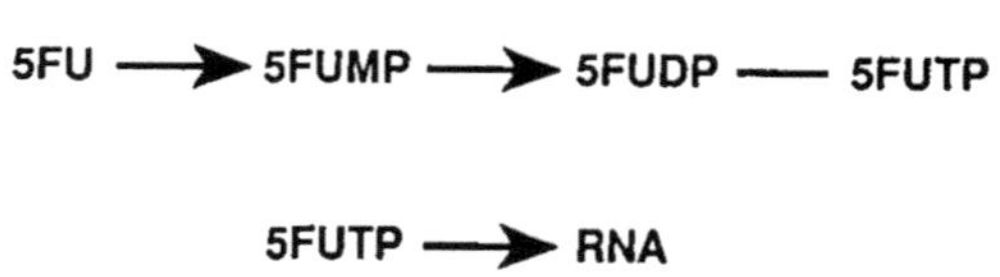

Figure 13. Pathway of conversion of 5FU into compounds affecting RNA.

soles of the feet and palms of the hands (the so-called hand-foot syndrome) have been reported, and these toxic side effects may be dose-limiting.

Purine Antimetabolic Agents

6-Mercaptopurine

6-Mercaptopurine (6MP) and 6-Thioguanine (6TG) were among the earliest of the antimetabolic agents introduced at the beginning of the 1950s. Their importance lies in the contribution they have made to the cure of childhood acute lymphoblastic leukemia, where their use combined with methotrexate as continuation therapy has been responsible for the high cure rate. Both 6MP and 6TG were found to induce remissions in acute lymphoblastic and myeloblastic leukemia. They were not effective in chronic leukemias or lymphoid malignancies. The exact manner in which 6MP achieves its effect is still unknown, although it must inhibit purine synthesis and interconversion. The same uncertainty exists for 6TG. The metabolic transformation of 6MP, though complex, is of note because of the effect that commonly prescribed xanthine oxidase inhibitors such as allopurinol may have on decreasing 6MP degradation. Doses of 6MP should be reduced by 50% when allopurinol is being given concurrently (Figure 14).

Pharmacokinetics show a considerable variation in bioavailability. There is great interpatient variation in absorption, and controversy over whether the drug is best absorbed in the morning before food, or at night. The half-time range for elimination of drug given intravenously varies from 20 minutes to an hour, and the drug is almost entirely eliminated by metabolism, very little being excreted. If the drug is given by mouth, there is considerable first pass metabolism in the liver, but the half-life seems to be about three hours.

6MP is now used mainly as continuation therapy in acute lymphoblastic leukemia. 6TG is preferred in induction regimens for acute non-lymphoblastic leukemia,

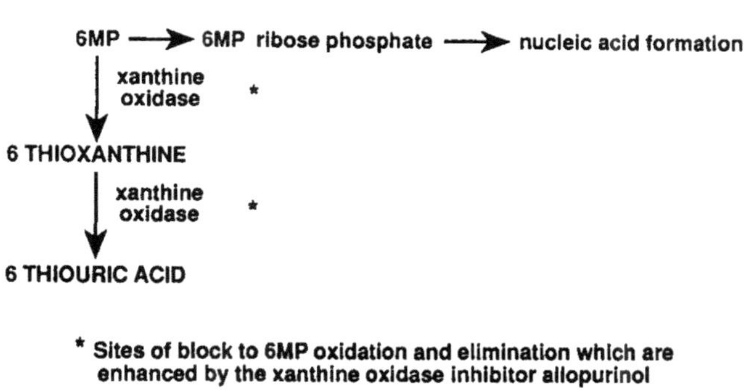

Figure 14. Pathway of 6-mercaptopurine (6MP), showing the reason for reducing 6MP dose when allopurinol is given.

and both have been used in the blast crisis of chronic myeloid leukemia. Myelosuppression is the main toxic side effect, but in some patients a mild cholecystatic jaundice has been reported. However, this is usually reversible on withdrawal of the 6MP or 6TG. Nevertheless, care should be taken when either of these drugs is being administered in conjunction with other hepatotoxic agents, such as doxorubicin.

Cytosine Arabinoside

Another example of an antimetabolite which is effective in the treatment of leukemia is cytosine arabinoside, ara-C. This compound was of great interest, and following its introduction it was found to have efficacy against a highly resistant form of leukemia—acute myelogenous leukemia—which had otherwise responded poorly to antimetabolites such as 6MP and methotrexate. With ara-C the response rates rose to more than 50%, and when it was combined with anthracyclines (see below) further improvement in remission induction occurred, and these two agents have become the mainstay of treatment of acute myelogenous leukemia.

Ara-C was originally isolated from a sponge, but is now produced synthetically. It is very similar in structure to deoxycytidine (Figure 15). The main difference is a hydroxyl group on the sugar, converting the latter from a deoxyribose to an arabinoside. The main event when ara-C enters the cytosol is its conversion to ara-CTP. This binds to DNA polymerase, stops DNA production, and hence arrests growth. Ara-C may also be incorporated into DNA, disturbing its synthesis by terminating DNA chain elongation. As can be seen from Figure 16, there are two

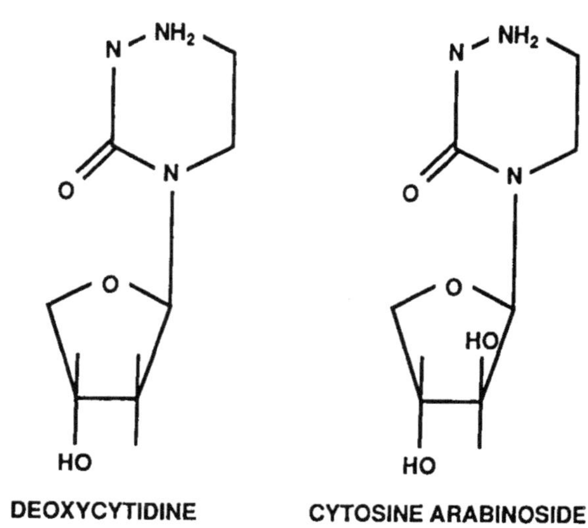

Figure 15. The similar structures of deoxycytidine and cytosine arabinoside.

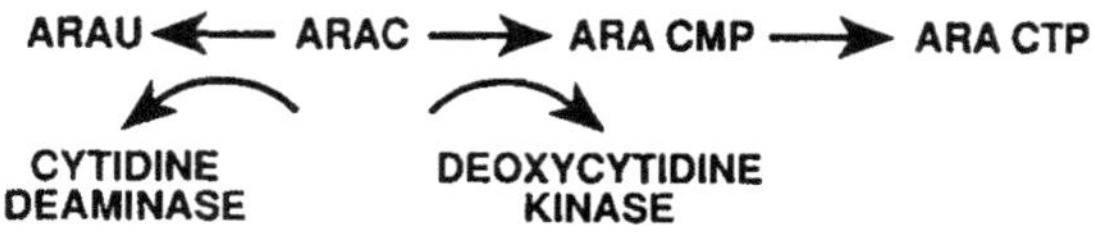

Figure 16. Pathways for the breakdown of cytosine arabinoside by the enzymes cytidine deaminase and deoxycytidine kinase.

mechanisms, both enzymic, which affect the amount of ara-CTP present, and hence the effectiveness of the drug. The kinases which produce ara-CTP, and the deaminases which destroy it, are vital to its activity.

Ara-C cannot be given by mouth, but is otherwise very versatile. It can be administered intravenously, intrathecally, and subcutaneously. This last route is becoming the route of choice for out-patient therapy, and since the half-life is so short, and the drug acts best on cells in the synthetic phase of the cell cycle, there are advantages in giving the drug by long-term subcutaneous infusion.

Studies on high-dose ara-C have shown that parenteral dosage of cytosine allows cytotoxic levels of the drug to be achieved in the cerebrospinal fluid. This has been helpful in prophylaxis and treatment of meningeal leukemia. Very high doses of ara-C have produced (in addition to the usual myelosuppression and gastrointestinal toxicity) specific effects on the central nervous system, including damage to the cerebellum with nystagmus, severe ataxia, and eventually (if the drug is not stopped) central nervous effects on the cortex with dementia and coma. Early detection of nystagmus is helpful in avoiding these serious toxicities. Another unusual effect is the acute conjunctivitis produced by ara-C. This can be prevented by giving steroid eye-drops.

After intensive investigation, the rationale for high-dose ara-C has been shown to be flawed, since the production of the effective ara-CTP is limited by enzymic process involving the kinases. Since these form a rate-limiting step in the metabolic pathway, increasing the dose of ara-C will not be followed by a greater antitumor effect.

PHARMACOLOGY OF DRUGS DERIVED FROM NATURAL SOURCES

A number of very potent cytotoxic drugs have been derived from natural sources such as streptomyces (doxorubicin, daunorubicin, actinomycin D, and bleomycin), plant products such as the vinca alkaloids from the Madagascar periwinkle, and epipodophyllotoxins from the mandrake plant. Many were found as the result of antitumor effects being noted when they were screened in the search for new antibiotics (for example). Among the most important are the anthracyclines. These have the basic formula shown in Figure 17. Relatively small structural changes distinguish doxorubicin, daunorubicin, and epirubicin, but their clinical activity and

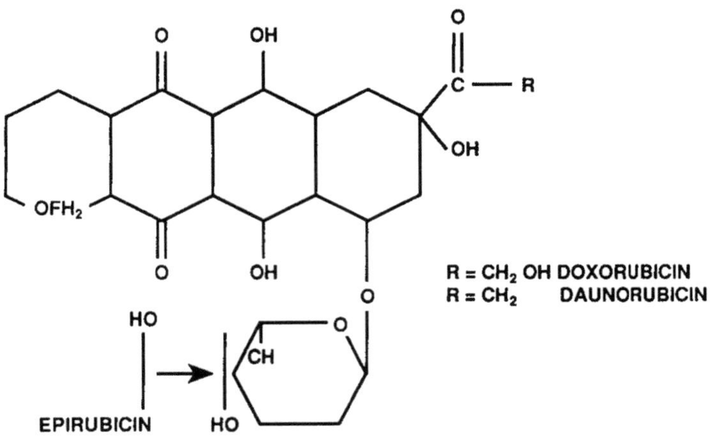

Figure 17. Structure of doxorubicin and daunorubicin.

to some extent their toxicities are widely different. Thus, while doxorubicin is effective in acute lymphoblastic and myeloblastic leukemia, it also has activity in solid tumors such as breast, ovary and lung cancer. It is used in lymphomas and in childhood malignancy. Daunorubicin, on the other hand, is a most effective agent against myeloblastic leukemia, and has little effect against other solid tumors. Both have serious cumulative cardiotoxicity, but this is less apparent in the analogue, epirubicin.

Although in clinical use for many years, there is uncertainty as to their mechanisms of action. The anthracyclines can be shown to intercalate with DNA. This mechanism cannot account for various features of their antitumor activity, and other actions have been proposed. Recently a mechanism whereby anthracyclines bind to an enzyme called topoisomerase II, which promotes strand breakage and resealing, has been described, and is thought to be increasingly important. A diagram of the action is shown in Figure 18.

Anthracyclines can cause topoisomerase II-mediated DNA changes, and these effects are seen at concentrations of anthracycline which are achievable in patients. Supporting evidence for this theory comes from the fact that when topoisomerase II enzyme levels fall in patient tumors, anthracycline activity is inhibited, and the tumor becomes resistant to their effect.

Another mode of action relates to the production of superoxide. Bacterial cell killing has long been known to be due to superoxide. Semiquinones, which form part of the anthracycline structure, are potent producers of superoxide. Superoxide can form the basis for the production of free radicals such as hydroxy groups, which are perhaps the most powerful cytotoxic moieties known. Unfortunately, in addition

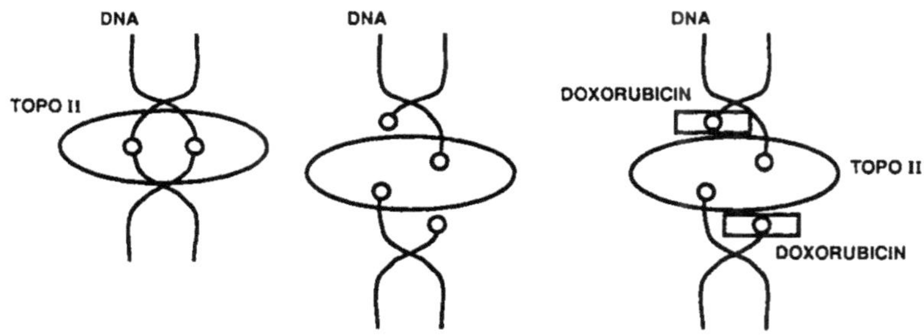

Figure 18. Diagram of the mechanism of action of topoisomerase II.

to their effect on the tumor cell, these free radicals, in conjunction with iron, have the ability to cause damage to the myoblast in the myocardium, and are thought to be related to the serious cumulative cardiotoxicity of both doxorubicin and daunorubicin.

Doxorubicin is one of the most potent sclerosing drugs known. Repeated long-term dosage, or administration as infusion, requires the use of a central venous line. This results in the drug being diluted, and reduces the incidence of extravasation and thrombosis. Doxorubicin has a short $t^1\!/_2\alpha$ of about 10 minutes. $T^1\!/_2\beta$ is three hours, but the $t^1\!/_2\gamma$ is very long, at more than 30 hours. This has a clinical implication, because the use of freezing caps to prevent the profound alopecia produced by the drug is not feasible when the drug circulates for such a long time. Bolus injection of doxorubicin results in high serum levels, and it appears that cardiotoxicity is related to these. Cardiotoxicity is reduced when the drug is given by infusion (Figure 19).

Many attempts have been made to reduce the cardiotoxicity of anthracyclines by giving other agents. Among the most useful so far developed is ICRF187, which has been shown to reduce the number of cardiac events in women on long-term anthracycline treatment for breast cancer (Figure 20). Other toxic side-effects are myelosuppression, mucositis, hair loss, and severe local injury on extravasation (as mentioned above).

Doxorubicin (adriamycin) is effective in a wide range of hematological and solid tumors. It has been a component in a number of well-established multidrug regimens for Hodgkin's disease (ABVD: adriamycin, bleomycin, vinblastine, and dacarbazine), and non-Hodgkin's lymphoma (CHOP: cyclophosphamide, hydroxydaunorubicin (= doxorubicin), oncovin, and prednisolone). Daunorubicin is preferred in remission induction regimens for acute myelogenous leukemia because it causes less mucositis, and many variations on DAT (daunorubicin, ara-C, and thioguanine) have been the most successful induction regimens in this form of leukemia.

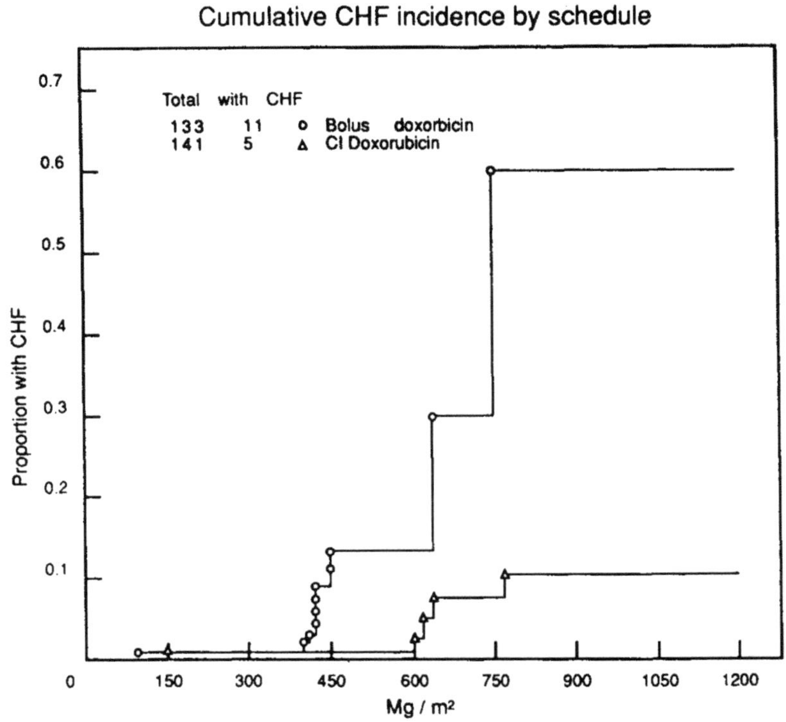

Cumulative incidence of congestive heart failure by dose of doxorubicin, comparing continuous infusion and 3 weekly bolus administration, p=0.004. [From Hortobagyi *et al.* 1989]

Figure 19. Reduction in cardiotoxicity when doxorubicin is given by infusion.

In breast cancer, CAF (ara-C, adriamycin and fluorouracil) has shown a significant survival advantage when compared with CMF (cyclophosphamide, methotrexate, and fluorouracil), and this has been confirmed in an overview analysis indicating the benefit of adding doxorubicin to breast cancer therapy.

Bleomycin

This agent is produced by *Streptomyces verticillus.* The bleomycins are a mixture of polypeptides of low molecular weight. They achieve their effect by DNA cleavage. The $t^{1/2}\alpha$ is quite short at 24 minutes, $t^{1/2}\beta$ is 2–4 hours. Over half the material is excreted in the urine. The rest is metabolized by hydrolases in the tissues. Bleomycin has remarkably little myelosuppressive activity. However, it does have other life-threatening toxicities, such as pulmonary fibrosis and acute hypersensitivity reactions. Severe skin reactions may also occur. The administration of radiotherapy at the same time as bleomycin is not advisable, as this induces

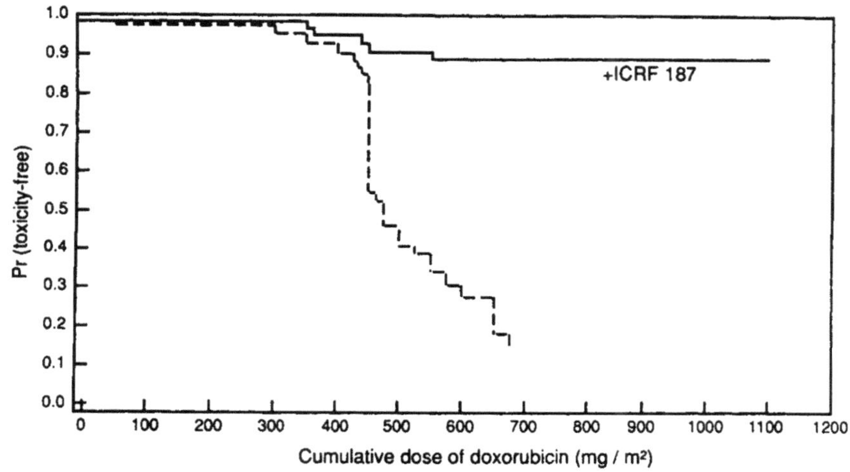

No. patients at risk	200	400	600	800	1000	1200
FDC 74	65	45	6	2		
FDC+ ICRF -187 76	64	45	30	17	11	4

After Speyer et al., 1988

Figure 20. Prevention of doxorubicin cardiotoxicity with ICRF 187.

radiation fibrosis. Bleomycin should be given with caution in the presence of renal dysfunction. It has been a valuable agent in programs for Hodgkin's disease and non-Hodgkin's lymphoma, has made a major contribution to curative therapies for testicular cancer, and has been of value in treating cancers of the head and neck.

Plant Alkaloids

The possibility that cancer might be cured by natural products has been of interest since medieval times. Salves containing alkaloids such as colchicine were used for superficial cancers, and there were reports of occasional success. Modern alkaloids of proven value are those derived from the Madagascar periwinkle (vincristine and vinblastine, see Figure 21), taxol (derived from the Western Yew tree bark, *Taxus brevifolia)*, and the epipodophyllotoxins (derived from the may apple or mandrake plant).

All these compounds have an action on tubulin, the highly complex 100 kD protein that forms the spindle in the mitosing cell. It is interesting to note that colchicine is currently used to produce so-called colchicine arrest in the preparation of cytogenetic samples, and to some extent this may be a visual representation of the action of some of these other agents.

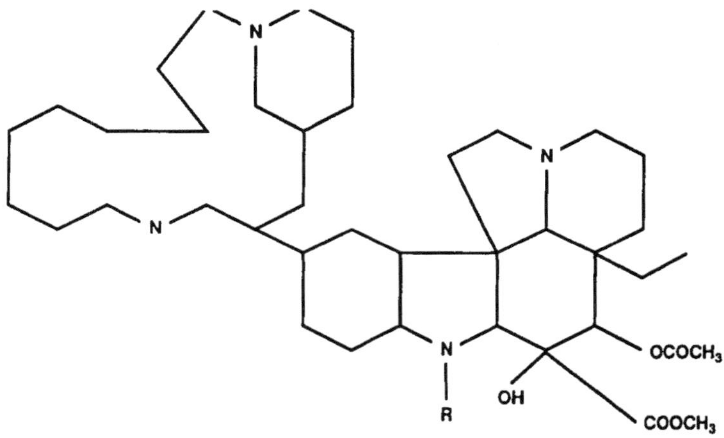

VINBLASTINE CH_3 SUBSTITUTED FOR R
VINCRISTINE CH_0 SUBSTITUTED FOR R

Figure 21. Structure of the vinca alkaloids.

The vinca alkaloids achieve their antitumor effect by binding to the microtubule and preventing its assembly; they have a very complex structure, as indicated in Figure 21.

Taxol enhances the production of tubulin, and this is certainly related to its activity, but whether this is because the tubulin produced is defective, or because the excessive material interferes with other cellular functions, is currently not known. Epipodophyllotoxins affect tubulin, but derive their major effect from DNA damage, strand-breaking and acting with topoisomerase II as described above.

Minor substitutions of a methyl or aldehyde group on the catharanthine moiety produce profound changes in the clinical spectrum of activity and toxicity. Vincristine has a broad spectrum of activity in acute lymphoblastic leukemia, Hodgkin's and non-Hodgkin's lymphomas, solid tumors of children and adults, and is relatively non-toxic to the bone marrow. Vinblastine is very effective in Hodgkin's disease, but has little effect in other malignancies, and in addition has a more profound myelosuppressive activity. Both compounds share a dose-related neurotoxicity which can be dose-limiting, and need to be given carefully parenterally, as extravasation results in tissue necrosis. The drugs are metabolized in the liver.

Taxol

This is a recently introduced agent which has already been shown to have activity in ovarian cancer. It has been difficult to produce a soluble product, and one that is without a high rate of anaphylactic reaction. It is largely metabolized, with only about 5% excreted in the urine. It has a short half-life of 20 minutes, and a $t^1/_2\beta$ of 6–8 hours. Myelosuppression, alopecia, and sensory neuropathy are other toxic side effects.

Epipodophyllotoxins

Etoposide (VP-16) and teniposide (VP-23) have been available for the treatment of malignancy for some 20 years. Etoposide has become established as an agent for the treatment of acute leukemia, Hodgkin's disease, non-Hodgkin's lymphoma, testicular and other germ cell tumors, and small-cell lung cancer. Teniposide remains an investigational drug.

As mentioned above, DNA strand linkage and topoisomerase II enzyme mechanisms predominate in their activity. About 40% of the drug is excreted, and etoposide has a half-life of 6–8 hours. The drug can be given orally, but bioavailability is very variable. Studies have shown that when given intravenously in small-cell lung cancer, for example, a five-day schedule is the most effective way to administer etoposide. More recently, it has been shown that epipodophyllotoxins can cause breaks on chromosome 11, which is associated with a high incidence of secondary acute leukemias.

Platinum Analogues

These fascinating compounds bring the cycle of anticancer drug development full circle 125 years after Lisauer showed the effect of inorganic arsenic on tumors. Rosenberg observed that the fluid surrounding platinum electrodes inhibited the growth of bacteria due to the formation of cis-Pt $(II)(NH_3)_2Cl_2$. This compound was found to have antitumor activity and has the ability to alkylate guanine. Cis-platinum and its analogue, carboplatin (Figure 22) have established themselves as curative drugs for testicular cancer. They have good activity in small-cell carcinoma of the lung, bladder cancer, and ovarian cancer.

Both drugs have to be given intravenously, and while cis-platinum is mainly metabolized, the major remaining portion is excreted by the kidney. In the case of carboplatin, virtually 90% is excreted by the kidney. $T^1/_2\alpha$ is similar for both drugs, at 20–30 minutes; $t^1/_2\beta$ is about one hour, and $t^1/_2\gamma$ is approximately 24 hours in the case of cis-platinum, and up to 40 hours for carboplatin.

Both compounds are nephrotoxic, leading to renal failure and magnesium-losing nephropathy. This toxicity can be countered by increasing fluid intake during the course of therapy, and by carefully monitoring creatinine and ethylenediaminetetraacetic acid clearance. Until the advent of the 5HT antiemetic group of drugs, nausea and vomiting, to a degree which often caused discontinuation of treatment, was a feature, but fortunately this can now be countered. Peripheral neuropathy and auditory impairment also occur, and are dose-related. While

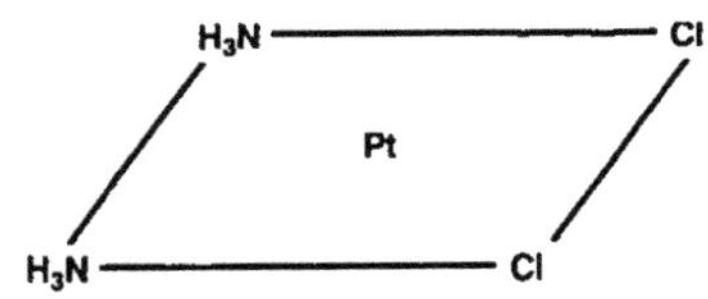

CISPLATIN

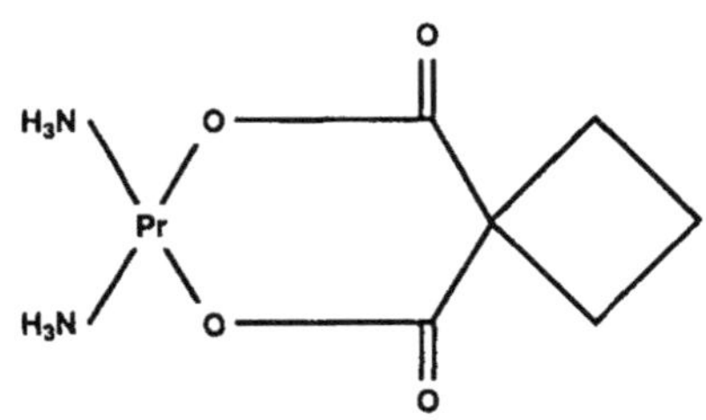

CARBOPLATIN

Figure 22. Structure of cis-platin and carboplatin.

cis-platinum is relatively nontoxic to the marrow, carboplatin is myelosuppressive and produces neutropenia.

The survey of anticancer drugs given above is far from comprehensive. This introduction is not intended to produce a complete reference for the chemotherapist, but to illustrate the main classes of anticancer drugs, and the principles of structure, activity, pharmacokinetics, and toxicity of the main members of each class. With the information given the next part of the chapter, describing how these drugs may be employed effectively, will (it is hoped) be more readily understandable.

THE USE OF DRUGS FOR CANCER TREATMENT

As will have been evident from the previous section, the cancer physician is now provided with a powerful armamentarium of drugs with which to treat cancer. Each has a special range of activity, pharmacokinetic properties, and toxicity, and just as the surgeon must be skillful in the use of the various instruments at his disposal in the operating theater, so the medical or radiation oncologist must know how to handle each of these agents.

Objectives

It is important to identify the objectives of chemotherapy, used either alone or in combination with surgery or radiotherapy. Cure may be a realistic aim in such diseases as leukemia, lymphoma, testicular tumors, choriocarcinoma, and pediatric tumors. Excellent palliation may be feasible in low-grade lymphomas, myeloma, genitourinary cancer, and gastrointestinal cancer, even when overt disease is present, but it is probably unreasonable to expect response in (for example) widely disseminated melanoma or renal carcinoma.

Over the past decade or so, cancer physicians have introduced a number of terms to describe their aims. Remission induction (usually applied to acute leukemia) indicates the elimination of all symptoms and signs of the disease, restoration of a normal blood picture, and a bone marrow with no evidence of leukemic blasts. The concept of complete remission or complete response is a vital one in cancer, for unless this is achieved, cure is very unlikely to occur. Consolidation is a term applied to a second period of intensive therapy, when the patient is in complete remission, in an attempt to reduce still further any undetectable residual disease. Maintenance therapy (probably better described as continuation treatment, and now largely only employed in acute lymphoblastic leukemia in children) describes continuous or pulsed doses of antileukemic agents given over a long period in order to finally eliminate any possible residual disease. So-called adjuvant chemotherapy (a term used more often in solid tumors) is based on similar principles, and follows the recognition that even pathologically localized tumors may have already spread in the form of micrometastases. These micrometastases, which are small, well-vascularized, and particularly susceptible to chemotherapeutic agents, are not detectable, but have been shown, in such childhood tumors as Wilms' tumor, to be eliminated by courses of chemotherapy.

Finally, chemotherapy can have a role in palliation. Here the intent is certainly not curative, and may not even primarily aim to extend survival, but is largely directed at the control of symptoms, though added useful life may be incidentally achieved.

Diagnosis

It is important to be absolutely certain of the diagnosis and of subdivisions of disease. This entails sophisticated techniques which may be highly relevant to any decision on therapy. One example would be making the distinction between small-cell lung cancer and non-small-cell lung cancer. In the former, currently available chemotherapy can be effective and relatively nontoxic, whereas squamous carcinoma is (on the whole) resistant. Subgroups of lymphoma are divided into high- and low-grade, and these require very different chemotherapeutic agents and approaches to treatment: while cure is possible in the high-grade lymphoma, in the low-grade condition it is still subject to experimental approaches. Even more refined analysis of lymphomas, with a distinction between T and B cell versions, has shown the beneficial effect of cytosine arabinoside in the former, and of alkylating agents in the latter.

Table 1. Definition of the Various Responses to Anticancer Drugs

Response	*Criteria*
Complete response	Complete disappearance of all demonstrable disease
Partial response	More than 50% reduction in the sum of the products of the longest perpendicular diameters of the tumor, with no disease progression elsewhere
No response	No change or less than 50% reduction
Progression	Increase in size of the tumor at any site

Staging

Extent of disease will affect planning of therapy with anticancer drugs. The approach to localized disease will be very different from that used where metastasis has occurred. While adjuvant chemotherapy programs, which may be associated with either short- or long-term effects, are justifiable in the case of localized disease where an improvement in cure rate is possible, a very different approach is needed in metastatic disease. If there is a realistic chance of cure, repeated chemotherapy with a reduction in the tumor burden to zero, is the only way to ensure this. No patient with cancer has ever been cured unless a complete response was achieved. It might be helpful to define responses, and these are shown in Table 1.

In order to achieve complete response, drugs to which the tumor is sensitive have to be given sequentially, with an interval to allow recovery of specially sensitive organs such as the bone marrow or gastrointestinal tract. If hematological toxicity is the dose-limiting feature, this can be circumvented by autologous, allogeneic, or peripheral blood progenitor cell transplantation, as described later. In most cases, however, a combination of drugs, each with some activity against the tumor, and (if possible) with differing toxic side-effects, are administered. There should be no pharmacokinetic incompatibilities, and drugs should not inhibit the pharmacodynamic action of other drugs. For example, the use of 5FU before methotrexate prevents the antifolate activity of methotrexate. It is an advantage if the drugs are synergistic; for example, leucovorin enhancing 5FU activity by inhibiting thymidilate synthesis is a combination shown to have practical results in treating solid tumors in the clinic.

Drug Resistance

This is dealt with in more detail elsewhere, but must form an important part of decision making on cancer chemotherapy. An important principle underlying the combination of drugs is that agents should be non-cross-resistant. The Goldie–Coldman hypothesis emphasizes the increasing risk of a resistant mutation occurring with each cell cycle, hence the importance of eliminating cells as quickly as possible. Cells resistant to one agent are killed by the other, and that particular clone is then destroyed. Well-recognized mechanisms of resistance are now identified; multidrug resistance needs a special approach to inhibit the activity of p170 glycoprotein, and the experimental use of verapamil and other blocking agents continues. Altered drug binding to topoisomerase II would be another avenue for research.

Combination With Other Treatment Modalities

Radiotherapy is effective against many localized tumors, and outstanding results (particularly in children's tumors) have accompanied the judicious combination of the two modalities. Agents that enhance the radiation reaction should be avoided during combined treatment, e.g., actinomycin D, 5FU, doxorubicin, and bleomycin. In the case of the latter two, irradiation of the heart or lungs concurrently may be very damaging. The role of chemotherapy combined with interferon is currently under investigation. Conflicting results have been achieved with these therapeutic agents used as induction therapy, but there seems reason to believe that continuation or maintenance therapy with interferon may be of value in some B cell malignancies.

New Principles

The major problem with all chemotherapeutic agents is their lack of specificity. All drugs, to a greater or lesser extent, damage normal tissues, and chemotherapy relies on the efficiency of normal tissue repair mechanisms, to the extent that these recover more quickly than tumors. Previous attempts to act on tissue rich in phosphatases and other enzymes (viz. the development of cyclophosphamide) have been unsuccessful.

In an effort to overcome this lack of specificity, attempts have been made to liberate drugs *in situ*. The antibody directed enzyme prodrug therapy (ADEPT) program administers an antibody enzyme conjugate which selectively combines with the tumor. An inactive prodrug is then given, which is activated by the enzyme at the tumor site. Experimental studies involving a benzoic acid mustine prodrug activated by carboxypeptidase have been carried out in culture and animal xenografts. Preliminary studies in patients with adenocarcinoma of the colon have shown this approach to be feasible, and minor responses have been noted. To be successful, this principle would need highly specific tumor localization of a prodrug active against the tumor type. Any drug liberated would need to be confined to the area of liberation, and not allowed to escape to damage normal tissues. This is an interesting concept deserving of further investigation.

DRUG TOXICITY

The toxicity of individual drugs has been described above, and most anticancer drugs share common toxicities, damaging the rapidly dividing cells of the bone marrow, gut lining, gonads, and hair follicles. Anticancer drugs may also have toxicities specific either to a particular group of drugs or even to one member of a family of agents. The timing of toxicity is crucial. This enables the clinician to be alert to a possible problem and to deal with it. It should be emphasized that most immediate or early toxicities are now well-recognized, but a delayed or late toxic effect has often been unexpected, and there may well be late effects which are not yet recognized. Nobody has yet lived for threescore years and ten following a cure with chemotherapy.

Immediate toxicity can be defined as occurring within minutes or hours. In the case of anaphylactic shock, the availability of epinephrine and hydrocortisone may be vital. Common immediate toxicity includes local tissue necrosis from extravasation, phlebitis, hyperuricemia, renal failure, or tumor lysis syndrome. Skin rashes are often seen shortly after administering drugs. Specific early toxicity includes hemorrhagic cystitis in poorly-hydrated patients not given mesna before treatment with cyclophosphamide or ifosphamide, or radiation reactions with doxorubicin and actinomycin D therapy. Occasionally fever and chills are seen with the administration of bleomycin.

Early toxicity (occurring within days or weeks) commonly includes leukopenia, thrombocytopenia, stomatitis, diarrhea, and alopecia. Specific toxicities seen during this time include paralytic ileus (in the case of vinca alkaloid therapy), pulmonary fibrosis (bleomycin), ototoxicity (platinum compounds), neuropathies (vinca alkaloids and platinum compounds), and cerebellar ataxia with cytosine arabinoside and 5FU.

Delayed or late toxicities, seen months or years after therapy, commonly include anemia, hepatocellular damage, azospermia and (after some time) either acute leukemia or solid tumors as second malignant growths. Specific delayed toxicities are heart failure resulting from excessive doses of doxorubicin or daunorubicin, encephalopathy from prolonged high-dose methotrexate, or learning disability with prophylactic cranial irradiation and chemotherapy.

DRUG RESISTANCE

It is generally agreed that biochemical resistance to chemotherapy is a major cause of failure to cure cancer. If resistance to the agents commonly used at present could be eliminated completely, it would be possible to cure cancer now. This explains the importance of drug resistance as a topic.

Tumors are either initially resistant or insensitive to chemotherapeutic agents or, after a period of treatment, acquire resistance. Examples of intrinsically resistant tumors are melanoma, renal cell carcinoma, and non-small-cell lung cancer. Acquired resistance is seen in breast cancer and small-cell lung cancer. Although this is a useful guide in the clinic, it may conceal the important principle that all tissues are initially sensitive.

Goldie and Coldman produced the hypothesis that the more cells there were in a tumor, the greater the chance that some would be resistant to chemotherapy. Since the progeny of these cells would divide while sensitive cells were eliminated, eventually the tumor would consist wholly of resistant cells. With every doubling of the tumor cell population, the number of resistant cells would increase. However, some tumors shed a large number of dead cells, so that their growth is very slow, and size bears no relationship to the number of cell divisions they have been through. Many very small colonic tumors may have gone through a thousand cell divisions by the time they are apparent clinically. Inevitably these tumors would largely consist of resistant cells. A rapidly-growing tissue such as a small-cell lung cancer loses relatively few cells, and, having gone through far fewer cell divisions, is less likely to contain a large resistant population. Therefore, the response to chemotherapy in this tumor might appear excellent.

Although Goldie and Coldman thought resistance was acquired in one step, it seems likely (from what is now known about genetically based changes) that drug resistance is acquired in stages. It is also apparent that there are many ways in which a cell can become resistant.

Mechanisms of Cell Resistance

This section will be concerned with the mechanisms of intrinsic drug resistance. A large number of mechanisms result in chemotherapeutic drugs being less effective than they ought. Some of these are specific to a particular drug (e.g., gene amplification resulting in an increase in the enzyme dihydrofolate reductase and producing methotrexate resistance). Others, such as the enhancement of the p-glycoprotein 170, result in a number of drugs being eliminated very rapidly from the cell (e.g., the vinca alkaloids, antitumor antibiotics, and etoposide) and therefore being unable to achieve their effect. The main reasons for drug resistance are:

1. The drug is actively pumped out of the cell as it enters (as with p170 glycoprotein enhancement).
2. The drug fails to cross the cell membrane adequately.
3. Having entered the cell, the drug fails to be activated.
4. Having been activated, it is rapidly de-activated before it can become effective.
5. The damage it produces is rapidly repaired.
6. The drug target is rapidly increased so that the metabolic block is bypassed.
7. Alternative biochemical pathways are opened.
8. Decrease in topoisomerase aids DNA breaks.

Some of these mechanisms are quite specific, but a considerable number of commonly used chemotherapeutic agents are affected by one or more.

Induction of Drug Resistance

Since drug resistance is conveyed from generation to generation, it must be genetically based. Further evidence is provided by the fact that transfection of the amplified gene into a nonresistant cell conveys resistance. As examples of how drug resistance may be induced, consideration will be given to a single drug, methotrexate, and then the problem of multidrug resistance will be discussed.

Resistance to Methotrexate

There are several methods whereby cells become resistant to methotrexate. It may pass through the cell membrane by passive diffusion or by means of an energy-dependent mechanism. The drug that passes through the membrane of the cell is polyglutamated, and once this occurs cannot diffuse out. Mutations leading to impaired active transport, or failure of polyglutamation, will lead to decreased uptake, and hence the cell's resistance. Methotrexate may pass into the cell, but there meet variants of the enzyme dihydrofolate which have arisen, again, as the result of mutation. These DHFR variants do not bind so avidly with methotrexate, and hence can still result in enough thymidilate synthesis to allow the cell to divide. The most common mechanism is the amplification of the gene responsible for DHFR production. This occurs in small steps, eventually leading to hundreds of copies of the DHFR gene, and consequently to very high levels of drug resistance. This has been proven in human tumors treated with methotrexate.

Multidrug Resistance

It has been noticed that resistance to one chemotherapeutic agent conferred resistance to others; thus, Chinese hamster ovary cells selected for resistance to colchicine were found to also be resistant to other naturally occurring anticancer agents, including antitumor antibiotics and the plant alkaloids. Antimetabolites and alkylating agents are usually not affected.

Cells which show multidrug resistance were found to have a larger than normal amount of a p-glycoprotein of 170 kD in their membrane, and it was also found that the genes responsible for the production of this material, *mdr*-1, were amplified. Proof of their activity has come from transfection studies in which the *mdr*-1 gene, transfected to another cell, induces multidrug resistance. A model of the p-glycoprotein suggests how the energy-dependent pump may act both by preventing the ingress of the cytotoxic drugs listed in the table, and by speeding their egress.

Topoisomerase

The presence of topoisomerase enzymes (in particular, topoisomerase II) is essential for the effective action of such anticancer drugs as doxorubicin, amsacrine and etoposide. When reduced or absent, the chemotherapeutic agent cannot bind to the DNA strand and prevent religation of the DNA and repair.

Glutathione

This is a sulphydryl compound which can react with reactive agents produced by some cytotoxic drugs such as alkylating agents. It is particularly effective in neutralizing the active groups on alkylating agents, thus inducing resistance. Although glutathione levels tend to be increased in resistant cells, this is not always the case, and the relationship is certainly not as clear as those described for other mechanisms of resistance.

SECTION II: EFFECTS OF DRUGS ON CELLS

Tumors are heterogeneous with regard to the proportion of cells that are capable of self-renewal, so-called stem cells. Tissues capable of self-renewal, e.g., bone marrow cells and the cells lining the intestine, contain a relatively small number of stem cells. On receipt of a signal or signals that indicate a requirement for proliferation, the progeny of the stem cells undergo clonal expansion and subsequent maturation and differentiation, without further cell division. The latter concept is illustrated by the differentiation of metamyelocytes into mature neutrophils and the maturation of reticulocytes into red cells, in parallel with the synthesis of hemoglobin. In situations where treatment is being given with curative intent, for example, remission induction and consolidation therapy for acute myelogenous leukemia, the aim of the treatment is to eliminate the clonogenic cell, hence the aim is lethality for stem cells. Various assays are in use to assess the efficacy of killing. They are described briefly below.

Colony Forming Assays

These actually measure the number of cells surviving a given treatment, rather than the number killed. Assessment of cell survival depends on demonstrating that after treatment, be it chemotherapy or radiotherapy, the cell in question can still produce a colony, i.e., a large cluster of progeny. In practice, the assay is carried out as follows:

Cells growing as a single layer in culture are irradiated or exposed to a drug.

↓

A single cell suspension is made and the cells are counted.

↓

Serial dilutions are made and plated in tissue culture.

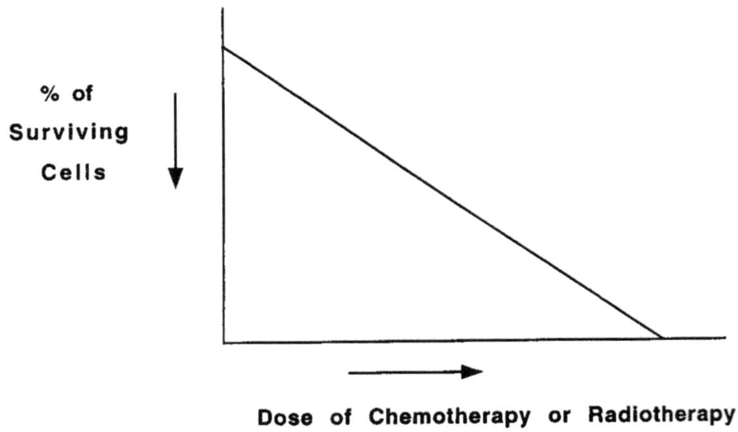

Figure 23. Relationship between percentage of surviving cells and dose of chemotherapy or radiotherapy.

↓

The culture is incubated at 37° C.

↓

The surviving cells proliferate and form colonies.

↓

The number of colonies is counted.

Only surviving cells will form true colonies, i.e., clusters of 50 or more than 50 cells, although killed cells may form "abortive" colonies.

$$\text{The plating efficiency (PE)} = \frac{\text{number of colonies}}{\text{number of cells plated}}$$

A control experiment is also carried out using untreated cells. The ratio of PE treated cells: PE control cells = the fraction of cells surviving the treatment. The experiment can also be carried out using a range of radiation or drug doses and a survival curve derived; survival decreases as the radiotherapy dose or drug dose increases (see Figure 23).

In Vivo/In Vitro Assay

An alternative method is to take a biopsy from a tumor growing in an experimental animal, or a biopsy from a patient (e.g., peripheral blood from a patient with acute myeloblastic leukemia) before and after treatment, and then proceed as above.

A cell suspension and serial dilutions are made and the cells plated in liquid or semi-solid media. The number of colonies before and after treatment (or derived from treated and control animals) is counted and cell survival calculated as above.

In Vivo Assays

These differ from the previous two methods in that treated (irradiated and control) cells are transplanted into mice. The principle of the assay is shown below:

A tumor growing in an experimental animal is irradiated.

↓

The tumor is removed and a single cell suspension is prepared.

↓

The cells are counted and serial dilutions prepared.

↓

Cells are injected subcutaneously or intramuscularly into groups of recipient mice.

↓

The animals are observed and the percentage that develop tumors is plotted against the number of cells injected.

↓

The number of cells required for 50% tumor "take", i.e., TD_{50} is determined.

$$\text{Cell Survival} = \frac{TD_{50}\ \text{control cells}}{TD_{50}\ \text{treated cells}}$$

Spleen Colony Assay

This assay has been used, in particular, to assess the effect of radiotherapy or chemotherapy on bone marrow stem cells. Once more, the principle is based upon bone marrow being transplanted into syngeneic recipient animals. The efficacy of a drug depends on two things: the concentration used and the duration of exposure. Therefore, survival is usually assessed in relation to drug concentration (with a constant duration of exposure) and duration of exposure (at a constant concentration). The assay is summarized below:

Treated and control bone marrow is injected intravenously into syngeneic mice that have been lethally irradiated.

↓

Seven to 10 days later the mice are killed and the number of colonies in the spleens of recipient animals is counted.

↓

$$\text{Cell survival} = \frac{\text{No. of colonies derived from treated cells}}{\text{No. of colonies derived from control cells}}$$

Survival Curves

As mentioned above, in general, the proportion of cells surviving chemotherapy or radiotherapy decreases as the concentration of drug or irradiation dose increases. To a certain extent, the rate at which killing occurs, i.e., the slope of the curve, depends on the proliferative characteristics of the cells in question, i.e., the rate of proliferation. Cells are inherently either slowly or rapidly dividing. Most drugs are most active against rapidly dividing cells; killing is greater if the cells are dividing quickly rather than slowly. Drugs, in turn, are either cell cycle-specific (cytosine arabinoside) or not. For cell cycle-specific drugs increasing the concentration of drug still further will have no further effect because all the cells in that phase of the cell cycle will have been killed.

Prediction of Response

It takes a very long time for a chemical compound considered to have potential antitumor activity to reach clinical use, and only a tiny proportion of drugs that are synthesized ever enter clinical trials. The first stage in the process involves *in vitro* testing against cell lines derived from human tumors, and screening in transplantable tumors in mice. Drugs deemed to be active against cell lines are then tested in human cancer xenografts as defined below.

Use of Xenografts

A piece of viable tumor from a patient with cancer is injected into an animal. Immune-deficient mice have been used, either congenitally athymic (nude) mice or immune-suppressed mice which will not reject a graft from a different species. Neither is ideal, but this is the closest experimental approximation to the human situation. Xenograft studies are therefore useful in the research setting, but it has not thus far been possible to individualize treatment on the basis of xenograft data.

Eventually drugs are tested at the Phase I, II and III level. Phase I studies are where the maximal tolerated dose is determined. Phase II studies are where the efficacy in a given tumor type is evaluated. Phase III studies are where efficacy in comparison with standard treatment in a randomized study is evaluated.

The ultimate logical step would obviously be an assay that would accurately predict response for an individual patient, leading to selection of the optimal treatment for that individual on the basis of *in vitro* sensitivity to a drug or combination of drugs, rather than empirical administration of treatment, which is what normally happens. Several attempts have been made to achieve this Utopian goal, with varying degrees of success. In general, *in vitro* demonstration of resistance has correlated with lack of efficacy *in vivo*. However, in terms of predicting responsiveness, the situation is much less satisfactory. Most of the assays that have been used have been based on estimating colony growth in semisolid agar and enriched media; thus, once more, these are clonogenic assays. However, such attempts to use *in vitro* assays to predict response to chemotherapy have been bedeviled by a number of problems. Virtually all depend on starting with a single cell suspension which is itself difficult to derive from a solid biopsy specimen without damaging the cells. Another pragmatic problem has been that often patients are in urgent need of treatment and there is no time to wait for the results of such an *in vitro* assay. A further difficulty is the fact that the majority of modern chemotherapy is given as combinations of drugs rather than single agents; such combinations are difficult to reproduce *in vitro*. Furthermore, results of experiments using single agents do not necessarily correlate with response to the same drugs given as a combination. In addition, not all tumor cells grow in culture and only some sub-populations within a tumor may grow. Finally, the conditions of the assay, i.e., *in vitro* concentration of drug and time of exposure, may be very different from the dose and schedule of administration in a patient.

Efforts to develop more accurate assays continue and these will need to be validated in prospective studies. At the present time, however, they have not improved survival of patients, which is ultimately the objective.

SUMMARY

The modern era of cancer chemotherapy is not yet half a century old, yet the range of drugs available, the multiplicity of their actions, and their ability to cure diseases, which at the time of their introduction were quite incurable, has been astonishing. While combinations of effective drugs have cured children with acute lymphoblastic leukemia, Hodgkin's disease in both children and adults, and testicular tumors in young men, there has (until recently) been a perception that little progress has been made in other solid tumors. It is, however, being admitted that in breast cancer, ovarian cancer, and other solid tumors, the duration and quality of survival may be improved, even if cure is not possible. Knowledge of how drugs act, how their toxicity can be countered, and how tumors become resistant to them, as has been outlined in this chapter, will enable further progress to be made.

REFERENCES

Bagshawe, K.D. (1992). Antibody directed enzyme prodrug therapy (ADEPT). In: Accomplishments in Cancer Research 1991 (Fortner, J.G. & Rhoads, J.E., eds.), pp. 154–169, J. B. Lippincott Company, Philadelphia.

Burchenall, J.H. (1977). The historical development of cancer chemotherapy. Sem. Oncology 4, 135–143.

Chabner, B.A. & Collins, J.M. (eds.) (1990). In: Cancer Chemotherapy-Principles and Practice. J. B. Lippincott Company, Philadelphia.

Ehrlichman, C. (1993). The pharmacology of anticancer drugs. In: The Basic Science of Oncology (Tamarnoch, I.F. & Hill, R.P., eds.), 2nd ed. Pergamon Press, Oxford.

Tannock, I.F. (1978). Cell kinetics and chemotherapy: A critical review. Cancer Treatment Reports 62, 1117–1133.

Van Hoff, D.D. (1990). He's not going to talk about *in vitro* predictive assays again, is he? J. Natl. Cancer Inst. 82, 96–101.

Workman, P. & Graham, M.A. (1993). In: Pharmacokinetics and Cancer Chemotherapy. Vol. 17, Cold Spring Harbor Laboratory Press, Cold Spring, NY.

INDEX

www.ingramcontent.com/pod-product-compliance
Ingram Content Group UK Ltd.
Pitfield, Milton Keynes, MK11 3LW, UK
UKHW020234290926
15470UKWH00002B/540

9 780762 301461